D0349381

THE TECHNIQUE
of
TELEVISION PRODUCTION

THE TECHNIQUE OF
TELEVISION PRODUCTION

By
GERALD MILLERSON

FOCAL PRESS
London and New York

First Edition 1961
Second Edition 1963
Third Edition 1964
Fourth Edition 1966
Fifth Edition 1967
Sixth Revised and Enlarged Edition 1968 (USA)
Seventh Edition 1969
Eighth Revised Edition 1970
Ninth Revised Edition 1972

Illustrated by the Author

Printed in Great Britain by
Richard Clay (The Chaucer Press), Ltd,
Bungay, Suffolk

CONTENTS

5

PREFACE

WHEN this book was originally written it sought to present the reader with a comprehensive study of the mechanics, techniques and aesthetics of television studio production. Many of the facilities and procedures it outlined were so new that they were experimental, confined perhaps to a few large organizations. Now many of them have taken their part in the operational scene of TV production, and we encounter further new innovations.

This typifies television. Equipment develops, techniques (and fashions) change. Once most productions were transmitted "live"; the viewer saw events exactly as they happened—for better or worse. But sheer economics have meant that television organizations must now record and market many of their products as well as meet varying network needs. Recording represents greater scheduling convenience; a greater opportunity to polish and supplement. So video-recording has grown.

True film-making is a relatively lengthy, expensive process, even when film-cameras use TV viewfinders to facilitate Director-monitoring. So TV studio production has continued to explore its own particular, continuous-action opportunities, augmented with location film and mobile video-tape recordings. Larger productions have tended to work on a scene-by-scene basis, retaking errors, tightening editing, varying treatment; the whole being subsequently re-edited in the final telerecording. But essentially, they remain *television* productions, rather than economy-budget "pseudo-films".

Fortunately for us all, a firm foundation of knowledge builds a bridge to all potential TV techniques—whether one seeks "spontaneous inspiration" or "calculated exploration". Various of the author's hypotheses appear to have stood the test of time, and helped to suggest persuasive opportunities the medium can offer. Television all too readily lends itself to dull routine treatments, and this, above all, the real professional avoids in a competitive world.

During the last decade, television has seen many interesting developments. Technical advances have been numerous, programme channels have extended and stations multiplied, colour TV has become established worldwide, television has brought an increasingly valuable teaching tool to the educator. Closed circuit TV is a familiar facility in innumerable diverse fields. Video-recorded package programmes are developing.

The situation in network studios too has changed. Economics, fashion, and technical evolution have influenced the current format. The use of 35 mm. film has diminished in television, and that of improved 16 mm. grown astronomically. Colour quality that earlier seemed unattainable has become routine. The advent of improved camera-tubes, high-grade video recorders, chroma-key, electronic lighting consoles, and the many less evident innovations, have modified both the productional approaches and the medium's apparent potentialities. And yet the future developments for television production are far from obvious.

Thanks to the mobility of the 16 mm. camera, an increasing proportion of hitherto studio-staged material has been shot on location; interviews, documentaries, song, dance, even "mini-feature" films have utilized the freedom of location shooting. Certain TV networks have restricted their studio productions largely to routines of soap operas, panel games, talks, and news telecasts, to augment film production and presentation, and TV remotes.

The future for TV networking must surely be beyond these narrow usages. Electronic opportunities are vast. Mobile video-tape, computerised TV editing techniques, and the opportunities for electronic matting, offer considerable economies and elaborations. Chroma-key of advanced design permits studio action to be integrated undetectably into designs, paintings, models, photographs, etc. This enables uneconomic but customary staging practices to be entirely reconceived. Instead of extensive scenic effort (with its construction, transport, erection, revamping, storage, and disposal problems, etc.), suitable backgrounds may be created to suit specific camera treatment. (A scheme devised by the author would permit photo-libraries of entire environments to largely replace built studio scenery. With it, extensive production treatment could be formulated and pre-coded before even entering the studio.)

It is only when we pause to take stock of such developments, that it becomes evident how stable and valuable the underlying production principles themselves have proved to be. It is these principles we shall study together here in this book, for whatever the transitional changes in usage, the fabric of the medium remains inherently constant.

Readers interested or engaged in the many allied fields of television such as stage, film, radio, photography, graphic arts, should find our studies here usefully interesting, for it is from these sources that television itself has grown.

12

1

THE TELEVISION STUDIO CENTRE

THE modern television studio takes many forms, from the small local station, with its film scanner and single camera channel, to mammoth studio centres with every technical facility. Something of the spirit of television production will certainly pervade them all. An indescribable air of expectant tension communicates itself to performers and technicians alike; whether the project is to be recorded for subsequent transmission or is being received at that instant in the home circle itself.

It is important for us to realize something of this urgency that forms the background of production. For later we shall be talking over its theoretical ideals, and it is only when we relate these ideals to the relentless immediacy of practical production conditions that we shall begin to see programming in its true perspective.

The television studio

Let us pay a visit in imagination to a typical large television studio. An indicator shows that they are on closed-circuit rehearsal. So, although all the studio equipment is "live", its output is only being monitored locally. At scheduled time, picture and sound will be fed to the master control room, and thence to the transmitters.

Spaced around the studio we see the set-designer's art: three-walled "rooms", part of a street scene, a summer garden, perhaps. But next to our surprise at the colourful realism of these settings, the first impression is one of endless lights. Suspended, clamped to sets, on floor-stands ... it is hard to appreciate that each has been placed and angled with careful precision, to fulfil a particular purpose.

Despite the size of the studio, and the numerous people around, everywhere is surprisingly quiet. There is little reverberation, thanks to the dampening effect of the acoustically-treated walls. Action is concentrated at the moment in a country house interior setting. The microphone at the end of the sound boom's telescopic arm follows the actors as they walk about the "room". A camera

13

moves silently over the specially levelled floor—where the slightest bump would judder the picture. Its long cable snakes away behind it, to a plug-point on the studio wall. Through the headsets that everybody wears, the voices of the programme and technical directors co-ordinate the manoeuvres of the camera, sound and stage crews.

The windows of the "country house" lead out on to a "sunny garden". Soft "bird-song" comes from a nearby foldback loudspeaker. Suddenly a spot-effects man sounds the horn of an "approaching car", and the floor-manager cues the newcomer to enter the garden. As he comes into shot the players in the room walk out to meet him. The cameras we have been watching pull away quickly and move to the next set; another camera and sound-boom have taken over from them in the garden set.

"Hold it, please!" comes over the talk-back loudspeaker—the action stops. The director's voice from the production control room (in this case behind the large glass panel set in the studio wall) explains how a slight change in the actors' positions would give him a better shot. They walk through their revised moves, he is satisfied with the alteration, and the floor is crayoned or taped with toe marks and camera location.

The director, using the production talkback system, asks a cameraman if he has room to pull back a little. The cameraman shakes the camera's head, giving a visual "no" with his picture; and another nearby camera pans over to show the director how the first camera is placed. By this remote control he can see the floor problems for himself, and rearrange his camera movements accordingly.

The floor-manager receives a go-ahead from the director over his radio-headset; so, calling for silence, he signals the action to carry on. The floor-manager is the director's contact man on the studio floor, and checks staging, action and performers on his behalf.

The game of make-believe springs back into being as if no mere technicalities had interrupted its progress. The rehearsal continues. ...

The production control room

The ideal arrangement of the control room is still a matter of opinion. At some studio centres we find all the necessary equipment and personnel housed in a single large room; economical, and facilitating close co-operation. But these separate activities may interfere with each other unduly, and some networks prefer to have isolated rooms for production control; video control; sound control. The centre we are visiting has this arrangement.

Through the window of the production control room we can see the studio below. The room's dominating feature is the row of picture monitors. These preview screens give a continuous view of what the three or more studio cameras and other video sources are seeing. Any shot can be switched or faded over to the transmission channel, when it will be seen on the master monitor, ready for transmission. Over a loudspeaker, programme sound is heard.

Facing these monitors sit the production director and his assistant (who checks timing and calls out shot numbers). Over the desk talk-back microphone he is instructing the studio crew, guiding their moves, warning them of action to come. He may operate the buttons and faders for video switching himself, but most networks consider the programme director too preoccupied with the many other aspects of production, and delegate this job to another person. This may be a specialist switcher (vision mixer), or the technical director.

The technical director is in charge of the technical operational staff on the show. As well as being a technical adviser, his duties may include those of :

Camera director—actively assisting in staging and treatment.
Switcher—doing his own vision mixing.
Or supervisory engineer—looking after the engineering aspects of production; acting as undercover assistant; aligning effects shots; checking source availability, etc.

Through his own talk-back circuits he can talk with individual members of his crew.

A number of other specialists are here checking their respective contributions to the production:

The lighting director, who has planned, arranged and supervised the studio lighting, and may be responsible for video quality.
The set designer, responsible for the scenic treatment.
A make-up artist, assessing the make-up requirements.

From there on, the population is extensive but variable. A host of personages, including the producer, assistant directors, production manager, sponsor, script girl, wardrobe ... and so *ad infinitum*.

The vision control room

Technical advances have resulted in television cameras of increasingly sophisticated and compact design. For remote location work, hand-held man-pack cameras enable the cameraman to become virtually a self-contained unit, transmitting his sound and picture to a convenient pick-out point. In the studio, however, heavier duty camera designs are generally used. Usually mounted on a pedestal or rolling tripod, each television camera's long two-way multi-core cable is plugged via a studio wall point to its distant camera control unit (CCU). There the major part of the camera channel's electronics provide the electrical supplies, pulses, waveforms, etc., required to generate and amplify the picture. After electronic adjustment, the respective picture passes to the video switching console, where each camera's output can be selected for transmission.

The studio cameras' group of CCUs may be housed in the vision control room itself, in a nearby apparatus room, or a remote complex where equipment for several studios is communally situated. Circuit controls on the camera control units allow picture quality to be adjusted. This process was originally done by operational video engineers seated at each CCU. Later, as equipment developed, one man became able to control two units' outputs—an arrangement still used in certain applications. Subsequently, networks introduced the concept of a vision desk at which all the various operational controls could be remotely adjusted. Here a single vision control operator or video engineer continually modifies the brightness, tonal and colour quality of all picture sources, matching them for optimum effect.

To enable him to do so, a row of preview picture monitors continually displays each video source, while others show the resultant studio output picture. By comparative switching, and by examination of nearby waveform monitors (showing an electrical "graph" of the video signals) the operator achieves pictorial continuity in close liaison with the lighting director's interpretation.

The vision control or video operator has a deceptively self-effacing job. As he processes the pictures he combines technical and operational skills and artistic sensitivity. While correcting and improving the picture quality, he is continually watching for technical faults and defects. Without him, the viewer would be distracted by changes in picture brightness and contrast, over-aware of picture transitions, encounter unclear or ambiguous reproduction.

Where "one man vision control" is employed, we find that the main video controls for each camera are juxtaposed to permit rapid adjustment. The ingenious master-control knob for each camera channel can tilt backwards and forwards (lens aperture variation to correct exposure). Rotation changes the picture's "black level" (sit),

16

lightening or darkening all picture tones. The main effect of "sitting down" is to merge the shot's darkest tones to black, while "sitting up" thins out relative tonal density.

Pressure on the camera's master-control knob provides comparative switching between that channel's picture, and the main studio output picture selected by the switcher (vision mixer) in the production control room. This instant examination on the same picture monitor enables the video operator to compare and match brightness, tonal, and colour quality.

Colour quality is controllable by further instrumentation, which may contain a vari-position joystick ("paintpot") moving the overall colour bias towards any selected hue, or pre-set adjustments (gain and gamma control of the RGB signals) known as TARIF, or a colour balance control panel permitting individual channel parameters to be varied (i.e. gain and black-level controls for each camera's red, green and blue channel).

In order to make one-man vision control practicable, extremely close liaison with the lighting director is imperative. This can be facilitated by arranging adjacent video control and lighting consoles.

The sound control room

Adjoining the production control room through a communicating glass window is the sound control room. Here the sound mixer (audio control man) sits before a large audio control console. His attention is divided mostly between the flickering needle of his volume indicator and his picture monitor. Adjusting the volume and quality of the programme sound to match with this picture, he guides (by his private-wire talk-back circuit) the sound men down on the studio floor.

Thus he can warn boom operators against dipping into shot, casting visible boom shadows, and similar hazards, while assisting them in achieving sound perspective to suit the transmitted picture.

Nearby, another technician is operating a row of disc replay turntables that are supplying the bird-song effects we heard earlier over the studio foldback loudspeaker. Tape recorders, racks of audio amplifiers, patchboards, and an electronic reverberation unit, complete the general set-up.

This, very briefly, is the production scene. In essence, it is the same everywhere; in detail, the elaboration and layout change from one studio to the next. Let us look more closely at the fabric of television itself—the television picture, and how it is formed.

2

THE TELEVISION PICTURE

ONCE the initial novelty of being able to convey pictures by television has worn off, we naturally tend to enquire whether we cannot obtain a more faithful image. Improvements are certainly possible.

The ideal and the attainable

We can have better definition (more picture lines), reasonably naturalistic colour, a greater brightness range, stereoscopic pictures, stereophonic sound. But even supposing that they were ultimately desirable, technological limitations and economics would delay their realization.

Fortunately the human mind is extremely tolerant, and can be satisfied very largely with substitutes falling quite wide of perfection. Sometimes the audience needs to be conditioned or trained to accept them; but accepted they invariably are.

But in simplifying our presentation of the external world down to a smallish, flat, coloured or monochrome picture, remember we are necessarily only *representing* natural features. We are not showing life. We are only displaying an interpretation, a symbolized substitute. That is all the motion picture does. Hence the need for art and technique when handling these media, as we shall see.

How television works

Considering its technical complexity, television is surprisingly simple fundamentally. An image of the studio scene is focused on to a small screen inside the camera-tube. This screen has a surface of thousands of tiny light-sensitive dots (elements). According to how much light is falling upon it from one particular part of the scene, each dot becomes electrically charged. Over this screen (the photo-cathode or target), we therefore have a charge pattern built up, proportional at each dot, to the respective light and shade in the scene.

A small gun in the camera-tube generates a continuous beam of electrical particles (electrons). This fine inertialess "pointer" explores the charge pattern on the camera-tube screen. "Reading" over it in a series of lines, the beam neutralizes each picture element in its path, so producing a varying electric current. This current is proportional in turn, to the charges and hence to the pattern of light and shade in the televised scene.

The current, known as the video or picture signal, is subsequently amplified and passed to the video switching console (vision mixing desk), as we have seen.

As each element is scanned and gives up its information, it becomes "wiped clean" and can therefore respond to any new light image it may receive. This charge-forming and systematic "reading" is a rapid, continuous process. The complete picture or frame is explored twenty-five or thirty times each second; the number of lines used depending upon the system's standards, e.g. 525, 625, 819 lines. The spot's scanning-rate being too fast for our eye to follow,

Fig. 2.1. Pt.1.

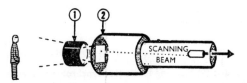

THE PICTURE SIGNAL. The camera lens (1) focuses the scene on to a light-sensitive surface (2) in the camera-tube where a corresponding pattern of electrical charges is formed. This is rhythmically scanned by an electron beam in a series of lines. The variations it "sees" as it scans each detail in turn are transformed into a continuous electrical video signal. The brighter the detail, the stronger the video at that point.

Fig. 2.1. Pt.2.

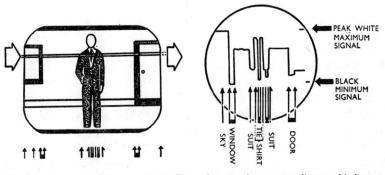

Left: A typical scanned line.   *Right:* The video signal corresponding to this line.

19

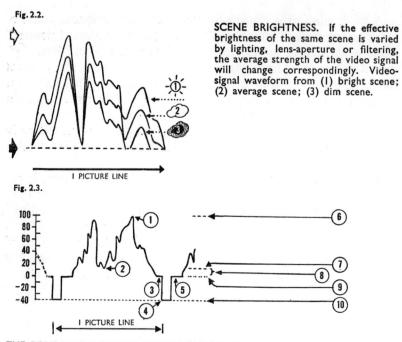

Fig. 2.2.

SCENE BRIGHTNESS. If the effective brightness of the same scene is varied by lighting, lens-aperture or filtering, the average strength of the video signal will change correspondingly. Video-signal waveform from (1) bright scene; (2) average scene; (3) dim scene.

I PICTURE LINE

Fig. 2.3.

I PICTURE LINE

THE COMPONENT PARTS OF THE VIDEO SIGNAL. (1) a peak white (peak modulation); (2) a peak black; (3) front porch; (4) horizontal (line) synchronizing pulse; (5) back porch; (6) white level (reference white); (7) black level (reference black); (8) set-up (pedestal); (9) blanking level; (10) synchronizing level.

we see only the resultant effect: a rectangle of fine horizontal lines.

To reduce flicker problems, the beam is made to read the odd lines (odd field) of the image first (i.e. lines 1, 3, 5, ...) and then return to scan the even lines between them (i.e. lines 2, 4, 6, ...). This interlacing method of scanning effectively saves "etherspace", allowing more stations to work in any band.

Transforming the video signal back into a picture is again a comparatively simple process. Upon the inside of the receiver picture tube's (kinescope) face is a coating of fluorescent powder. This has the property of glowing when electrons impinge upon it. A gun in the picture-tube (like that in the camera-tube) produces such a stream of electrons, and this is made to scan over the powdered screen in a regular series of sweeps, in synchronism with the exploratory beam in the studio camera. Horizontal and vertical 'sync' pulses keep the respective beams in step.

As the picture-tube's beam scans, the screen brightens in its path. A constant-strength beam would trace an even rectangle (raster) of fine horizontal lines. But if its strength is varied, spot brightness varies correspondingly. So, by applying the video signal to regulate the picture-tube's beam, a pattern of light and shade can be built up on the screen's phosphor, corresponding to the light distribution in the studio scene.

This picture dies away rapidly and is replaced, but the rate of the continually scanning beam is too rapid for the eye to follow, and we see a clear, unflickering complete image.

For colour television, the face of the picture-tube is covered with systematic dot patterns of three phosphors which glow red, green and blue respectively, when excited by their corresponding electron beams. These primary colours blend to produce a wide range of hues (see Chapter 21).

TABLE 2.1

INTERNATIONAL TELEVISION STANDARDS

No. of scanning lines in system:	525 American F.C.C.	625 European	819 C.C.I.R.
No. of lines seen in viewed picture (the rest lost in blanking periods)	483 to 499	563 to 587	737
Frames per sec.	30	25	25
Fields per sec.	60	50	50
Scanning rate (line frequency) per sec.	15,750	15,625	20,475
Vision band width	4·2 MHz	5 MHz*	10 MHz
Sense of modulation	—	—	+
Sound modulation	F.M.	F.M.	A.M.
Picture shape (aspect ratio)	In all systems		
Interlaced pictures			

* Certain countries 5·5 MHz and 6 MHz.

3

THE TELEVISION CAMERA

THE television camera is a hybrid of the photographic and the electronic worlds, and we can appreciate its features best if we take separately:—

1. Physical characteristics—basic mechanics and optics.
2. Electronic performance—how electronic characteristics limit the quality of the televised picture.

The understanding of these principles can help us considerably in determining staging and production problems, although most equipment will only be available in the largest organizations.

Physical characteristics

Camera mountings

Most television cameras are too large and weighty to be hand-held, but even where this is practicable (as in the Vidicon), camera-shake and an unsteady picture are hard to avoid. Furthermore, an unaided human operator is highly unlikely to be able to carry out smoothly and unobtrusively the various moves, elevations and contortions required in modern studio production.

Fig. 3.1.

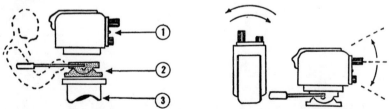

CAMERA HEAD—MOUNTING AND MOVEMENT. *Left:* The camera head (1) is attached via the panning head (2) to the mounting (3). *Right:* The panning head provides lateral panning and vertical tilting which can be restrained or stopped completely by variable frictional pressure.

Fig. 3.2.

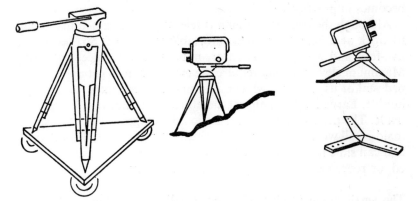

THE TRIPOD. *Left:* A simple 3-legged stand with independently extensible legs. Height is pre-set and not readily readjusted. It is light and portable but static, unless mounted on a castored base (rolling tripod), permitting limited wheeling.

Centre: A tripod enables the camera to be set up on sloping or uneven ground (unlike other types of mounting).

Right: The low tripod provides ground shots. A tripod is sometimes placed on a metal triangle or crowsfoot (*bottom*) to prevent its legs from slipping.

To provide such flexibility, various types of camera mounting have been evolved in motion picture making. These have been largely inherited by television and developed for its special needs.

The pedestal. A one-man camera mounting, the pedestal has high manoeuvrability. It has panning/tilting action, its height can be adjusted from about 3-5 ft., and it can easily be pushed around the floor. Changes in column height or pedestal movement can, within limits, be carried out while on shot. But remembering that the cameraman then has to focus and compose the picture and control the camera head while pushing and guiding the mounting, one cannot expect the smoothest or subtlest of camera-work. At eye-level,

Fig. 3.3.

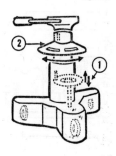

THE PEDESTAL. Camera height is adjustable from 3 to 5 ft. by handcranking, counterbalancing, pneumatic or hydraulic control (1), according to design. The three linked base-wheels are freely steerable by the steering-ring (2) or can be set in pre-selected locked positions. The whole unit weighs 350 to 500 lbs.

this can be a difficult feat; at maximum or minimum heights it becomes impracticable.

Although the pedestal column is telescopic, the cameraman tends to work at a height that is comfortable to himself. So, between a five-foot and a six-foot operator, noticeable differences can arise. Maximum elevation may require the cameraman to get up on to the pedestal or use a stool; while minimum elevation may see him bent double. Large height-changes, therefore, cannot always be speedily made. The pedestal has several basic advantages, and is the studio maid-of-all work. Comparatively cheap, it requires little floor space, and can sidle into odd corners and confined spaces. It can be pushed, or reorientated very rapidly on its own axis.

The small crane (dolly, velocitor, small boom). The small crane enables many elaborate camera movements to be carried out with precision. Although it is much bulkier than the pedestal (around 800 lbs. weight), and needs room to reposition, its flexibility is high. Where the cameraman forsakes his seat to kneel, stand or squat on the mounting, he can obtain even greater variation. But perhaps the most essential requisite is a level, even floor surface, to permit smooth tracking.

Assisting the cameraman is an operator who pushes and guides (tracks) the mounting. Where the crane-arm is to be raised/lowered or slewed, he may also carry out these adjustments. Alternatively, a third person (the second operator) may sit on a detachable seat, for this purpose.

The camera-crew's movements will be co-ordinated by finger signals given by the cameraman's panning-handle support hand—since his other hand will, in normal operation, be occupied with the focus adjustment.

Fig. 3.4.

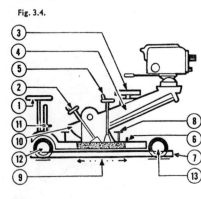

THE SMALL CRANE. Approximate dimensions: base size, 3 by 5 ft.; overall area 4 by 9 ft. (1) Tiller handle or steering wheel; (2) wheel to raise/lower crane-arm (can be locked off); (3) cameraman's seat; (4) crane-arm (jib tongue); (5) wheel to rotate turntable and crane-arm; (6) possible 2nd operator's seat (operating 2 and 5); (7) detachable cable-guard prevents floor cables from fouling wheels; (8) turntable locking screw; (9) turntable; (10) geared sector for crane-arm height adjustment; (11) screw-jacks; (12) rear wheels—guided by tiller 1; (13) front wheels—fixed as shown or at right angles to this direction.

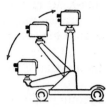
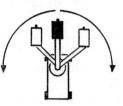
Fig. 3.5.

SMALL CRANE MANOEUVRES. *Top left:* Craning (booming) up — Maximum lens height about 7 ft. Craning (booming) down —Minimum lens height about 3 ft.

Top centre: Lateral craning (slewing) left and right.

Top right: Panning off the forward axis of the crane-arm must usually be restricted to ± 20° if the centrally-seated cameraman is to be able to see into his viewfinder. Wide panning movements are risky.

Right: Two methods of trucking sideways are shown (heavy arrows) while the operation of tongueing towards (in) and and away from (out) the scene is illustrated (*left,* light arrows).

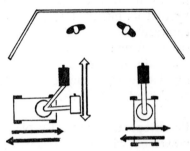

The crane (or boom). In principle, the crane resembles the small boom. The Motion Picture Research Council crane is to be found in the studios of most large television networks (Fig. 3. 6). Because it needs even more floor-space than the standard dolly and requires a three or four-man crew, we shall normally only find it in productions that can make full use of its flexibility in production numbers, mass spectacle, and similar large-area displays.

Fig. 3.6.

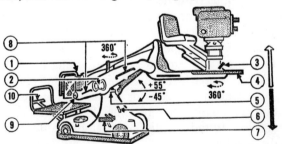

THE CRANE. Maximum height, about 9 ft.; minimum height, about 1½ ft.; weight complete, 3640 lbs. plus; base, 6 ft. x 3½ ft.; overall length, 14½ ft. maximum; maximum speed, 5 m.p.h.; Electrically driven. (1) Counter-balance weights; (2) column elevation lock—prevents boom-arm moving when mounting is idle; (3) foot brake, for camera platform rotation; (4) rotatable platform carrying camera, panning-head, seat and cameraman. Rotates electrically, or by foot movement; (5) hydraulically raised central pedestal, 15 ins. movement; (6) pedestal hand pump and release valve; (7) brake and forward/reverse movement levers; (8) support handle to steady swings of boom; (9) fine counterweight adjustment; (10) steered by Tracker operator.

As with the small crane, the camera and panning-head remain level, irrespective of crane-arm angle. In the crane, the cameraman's seat is an integral part of this mechanism. And because his seat is concentrically mounted with it, he can rotate the camera head and himself through 360°, by walking his feet round the circular platform on which they rest.

Owing to its size, weight and balance, the crane requires highly-skilled operation if it is to achieve smooth, safe movement. A misjudgment can cause the arm to overswing, or the dolly to plunge beyond its intended position. Considering that the two-ton crane can reach a top speed of about 5 m.p.h., and the really large cranes (that elevate to 22 feet) can attain 25 m.p.h., the crew's task within a set-strewn studio is no easy matter. The cameraman, needless to say, has to have complete confidence in his crew's judgment—and a good head for heights.

Power-operated dolly. By placing remote controls for all camera movement within easy reach of the cameraman, one could avoid the need for assistants, and overcome co-ordinating problems. Where a camera is panning, slewing and tracking simultaneously, for example, this can demand almost telepathic team-work. Foot-pedals, hand twist-grips, etc., would allow absolute mechanical control by the cameraman.

But there would be snags. He would lack the guidance and vigilance of his crew. Already he has to remember action, subject movement, shot sequences; to focus and compose and to change lenses. For long continuous sequences this requires appreciable concentration. With each extra control he has more to remember.

There are vital physical problems, too. With his head tucked into his viewfinder, he is in no position to avoid obstacles in his path. He is liable to depress the camera on to anything beneath the crane-arm's overhang. Similarly, he might hit something above him. If, for instance, he is tracking backwards through an archway at too high an elevation, he might severely injure himself, as more than one cameraman has found to his cost. Under a second operator's control, this is avoidable.

Mechanization can alleviate the tedious manual effort of dolly operation, so power-operated dollies have found increasing popularity in large organizations. Electrical control must, however, provide the precision that manual operation allows, if subtle camera treatment is to be achieved. Also, foolproof devices must exist to prevent any sort of runaway by the apparatus itself.

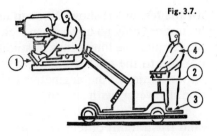

Fig. 3.7.

Low dollies ("creepers"). Basically a steerable wheeled platform some 6 in. from the floor. The camera head and seat are fitted on a turntable at the front, placing the lens centre at about 2 ft. Where long periods of low-angle shooting are required, this may provide a more convenient solution than mirror-periscopes (Fig. 20.16, Pt. 3).

Unorthodox mountings. Providing a camera head is positioned securely and suitably, there is virtually no limit to the arrangements that may be made for its mounting. By means of a clamp on the panning-head, cameras have been mounted on tubular steel barrels, gantries, cradles, hoists, rubber thongs, etc. The cameraman's size and weight, however, have usually to be considered. Against the delights of these unorthodox mounting methods, we have to weigh the disadvantages of restricted camera head movement, camera isolation and availability. Often an indirect approach to the situation (e.g. using mirrors, pre-filmed inserts, etc.) ultimately proves more convenient. Hydraulic platforms, helicopters, all have their uses.

Studio camera facilities. Few television studios have more than six cameras. Even large network productions often manage with four. Two small cranes (or power-operated dollies) and two pedestal cameras will meet the reasonable demands of even an elaborate drama production. Many small studios get by with rolling-tripods alone. Not only economics but studio space will determine how elaborate a mounting can be usefully employed under average production conditions. The more complicated or cramped the settings, or the smaller the studio, the fewer the opportunities for cranes. Where a small crane (or two) is part of the camera facilities, it will normally carry the bulk of the show, being supported by the pedestal cameras. Some prefer nothing but highly mobile pedestals.

The director has always to suit his production treatment to the available facilities. If he wants a shot looking from the head of a staircase, he may find that the camera concerned has to remain there

27

for the rest of the show. Nor can he envisage high crane shots if his mountings only go up to seven feet. These matters are obvious when we examine the limitations of studio mechanics. But they are not so obvious to the critical viewer, who sees what appear to be missed opportunities or unimaginative camera treatment during a production. However, with substitution and subterfuge, one can overcome a great many of these limitations, and here lies much of the challenge (and frustration) of television production.

Remotely controlled cameras. Camera operation, like most mechanical processes, can be locally or remotely controlled. Many adjustments lend themselves readily to either arrangement. Taken to its extreme, we could have unmanned robot dollies, controlled by operators in the production control room. Or complete automation!

A certain amount of remote selection relieves the cameraman from distracting mechanics. But most control is best left in his hands. In remote telecasts where it is undesirable or impracticable to have a cameraman beside the camera, distant operation may on occasion

Fig. 3.8.

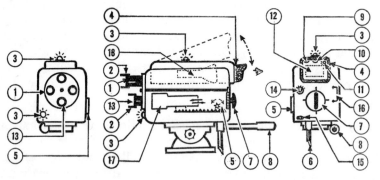

THE CAMERA HEAD. *Left:* Front. *Centre:* Side. *Right:* Back. (1) Lens turret, with 3 to 6 lenses according to design, rotated by rear handle (7) or remote switching. Replaceable by single zoom lens; (2) interchangeable camera lenses with lens hoods; (3) camera tally light—indicates when camera is transmitting; (4) viewfinder—can often be tilted to facilitate viewing; (5) focus control—moves camera-tube behind fixed lens (coarseness varies with lens-angle); (6) cable to Camera Control Unit, 400-1000 ft. carrying power supplies to camera, video output, talk-back, etc.; (7) lens change control, turns turret (1); (8) panning handle to guide camera head; (9) lens aperture indicator (inside viewfinder); (10) lens indicator (inside viewfinder) showing lens in use; (11) miniature tally light, visible to cameraman when looking into viewfinder; (12) small viewfinder kinescope showing TV picture (optically magnified) seen through taking lens. Usually shows slightly greater area than actually transmitted, allowing cameraman to see outside picture area; (13) taking lens—its position differs with equipment design; (14) lens aperture control (may be remotely controlled); (15) microphone headset jack-point—cameraman can communicate with directors, technicians, etc.; (16) clip for camera cue-cards; (17) camera tube; (18) viewfinder tube.

offer a convenient solution. But its value in the comprehensive television studio is arguable, except for news and continuity use.

The camera lens

The camera lens is actually made up of a series of individual lenses (elements) mounted together in a tubular metal housing. This screws into the camera's lens turret. According to the type of lens system required, the separate optical properties of these elements are combined; and the aberrations of simpler lens systems reduced.

Although we need not concern ourselves here with the technicalities of lens design, a basic knowledge of how lenses behave will give us a better idea of how to use them.

Focal length and lens angle. In television production we shall hear lenses identified in two ways.

By their *focal length*—long focus; short focus; telephoto lenses; a "so many inch (or millimetre)" lens,
or by their *lens angle*—wide angle; narrow angle lenses; lenses of "so many degrees".

There is nothing mysterious about either of these methods. The focal length (F) of a lens is marked on the lens mount (e.g. F = 2 ins.) and is measured in inches or millimetres.

LENS ANGLES. The lens covers a 4-sided pyramid, having one base dimension (V) three-quarters the length of the other (H). Anything outside it is unseen. We use the vertical angle of view when studying elevations (side views) of scenes. Most production organization is done on plans however, so the horizontal angle of view (plan view) is normally used.

Fig. 3.9.

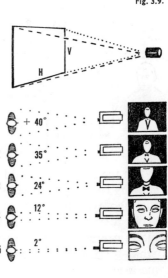

For a given camera position the narrower the lens angle, the closer the subject appears, but less of the scene is visible. The wider the angle the more distant the subject appears, but more is visible. *Right* are shown the approximate angular fields covered by (1) very wide-angle; (2) wide-angle; (3) normal; (4) narrow-angle; (5) very narrow-angle lenses. Reducing the lens angle "blows up" a proportionately smaller area of the centre of the shot, to fill the whole screen. Changing from one lens to another 3 times its angle will include 3 times more subject area in the picture, but with subjects 1/3 the size of the original. It is the equivalent of increasing the camera-subject distance by 3 times. Changing to a narrower angle has exactly the opposite effect.

Fig. 3.10. Pt.1.

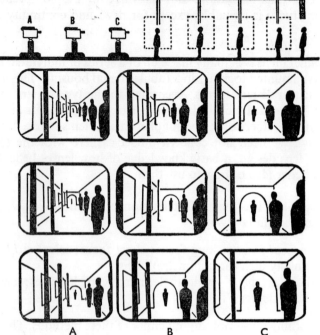

LENS ANGLES AND CAMERA MOVEMENT. The effect of varying camera-subject distance with various lens angles is shown above. *Top:* Wide-angle. *Centre:* Normal. *Bottom:* Narrow-angle.

Fig. 3.10. Pt.2.

This shows the effect of changing lens angle and altering camera distance to keep the main subject the same size: (1) narrow; (2) normal; (3) wide.

When the lens is its focal length away from the surface on which it is being focused (i.e. the light-sensitive screen of the camera-tube —the photographic film, etc.), very distant objects will appear sharply focused. Taking it further from that surface will cause subjects nearer the camera to come into sharp focus instead. That is what we are doing when focusing a camera; altering the lens/ camera-tube distance.

We are chiefly interested in how much of a scene a particular lens will show from a given distance away.

For any camera, the longer the focal length of the lens it uses, the narrower its coverage or angle of view. The shorter its F, the more it takes in from the same position. But a lens of "X-inch" focal length will not give the same shot with all size cameras. Used with a large camera-tube or film area, it will behave as a wider angle lens, while for a small area it will cover a comparatively narrow angle.

In short, focal length alone tells only part of the story. It is more convenient operationally to refer to a lens by its lens-angle or angle of view.

Perspective and its distortion. Some lenses have gained an undeserved reputation for distorting perspective. As we shall see, this idea is misleading, for distortion comes from inappropriate usage, rather than the lens itself.

It is only when the lens is incorrectly used relative to the picture viewing conditions that this distortion arises. We need to take care, then, in our choice of camera lens angles—unless of course we actually require distortion for specific effect.

One is apt to take the illusion of perspective very much for granted.

What, then, is natural perspective? In everyday life we shall accept pictorial perspective as normal when the relative sizes and distances of objects appear to diminish at natural rates.

A photograph's perspective will look natural to us when our viewing angle to the photograph equals that of the camera lens-angle used to shoot it.

So much for the effect. What can we do about the cause? How can such distortion be avoided? Clearly, if we use any camera lens-

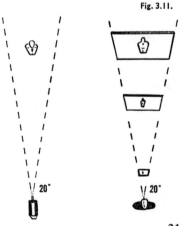

Fig. 3.11.

PERSPECTIVE DISTORTION. When we adjust our distance from a photograph so that our viewing angle is the same as that of the camera lens that shot it, the perspective looks natural. Viewing from too close or too far makes its perspective look unnatural. This is termed perspective distortion. *Left:* Camera taking pictures with 20° lens angle. *Right:* Viewer adjusts distance from large, medium or small picture to obtain 20° viewing angle.

31

Fig. 3.12.

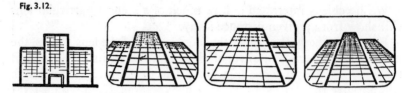

NATURAL PERSPECTIVE. The eye is accustomed to distortion of various forms. But perspective distortion can cause natural perspective to become so exaggerated that the eye can no longer accept it as natural. *Left:* A building may be of this shape in reality. *Centre left:* As the eye and a normal camera lens see the building from street level. *Centre-right:* Narrow-angle lens view. *Right:* Wide-angle lens view.

angle other than the audience's viewing angle (around 12° to 20°), distortion will occur.

Fortunately for the director, perspective-distortion is not too obvious for many kinds of scene. In fact, he can often turn it to his advantage to get pictorial illusions that would otherwise be impracticable or expensive to obtain. Perspective-distortion can create queer, unpleasant effects unwittingly, however. By outlining the two extreme types of distortion (for very narrow and very wide-angle lenses), we shall know what variation to expect as the camera lens-angle deviates from normal.

THE NARROW-ANGLE AND TELEPHOTO LENS (Long Focus). The forms of perspective-distortion introduced by narrow-angle lenses are a familiar feature in telecasts where the camera has to be positioned some way from the subject. Depth appears remarkably compressed; foreground to background distance seems shorter than it really is.

Fig. 3.13.

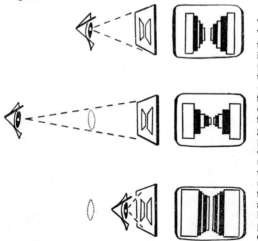

VIEWING ANGLE. *Top:* When the viewing angle equals the camera's lens angle our impression of the picture's perspective will be similar to that of the original scene. *Centre:* Viewing the picture from too far away or using a camera lens of too wide angle for normal viewing conditions gives a false impression of perspective. Depth and distance are exaggerated; distant subjects unduly small. *Bottom:* Viewing the picture too closely or using a too narrow-angle lens for normal viewing conditions again gives a false impression. Depth and distance are compressed; distant subjects unduly large.

Subjects look squashed or "cut-out". People take an interminable time to get to and from the camera. Distant objects are shown disproportionately large.

In extreme cases this compression creates most bizarre effects:

Seen end-on, an Atlantic liner can look only a few feet long.

At the race track, horses gallop vigorously, but appear to cover little ground.

Flat, cramped figures of a rowing eight squat in a boat only a foot or two in length.

On location such distortion may have to be tolerated where close-shots can only be obtained by narrow-angle (long focus) lenses. But in the studio, where closer camera positions are practical, there is little excuse for it.

Furthermore, a camera using a narrow-angle lens does not handle smoothly. Even slight camera head or dolly manoeuvres are liable to jerk.

Not only settings, but people too, can appear distorted with a narrow-angle lens. Close-ups show flattened faces, flat noses, flat chests, reduced bone-structure, and similarly misshapen caricatured features.

THE WIDE-ANGLE LENS (Short Focus). Here perspective-distortion causes depth exaggeration. Foreground/background distance appears greater than it is. Such distortion can usefully make smallish settings look large. For spectaculars, or crowded studios, this ruse can be extremely useful.

Other features of this type of distortion are less acceptable. In elevated shots people are liable to develop bulbous heads, stumpy legs, and enlarged chests. In close shots we see protruding noses and exaggerated facial contours.

In extreme cases, the very wide-angle lens will distort verticals, so that uprights at the sides of the frame appear curved. On panning shots, verticals then appear to bend as subjects approach the border of the frame.

Perspective changes will occur wherever lens-angles deviate from normal, and when cutting between cameras using differing lens-angles. Sizes and proportions must vary, especially where the angles are considerably different and perspective lines are strong.

Luckily, the eye's tolerance, and past audience-conditioning, have made these variations in perspective less distracting than we might otherwise expect, but the skilled director avoids needless distortion.

33

Fig. 3.14.

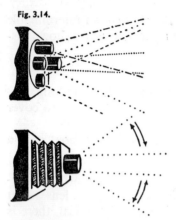

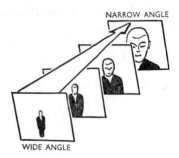

NARROW ANGLE

WIDE ANGLE

THE ZOOM LENS. A special lens system with continuously variable focal length and hence an adjustable angle of view. However, the internal complexity of a zoom lens makes it bulkier than a turret assembly, and optical design problems may result in inferior overall performance. For colour cameras, the zoom lens overcomes complex optical matching problems, and is universally fitted.

Top left: Instead of turning the turret to set the lens angle from a fixed selection (e.g. 10°, 18°, 24°, and 35°) the zoom angle (*bottom left*) can be adjusted anywhere in the zooming range. Typical lens angle variations of 1:5 (5°–25°) and 1:10 (5°–50°) are found.

Right: The lens is usually set at some pre-selected angle. Changing the angle in shot however, produces a zooming effect which has certain productional value.

The need for focus adjustment.

Many simple photographic and motion-picture cameras have no focusing mechanism. And yet, at the push of a button, they produce acceptably sharp pictures. Why then does the professional cameraman become so concerned with focus-following, depth of field and the rest of the focus jargon? On the face of it, we seem to have no need to focus these amateur cameras.

The answer is a slightly technical one and concerns a characteristic common to all camera lenses.

We can adjust any camera lens system so that it is focused on its *hyperfocal plane.* This done, all subjects between half that distance and the furthermost distance (infinity) will be in acceptable focus. In the simple fixed-focus camera, the manufacturer has set the lens there and left it.

We can adjust any focusable camera similarly. Just set it to the hyperfocal distance and leave it. Most of the scene will be reasonably sharp.

This hyperfocal distance (H) is found from:

$$\frac{\text{Focal length (F)}^2}{\text{lens aperture (f)} \times \text{circle of confusion}}$$

TABLE 3.1

HYPERFOCAL DISTANCES (in feet)

Focal length (inches) (F)	f2	f3	f3.5	f4	f4.5	f5.6	f6.3	f8	f11	f16	f22
1	42	28	24	21	18½	15	13	10½	7½	5	4
1½	63	42	36	31	28	22	20	16	11	8	6
2	83	55	48	42	37	30	27	21	15	10½	7½
2½	104	69	59	52	46	37	33	26	19	13	9½
3	125	83	71	62	56	45	40	31	23	16	11½
3½	146	97	83	73	65	52	47	36	26½	18	13
4	167	111	95	83	74	59	53	42	30	21	15
4½	187	125	107	94	83	67	60	47	34	23½	17
5	208	139	119	104	93	76	67	52	38	26	19
5½	229	153	131	115	102	82	73	57	42	29	21
6	250	167	143	125	111	89	80	62	46	31	23
6½	—	181	155	135	120	97	86	68	49	34	24½
7	—	194	167	146	130	104	93	73	53	36	26½
8	—	222	190	167	148	119	107	83	60	42	30
10	—	278	238	208	185	149	133	104	76	53	38

With a lens of focal length (F)—set to stop "f"—the scene will be acceptably sharp from ½ the "Hyperfocal Distance" to infinity.
Circle of confusion (e.g. $\frac{1}{200}$ in. for TV) defines minimum detail visible at correct viewing distance.

But this method is a compromise; most subjects are only reasonably sharp. This may be acceptable in a very small picture, or where we are not too critical, as in amateur snapshots.

A lens can only provide maximum sharpness at the distance to which it has been focused. Nearer and further than this distance, focus gradually falls off, until it becomes too blurred for us to accept as in focus.

The near and far distances within which subjects are still apparently sharp give us the depth of field (often erroneously called depth of focus—a different matter entirely).

How far focus can deviate from dead sharp before we notice it,

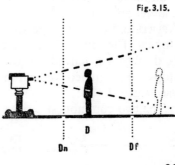

Fig. 3.15.

DEPTH OF FIELD. Dn, nearest distance in focus; Df, farthest distance; D, distance to which camera is focused.

Fig. 3.16.

LENS APERTURE. *Left:* Large aperture—small *f*-number e.g. *f* 1.9: shallow depth of field but less light needed. *Centre:* Medium aperture e.g. *f* 5.6. *Right:* Small aperture—large *f*-number e.g. *f* 16: increased depth of field but more light needed. Scene brightness must be enough for the smallest aperture used. The maximum aperture of a long focus lens is usually restricted by design (e.g. to *f* 4.5).

will depend upon the subject's detail, and upon the definition limits of our system.

Lens aperture (Stop). Within the lens-housing a variable circular diaphragm is fitted, known variously as the diaphragm, stop or iris. This device alters the effective diameter of the lens-opening over a wide range.

Changing the aperture does two quite separate things simultaneously, it adjusts the amount of light falling upon the camera-tube (or film), and alters the depth of field. This we shall consider later in Chapter 14.

If the camera-tube received excessive light from the scene, over-bright areas would be over-exposed in the reproduced picture. They would appear flaring and lack modelling (burnt-out). Under-exposure due to insufficient light would produce poor shadow gradation. The lens diaphragm enables us to regulate the exposure, to suit the scene-brightness and the subject tones, in front of the camera.

For bright scenes and light-toned subjects, one tends to reduce the aperture diameter (stop-down), while for dim surroundings we shall make the most of the available light, and open up the lens to a larger aperture.

APERTURE NUMBERS. Lens diaphragms are graduated in units called stops or f-numbers. The amount of light the lens lets through will be doubled (or halved) for each complete stop changed. Such large stops are often inconvenient though, and sub-divisional numbers are used instead.

The two commonest series of lens markings are:—

English   1.4/2/2.8/3.5/4/5.6/8/11/16/22/32
Continental 1.6/2.3/3.2/4.5/6.3/9/12.5/18/25/36

In practice we shall often see a mixture:—

$$3.5/4.5/5.6/6.3/8/11/16$$

To find the difference in light passed at two apertures we use the formula:—

$$\frac{(\text{first f No.})^2}{(\text{second f No})^2} = \text{light change}$$

Thus from f.4 to f.8 the image brightness changes to

$$\frac{(4^2)}{(8^2)} = \frac{(16)}{(64)} = 1/4$$

In other words, stopping the lens down from f.4 to f.8 will require four times the scene-brightness to expose the camera-tube (or film).

This we shall have to remember when discussing depth of field later. Stopping-down necessitates considerably brighter lighting.

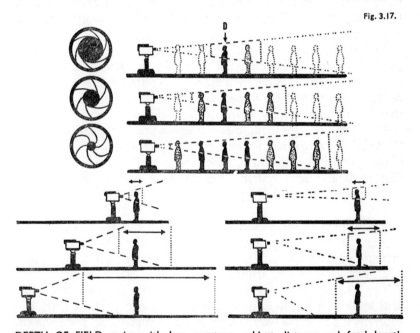

Fig. 3.17.

DEPTH OF FIELD varies with lens aperture, subject distance and focal length.

Top: For a given focal length and focusing distance (D), depth of field increases as lens aperture is reduced (stopped-down).

Bottom left: For a given focal length and aperture, depth of field increases as the camera gets further from the subject; but image size decreases.

Bottom right: For a given aperture and subject distance, depth of field increases as focal length decreases (lens-angle is widened); but image size decreases.

37

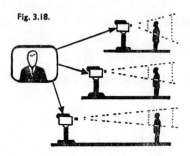

Fig. 3.18.

CONSTANT DEPTH OF FIELD. If the camera distance is adjusted for each lens angle to keep the same size shot, the depth of field will remain the same (for a given aperture).

Lenses are generally specified by their focal length (F), and their maximum aperture (f). The lower its minimum f-number, the faster the lens is said to be; for it can open up to let in more light under adverse conditions. However, lenses are not used wide open if possible. At those apertures (e.g. f.2) depth of field is restricted, and most lenses produce their best quality pictures at around f5.6 to f6.3. Larger or smaller openings than this critical aperture tend to reduce crispness and gradation in the picture.

Although all lenses working at the same stop should ideally be passing similar proportions of light, in practice this does not happen. Construction differences (e.g. the number of elements they contain) can vary their respective light-passing abilities, although their stop numbers may be measurably identical.

To obviate this problem, modern lenses are sometimes marked in 'T' numbers rather than f-numbers. These transmission numbers are obtained by measuring the relative amounts of light passing through that particular lens system. We can then make a more precise comparison between different lenses.

Supplementary lenses. Occasionally one finds that the available lens-angles are not quite suitable for a particular purpose. A slight increase or decrease may be wanted. This can be done by clipping additional lenses to the front of the present lens mountings.

The most common auxiliary is the *positive supplementary lens*, also known variously as a close-up, copying or portrait lens. This shortens the effective focal length of the main lens, so widening the angle. That can provide us with extra-wide-angle effects and facilitate big close-ups. The positive supplementary lens is easily recognized by holding it at arm's length, when we shall see a small inverted image in the lens.

The *negative supplementary lens* has the opposite action. This increases the focal length of the main lens, to which it has been clipped, so reducing the lens angle. It can be recognized by the tiny upright image seen in it at arm's length.

Where the supplementary lens is placed beyond the main lens' focal length, its effect is reversed. A negative lens now increases the lens-angle up to about $1\frac{1}{2}$ times.

Very close-shots. To fill the screen with a close-shot of a small subject we can:

Use a wide-angle lens close to the subject. (Narrower-angle lenses cannot normally focus close enough to get such large shots.) But the camera may then be so close that lighting is difficult; depth of field will be quite shallow; and wide-angle lens-distortions may be visible.

Use a positive-supplementary lens, which, by effectively widening the lens-angle, amounts to the previous method.

Use a lens extension-tube. This is a hollow metal cylinder about six inches long. Our lens screws into one end, the other fits into the lens-turret as normal. By this device the camera can now work with a narrower-angle lens further from the subject, and yet get higher magnification; so the subject can be handled or worked upon more easily. Distortion is less. But the focused depth is now very shallow, and considerably more light is necessary. Unless excessively long extension-tubes are used, none of the picture area will be masked off, and it becomes possible to fill the screen with a single tooth—when the occasion demands.

Colour values in monochrome

Like the human eye, the television camera and the photographic film do not respond equally well to all colours. They are more sensitive to some parts of the spectrum than to others. Colours to which they are particularly sensitive will be reproduced lighter than equally bright surfaces of other hues.

Most television camera-tubes and photographic materials used in the studio are manufactured with a *panchromatic* colour-response. That is, they can distinguish colours throughout the whole visible spectrum. This will seldom correspond accurately to our eye's response, but their monochromatic rendering of coloured subjects will appear natural enough.

There are camera-tubes and films which have an orthochromatic response, and so are blind towards the red/orange end of the spectrum. Reddish subjects are then reproduced abnormally dark. But we shall not usually meet this in the studio camera.

The camera-tube's colour-response affects many aspects of production:[1] make-up, set design, lighting, wardrobe. A poor red-response, for example, will cause lips to appear too dark; skin-tones blotchy and exaggerated. This will necessitate lighter shades of make-up. Where the red-response is excessive, faces appear pale and lips light. Make-up will then need to be darker (or more bluish) than normal. Although face and lip-tones are generally our quickest indication of a tube's colour response, the reproduction of all coloured surfaces will be affected.

Few surfaces have a pure spectral colour. Most will reflect an irregular section of the spectrum. So, from a blue surface, we may find a surprising proportion of red, yellow and green light reflected, were we to analyse the reflected surface colour spectrographically. Practically speaking, this means that variations over part of the camera's colour response can modify the reproduction of most other coloured surfaces.

Modern camera-tubes have a reasonably stable colour response, so that we are no longer likely to find bad changes in colour rendering on switching between cameras. But in tubes of different type or manufacture we can still encounter some variations at the red and/or blue ends of the spectrum.

Filters. Extensively used by the photographer and the motion picture cameraman, filters enable us to modify colour values or pictorial quality. They have a similar value in television, too, but currently find only limited use. Formed from special discs of optically-worked glass, or glass-mounted gelatine, they clip over the front of the camera lens.

Any filter introduces some light loss, and the working aperture has to be increased, or lighting augmented to compensate. This light attenuation is known as the filter's *factor*. A "times four" (x 4) filter, for example, requires four times the normal exposure.

Amongst the filters we are most likely to meet are:—

Neutral-density filters.
Colour filters: (i) corrective, (ii) contrast.[2]
Polarising filters.[2]
Special types[2] including: Graded filters, haze filter, night filter.
Diffusion discs.[2]

[1] An increase in the camera-tube's red-response will result in greater *sensitivity* also, due to the high red-content of incandescent lighting.
[2] See Chapter 20.

NEUTRAL-DENSITY FILTER. A grey-tinted, non-colour-discriminating filter. Where the scene is too bright for the working lens-aperture required, the overall light to the lens can be cut down with a neutral-density filter. Values range from 1/10th to 1/100th transmission. A 10% filter, for instance, will cut down the light nearly as much as reducing the lens aperture by three stops (e.g. f5.6 to f16).

Apart from occasions when sensitive cameras are used in sunny surroundings, and would otherwise be embarrassed by excessively high light levels, the neutral-density filter has other, less obvious applications.

We can open up a selected lens for a shallow depth of field (filtering to prevent over-exposure), while other lenses are stopped down for greater depth under similar lighting conditions.

Certain use has been made of integral graded neutral-density filter wheels to give continual light-control adjustment. The lens aperture itself can then be chosen for the required depth of field, and the rotary filter turned to give the correct tube exposure.

COLOUR FILTERS. Although colour filters cannot be used in a colour system without causing an overall tint, in a black and white system they enable us to modify tonal values of coloured subjects.

Corrective filters—are used with some camera-tubes to bring their response closer to the eye's. Where the tube's response to red is excessive, this may be held back with an appropriate green or blue filter. More often, a corrective filter is provided over the camera-tube face to compensate for any unwanted colour-cast in the lighting. Fluorescent light sources contain strong blue-violet radiation that can unbalance colour renderings. A Wratten No. 3 or No. 6 filter, by suppressing the blue end of the spectrum, can correct this.

Contrast filters—See Chapter 20.

The television pick-up tube's performance

From the mechanical and optical behaviour of our camera, let us look briefly at the pick-up tube itself. For its performance must influence staging practices in the television studio, as well as ultimate picture quality.

Probably the most universally used camera-tube today, the "*Image Orthicon*", has undergone many improvements over the years. Versions exist that are far less susceptible than hitherto to various spurious vision defects (see pages 48–50). At best its pictures can satisfy the most critical. It does, however, possess certain apparently inherent shortcomings that cause scenic brightness and contrast to affect the overall quality of tonal reproduction (viz.

41

redistribution effects, black-level variations). Moreover, darkest areas of the picture are speckled with marked "noise".

Still, the "perfect" camera-tube has yet to be created, and the Image Orthicon is very firmly entrenched for many years to come.

The advent of photo-conductive camera-tubes has provided compact equipment of simple, robust design. For smaller studios, where its movement-smearing tendencies are less evident, the "*Vidicon*" has proved a very stable economic camera. More recently introduced, the "*Plumbicon*" (low lag, higher sensitivity) has found direct application both for monochrome and the more exacting demands of colour television cameras.

Camera-tube sensitivity

All television cameras require a fair amount of light from the scene before they can operate efficiently. Given too little, pictures will be indistinct, smeary, lifeless, and scintillating with the grainy effect of picture-noise. In the studio there is little excuse for light-starvation, although during night-time scenes there may have to be a compromise between conflicting technical and mood requirements.

The image orthicon at its most sensitive can get acceptable pictures under light conditions too poor for the eye to see with comfort. This would permit extremely low light-levels in the studio. But such dim surroundings are depressing and uncomfortable to work in. Artistic lighting is difficult. We can no longer assess light values accurately by eye, for the camera tends to reveal slight shadows or spill-light, that are almost undetectable in the studio.

We could light to more convenient intensities, and then stop the lens down or use neutral-density filters to cut down the excess. That is often done. But small lens apertures can degrade lens performance and lead to unwanted depth of field. Filters can reduce picture definition.

Advances in camera-tube design have minimized various picture defects, at the same time reducing camera-tube sensitivity. Required light levels can vary considerably therefore, but typically 150 fc to 250 fc (1600 lux to 2700 lux) at $f4$ to $f8$ are common.

Picture quality
Picture detail (resolution; definition). When we talk of detail, we are really saying how readily we can distinguish between tiny adjacent areas of differing tones, or electronically speaking, how quickly the system can change from one tone to another during the scanning process.

Fig. 3.19.

PICTURE DETAIL. The ability of the television system to resolve small detail is checked with standard test charts. These contain gratings (*top*) or graduated resolution wedges (*bottom*). They are calibrated according to the number of lines they represent to fill the picture height. (e.g. 200 = 100 black, 100 white bars) or to the video frequency they produce in the video signal (e.g. 2 MHz, 3 MHz, etc.).

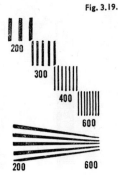

The maximum detail a particular television system can transmit is conditioned by the number of picture lines it uses. But more lines do not necessarily mean better reproduced detail. The standards of performance demanded of the whole video chain, from camera lens to home-receiver, become higher for increased picture lines. And it can happen that a nominally higher-definition system can provide picture detail that is no better—often poorer—than one using fewer lines.

But where the television equipment falls below that maximum, or a subject is not sharply focused, the reproduced picture will be to that extent unsharp. If the scene is detailless anyway, this lack of definition will, of course, be less noticeable.

Both camera and picture tubes will have highest definition near the centre, deteriorating towards the edges. Ideally, however, the resolution should not fall below the system's maximum.

Electronically speaking, detail resolution will fall off wherever video equipment suffers from a poor response to the higher frequencies; rather as sound reproduction begins to lose crispness as higher audio notes are filtered out. Certain video distortions, from phase-distortion to signal-reflections, can also be the reason for poor definition in a picture.

Optically speaking, there are several obvious causes for lack of clarity: dirty lenses, light-scatter in the lens, depth of field limitations.... And, of course, the camera may not be sharply focused!

Picture tonal gradation. Black and white are very arbitrary values. Our assessment varies with the circumstances in which we see them. White in dim surroundings may actually be reflecting less light than a black surface under brilliant illumination. In the television picture, white will be the screen brightness corresponding to the maximum

Fig. 3.20.

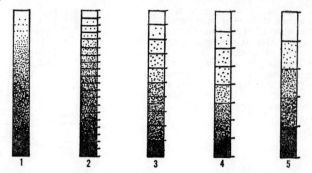

PICTURE TONAL GRADATION. Scenic and picture tones vary continuously from dark to light. But the tonal changes the eye can distinguish can be represented by a step-wedge, in which intermediate tones are grouped into steps, each proportionately lighter than the one below it. This tonal scale can be calibrated for reference purposes. A tonal wedge with too many steps would be confusing when trying to match a sample tone. Too coarse steps would give little idea whether the system has good half-tone reproduction. Any classification would be too simple to be useful. Although from the entire tonal range (1), a high quality TV system may distinguish as many as twenty different steps between black and white (2), the average picture monitor is often limited to around eight to ten tones (3), TV receivers and low-grade camera tubes may only just distinguish six to eight tones (4), while poorly adjusted receivers may not manage five (5).

video strength (picture-signal) the system can handle (usually the screen's maximum brightness, too). Black will—or should—correspond to the unlit face of the picture-tube.

Between these brightness limits lies a progressive range of intermediary tones (so-called half-tones). But our eyes cannot distinguish between any two tones (in daylight) unless one is some 2% brighter than its neighbour (doubling any light's intensity only seems to the eye to give a slight increase in brightness—the eye's sensitivity curve is logarithmic). So our eyes cannot detect over-subtle changes or differences between tones. They can be satisfied with what is, in fact, relatively crude half-tone reproduction.

Fig. 3.21.

REPRODUCED TONAL GRADATION. Fine half-tone rendition in the transmitted picture (*left*) is capable of being reproduced accurately, given suitable conditions (*centre*) or it may be degraded to a poor reproduction (*right*). But where the originating picture has inferior tonal rendition or definition, there is nothing that can be done at the receiver to improve it; and although noticeably poor on a high grade receiver, its shortcomings may hardly worsen the usual picture quality produced by a poor receiver.

Picture contrast. When discussing picture quality and staging problems, certain terms will frequently crop up in connection with tonal values. None of these is especially mysterious. Each helps us to analyse and assess tonal relationships. They take the guesswork out of picture creation.

CONTRAST RATIO is the term used when comparing the relative brightness of any surface, whether light, intermediate, or dark. Usually the tones being compared are neighbouring. If a face is four times as light as the background behind it, we will get good tonal

Fig.3.22.

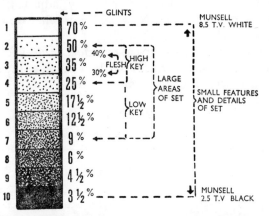

THE GREY SCALE. The figures on the left of the scale indicate grey steps: those on the right, the percentage relative light reflectance. Overall contrast range is about 20 : I, with each step $\sqrt{2}$ times the brightness of the next. The logarithmic scale looks linear to the eye.

Sample paint-cards and materials are compared on camera with a standard grey-scale chart. They can be tagged, and used to identify and interrelate tonal values. The final values of any surface will vary with: incident light intensity; shadowing; camera-tube exposure; gamma; video gain; black-level adjustment; surface angle to light and camera; surface texture; simultaneous contrast. Acceptable maximum contrasts: large adjacent areas 2 : I; large non-adjacent areas 5 : I; small adjacent areas 20 : I.

separation. There is a 4 : 1 contrast ratio between them. A low contrast ratio between them would signify that there was little difference between their brightnesses. Result? The face would merge into the background—the picture would lack "attack"—appear flat.

CONTRAST RANGE (subject-brightness range). Simply the contrast ratio between the lightest and the darkest tones in the scene or picture being discussed. The extreme contrast ratio: in a scene—the maximum the camera will have to handle; in a picture—the maximum contrasting tones in the picture being examined. The eye can accommodate a range of 100 : 1.

45

Photographic and television pictures cannot reproduce the extreme contrast ranges that we encounter in the world around us; certainly not with good half-tone rendition between these extremes. Estimates vary, but measurements suggest that for motion pictures— a maximum contrast between adjacent tones of 35 : 1 to 40 : 1 is attainable, and for the television screen—10 : 1 maximum between adjacent tones, 20 : 1 on high-grade monitors, 30 : 1 maximum between widely-spaced tones, 40 : 1 on high-grade monitors.

A contrast range of 30 : 1 is generally accepted as providing good picture quality photographically. Practically speaking, this is only attainable in television from a properly adjusted, high-grade receiver in a darkened room. Now we know what we can hope to achieve at best. What of the subject?

Let us look at a table of typical *reflectance values* for a selection of materials and painted surfaces (i.e. how much of the light falling upon them is reflected for us to see).

TABLE 3.11

TYPICAL REFLECTANCE VALUES

Percentage of light reflected   when falling upon

100	Ideal reflector.
98	Fresh snow. Magnesium carbonate.
90–70	White paint.
80–70	White china.
80–60	White paper
70–65	Light yellow paint.
50–40	Buff paint.
40–30	White skin tones. Light grey paint.
30–15	Orange paint. Green leaves.
30–10	Dark to medium grey paint.
25–10	Green paint.
20	Bronzed skin tones. Medium blue paint.
15–5	Red, brown paint.
10–5	Dark blue paint. Black paper.
5–1	Black paint.
0.3	Black silk velvet.

Actual reflectance values will depend upon the surface's finish, colour, etc., and upon the colour quality of the incident light.

From this list we can see contrast ranges that can arise. A black velvet dress against white paper could, theoretically, present a 260 : 1 contrast.

This assumes equal light falling on all surfaces. Where some areas are strongly lit, while others are in deep shadow, scenic contrast will be correspondingly higher. A 5 : 1 contrast between two surfaces could then become 50 : 1 or even 500 : 1.

Here is a situation of more than academic interest for those controlling the staging, lighting and picture quality. If the camera is presented with too high a subject contrast range, picture quality will be degraded, highlight and/or shadow detail may be lost, and picture blemishes arise electronically. The camera may only accept about 20 : 1. The picture-tube may not reproduce it.

Subject contrast must be kept down, unless tonal extremes do not matter, by limiting subject tones, or preventing high-contrast surfaces appearing in the same shot; by using even, shadowless lighting, so as not to accentuate existing contrast.

If contrast is kept to too narrow limits, the picture will look flat and lifeless. In practice one observes these precepts, but with some latitude, for too rigid control is impracticable and not always desirable.

Fig. 3.23.

TONAL GRADATION, AND GAMMA.
If the camera's output was strictly proportional to scenic tones, a continuous tone wedge would result in this video waveform (*right*).

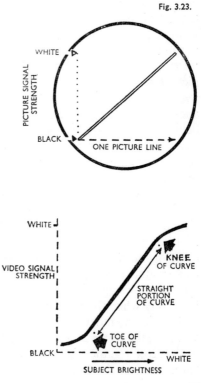

This waveform can be represented graphically (*below*). But the response of an actual camera-tube (or film emulsion) is not exactly proportional. It has a non-linear exposure curve (transfer characteristic) (*below right*). The toe causes "black crushing" so that darkest tones merge into black; exposing over the knee, or beyond video white-clipper limits, produces "white crushing", the lightest tones merging into white. Only within the straight part of the curve are relative tonal values reproduced accurately, so exposure is selected to place most scenic tones in this region.

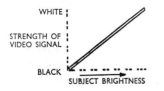

Fig. 3.24.

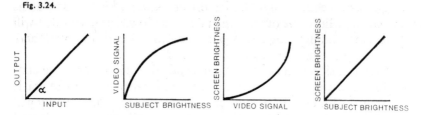

OVERALL GAMMA. From the angle of the slope, the curve's gamma can be deduced (tan α). *Left:* A camera-tube, video amplifier, or film emulsion, reproducing tones in proportion to those of the original scene, will have a gamma of one (unity). A low gamma device accepts a wide contrast range, but compresses it to fit reproduction limits, resulting in reduced tonal contrast (thin). A high gamma device accepts only limited subject contrast, this being expanded to fit reproduction limits, giving exaggerated tonal values.

Fortunately, the effect of a series of processes is multiplied, so that a *camera*-tube (*centre left*) having a low gamma combines with a *picture*-tube (*centre right*) having an inherently high gamma (about 2) to produce an overall reproduction of about unity gamma (*right*). Actually, a gamma of around 1.2 to 1.7 is usually chosen, in monochrome, to compensate for the absence of colour. In colour systems unity overall gamma is essential.

Spurious visual effects. A multitude of electronic defects can arise in the television picture. Some are created, or aggravated by unsuitable staging, lighting, video operation, etc. Others are inherent in the television system being used.

PICTURE NOISE (grain; snow). Close inspection of any television picture will show tiny, scintillating specks of light, fluctuating all over it. This is due to a spurious background of electrical disturbances, the video equivalent, in fact, of the background noise we hear behind phonograph records and A.M. radio broadcasts. In bad cases, the entire picture becomes covered with a "snowstorm". Noise is usually noticeable in television pictures, but will normally only become disturbing when the camera-tube is getting too little light, and video amplification has been increased to compensate.

BACKGROUND. One can sometimes detect in a picture a texture that remains constantly superimposed upon it, wherever the camera looks. It goes when we cut to another camera, of course. Taking the form of mottling, patches, or streaks of fixed irregular shape, this background is a shortcoming of the particular camera-tube. Background shows up worst when shooting even-toned areas of setting—especially when they become defocused.

STICKING (memory; picture retention). If the image orthicon camera-tube looks at a bright contrasting scene without movement for several minutes, a negative image of that scene will become "stuck-on" to its later shots. Aged camera-tubes are particularly

48

prone to this. Even where the foreground subject itself is active, if the camera is stationary, an image of the unmoving scenery will remain. Prolonged static captions, and visible light-sources in the scene, are the worst offenders.

Where there are no electronic orbital-scanning arrangements, the director's only solution is to have the cameraman weave slightly when on a sustained shot; to have cameras defocus when their shots are not being taken, and to avoid holding any still-shot overlong. Photoconductive camera tubes (e.g. vidicon, plumbicon) are still liable to "burnt-on" highlight images if held still overlong.

The image may remain for seconds or days. The cure is to shoot, defocused, at a bright, white surface until the image has disappeared.

STREAKING. These are clearly defined black or white smears extending across the picture, mostly aggravated by intense highlights or contrasty horizontal scenery lines. Stair-treads, horizontal stripes, palings, etc., of extreme contrast, are normally best avoided, especially when near a plain surface of the opposite tone.

DYNODE SPOTS. Appearing as white, defocused bright-spots or flares superimposed upon a picture, dynode spots are most noticeable in dark, low-key scenes and against dark backgrounds. A camera-tube design problem, they tend to be eliminated altogether only at the expense of picture definition.

HALOES (throw-off). A black aureole surrounding an over-brighthigh-contrast area, and obliterating the nearby picture; around light-sources (candle-flames, table-lamps) and reflections from shiny surfaces. Considerably reduced in certain image orthicon-tube design.

The primary solutions are to avoid high contrast; avoid (or "doctor") bright and shiny materials; keep studio lighting intensities high enough to make natural candle-flame intensities relatively less bright; use lower power light fittings in settings.

IMAGE ORTHICON GHOST. Under high-contrast conditions (e.g. light subjects against a dark background) we may see a ghostly image of the subject slightly displaced from the subject itself. (In colour cameras, similarly displaced coloured blobs can arise from "di-chroic reflections" in the camera's beam-splitting mirrors.)

CLOUDING. A form of electronic fogging, causing large dark areas to become mottled, appearing as milky-grey with light spots. Another high-contrast fault.

MESH PATTERNS (Moiré pattern). Owing to the camera-tube's scanning-beam pattern "beating" with the tube's target-mesh structure, a criss-cross pattern may appear in light scenic areas. Reducible by electronic adjustment, it cannot always be cleared without degrading picture resolution. An allied defect may take the form of flicker in highlights.

SHADING. Localized darkening or lightening over parts of the image orthicon's picture; reduced by electronic correction signals. Mainly seen when shooting plain light-toned surfaces.

In colour television shading appears as local hue variations.

REDISTRIBUTION. An image orthicon shortcoming in which light-toned picture areas modify the value of adjacent darker ones. The effect of such spurious changes range from "haloes" to shadow-thinning.

GEOMETRIC DISTORTION (non-linearity). A cramping or stretching-out, due to scanning circuits (or lens system), S curvature, "barrel", "pincushion" and trapezoidal distortions are sometimes found.

EDGE-EFFECT. Tonal contrast becomes artificially exaggerated at subject borders. This gives a slight ringing around the edges of subjects, apparently crispening definition. Aperture correction circuits also can produce a false black edging effect.

EXPOSURE-TIME. The image orthicon camera-tube has a relatively long exposure-time. It tends, therefore, to reproduce fast-moving objects as a blurred image; especially when short of light. This smearing or trailing is very clearly seen with the vidicon camera.

BLACK STABILITY. Ideally, a camera-tube should put out a picture signal that directly represents what it sees. It should not add spurious signals of its own. All its tones should rise from a stable black-level datum. Several kinds of camera-tubes do not have this stability. With the image orthicon there are large variations in the signal produced for a black subject area. Owing to electron-redistribution effects in the tube, we get no intrinsic black-level datum at any point in the picture. In other words, black wanders according to other lighter-toned areas appearing in the same shot. The vidicon, once adjusted, will put out a regular picture signal strength for black, as does the plumbicon. As in a colour camera variation of any colour channel affects overall fidelity, stability is essential.

Electronic clamping circuits reduce the wide variation in picture tones that an unsteady datum produces; but the cure is not a complete one.

Picture defects

Picture defects caused by the receiver

Most people watching a television programme take the picture and sound quality as they find it. If detail is hard to see, or tonal gradation poor, there is a tendency to lay the blame on the studio originating the programme. Sometimes they may be right, but the likelihood is that certain defects are originating in the transmission process, or within their own receiver. Even extremely costly high-grade monitors in the studio centre itself will have their vagaries, despite regular expert maintenance and skilled adjustment. Let us outline briefly the chief ways in which picture quality can become degraded in the received picture.

CONTRAST may be excessive, or insufficient (receiver adjustment, or weak reception).

BRIGHTNESS may be excessive—causing poor modelling in light tones, or poor focus; or it may be insufficient—causing poor tonal range; dim picture, unmodelled dark tones.

FOCUS may not be sharp overall, showing all picture lines clearly.

SHAPE AND SIZE. The picture may be too high or wide, its edges lost behind mask; non-linear.

INTERLACE. Picture lines may not be evenly spaced; pairing; coarsening picture lininess; lack of definition.

NOISE-LIMITER CIRCUITS may degrade highlight and light-toned modelling of surfaces.

BLACK-VARIATIONS. Picture black may not be constant. The darkness of the blackest tones in the picture may vary from shot to shot, with picture content. Black backgrounds become reproduced as dark grey. Studio lighting changes, or big tonal movement, cause the picture to become lighter or darker. Pictorial quality and mood vary. Studio lighting seems very variable; low-key scenes look poorly lit. These are symptoms of inferior receiver performance which are the bane of the creative artist.

TEARING. Jagged-edged detail; horizontal displacement of picture lines. Due to received interference; line-hold maladjustment; weak reception.

GHOSTS AND REFLECTIONS. Displaced images, ringing, etc., due to reflections as received signal bounces off nearby obstructions; to aerial imperfections, etc.

SNOW. Picture noise, arising from weak reception conditions.

INTERFERENCE PATTERNS. Stripes, herringbone patterns, splashes, spots, tearing, etc., due to electrical interference, instability, mistuning, etc.

COLOUR DEFECTS. Broadly, colour defects derive from *misregistration* (colour fringes); colour *shading*; exaggerated or reduced *saturation* due to incorrect contrast/brightness adjustment. *Hue controls* wrongly set, causing overall colour cast, or hue shift.

Inherent picture defects

A few picture defects are quite unavoidable. They are inherent in the television process. Where they prove too distracting, the director must either accept them or alter the shot concerned.

FLICKER. Pictures are televised at a rate of 30 (25 Britain and Europe) complete images per second. Effectively doubled by the interlacing process, this rate is normally too fast for the eye to perceive. Persistence of vision causes the pictures to appear continuous. However, when highlight-brightness is increased considerably, e.g. on a bright screen, we shall notice flicker in highlight areas.

An interlace-flicker effect between the odd and even sets of lines comprising the picture is noticeable in certain circumstances.

Flicker of a special kind is often visible on patterns made up of close, narrow, horizontal lines. Where clothing fabrics, brickwork, tiling, venetian blinds, etc., are seen at such a distance that their surface appears as close horizontal lines, *line-beating* or *strobing* will occur. It happens when the subject's lines coincide with, and fall outside, the picture's scanning-lines. A change in the length of shot, or slight defocusing, may cure or reduce the effect.

STROBOSCOPIC EFFECTS. Any fast rhythmical movement (e.g. a rotating wheel, pulsating light) will not be shown by television at its true speed. Because of the television scanning-process, a wheel rotating at 60 revs. per second (U.S.A.) or 50 revs. per second (Europe) will actually appear to stand still. Slightly faster speeds make it appear to move slowly clockwise; slightly slower gives counter-clockwise movement. The effect is similar to that seen in motion pictures, where the wheels of fast-moving vehicles turn slowly backwards.

INTERLACE CRAWL. As our eyes look up or down a television picture, the illusion of interlacing is destroyed. The line structure becomes coarser; vertical definition is halved. The effect comes from the relative speeds of our eye movements and the frame-scanning speed. When our eyes settle, or look over the picture horizontally, the interlacing reappears.

HORIZONTAL BREAK-UP. Whenever well-defined vertical or horizontal detail moves across the picture, it will break up. On close inspection, we see that this detail has disintegrated into two displaced sets of images, corresponding to the odd and even sets of scanning lines. This happens because of the inherent time-lag between these scanning periods (1/60 or 1/50 sec.).

4

PICTURE CONTROL

ATTRACTIVE picture quality is not easy to define. It does not necessarily imply being able to see all detail and tones in our subject clearly. A needle-sharp picture containing a wide range of tones can represent a technically accurate interpretation of the scene. Yet, if it does not have the persuasive strength to attract, direct and hold the viewer's interest, it has failed artistically.

Scenery and lighting can build up atmospheric effect, but whether this effect will be conveyed by the camera depends largely upon the awareness and skill of the video operator.

The technical need for picture control

The first step in all systems is the technical line-up of the camera equipment. Fundamentally, this involves placing test charts in front of the camera to check the camera channel's performance; adjusting the camera-tube's supplies to give optimum definition, tonal gradation, etc.; examining various electronic parameters. Details need only concern the video engineer. This done, the video operator has two sets of limits within which he must work. Those of the camera-tube and those of the video equipment.

The camera-tube's exposure limits

He must prevent the camera-tube from receiving too much or too little light from the scene. For this he has light control adjustment by introducing a neutral-density filter of suitable value to control the average light level, or by selecting an appropriate lens aperture.

Subject tones are only reproduced proportionally if they fall upon the straight portion of the camera-tube's exposure-curve, between its toe and knee (Fig. 3.23). Clearly, the longer this portion, the greater the tonal range that can be accommodated and reproduced.

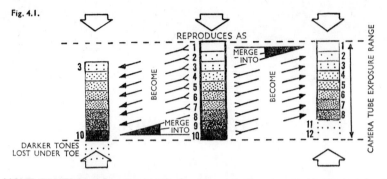

Fig. 4.1.

REPRODUCES AS

DARKER TONES
LOST UNDER TOE

CAMERA TUBE EXPOSURE RANGE

LIGHT CONTROL. By adjusting the lens' aperture (stop) and/or its neutral density filter, the overall brightness of the image on the camera-tube is altered.

Left: Under exposure moves scenic tones down the exposure curve and results in poor shadow gradation, lowest tones merging, considerable picture noise, picture definition deteriorating (trailing), tube blemishes (background) apparent, faces and light tones being reproduced darker (unless video gain increased). *Centre:* Correct exposure. *Right:* Over exposure results in lightest tones merging (crushing over the knee), light areas looking over-bright (flare), faces lacking modelling (plastic), picture defects (haloes, etc.) arising; *but* more detail being revealed in darkest scene tones 11, 12.

With the image orthicon camera-tube, the length of this straight portion varies with electrical line-up. Adjustment producing a good long straight portion can lead to noisy (grainy) pictures. A shorter linear region makes for noise-free pictures but coarser tonal rendition. Performance is a compromise, therefore.

A camera-tube's gamma can also vary with electrical line-up and with changes in relative proportions of tonal masses in shot.

For high-key scenes, the exposure-curve may become steeper (i.e. of higher gamma). Such changes are often referred to euphemistically as the tube's 'dynamic characteristic'.

Conversely, the photo-conductive camera-tubes possess a substantially linear, stable characteristic, without "toe" or "knee" regions. Circuits prevent highlights from peaking excessively in the video.

Video equipment limits

To prevent the camera's picture-signal from exceeding the electrical limits of the video equipment, clipper circuits are fitted. Several controls help the video operator to keep the televised picture within these limits: video gain—black-level (sit; set-up)—gamma adjustment (in monochrome systems).

If extreme subject tones are allowed to go beyond the camera-tube's or the video equipment's limits, they will be merged irrevocably into white and/or black masses.

The precise pictorial effect any of these adjustments will have, however, can only be estimated through skill and experience.

54

Fig. 4.2.

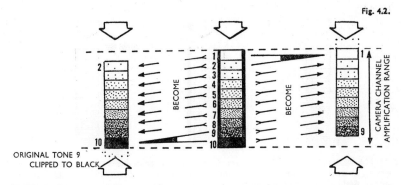

ORIGINAL TONE 9
CLIPPED TO BLACK

VIDEO GAIN. Adjusting amplification of the picture signal moves it up and down the camera-channel's transfer characteristic, between the white and black clipping limits.

Left: Insufficient amplification (undermodulation) results in pictures lacking "snap"; lightest tones too dim; lowest tones merging, (9) into (10) cut by the black clipper limit (but no camera-tube defects or picture-noise develop). *Centre:* Normal amplification. *Right:* Excess amplification (overmodulation) results in tones (1) and (2) being clipped to white by the white clipper limit, lightest tones merging, light areas looking over-bright, faces lacking modelling (but no camera-tube defects arise or lower subject-tones become visible. Picture-noise increases).

Picture Control is needed:

Technically—for pictures of good, consistent, technical quality.
Artistically—to interpret scenic and lighting treatment.

Opinions vary on the best method of picture control. There are two approaches: *fixed modulation* and *peaked modulation*.

Fig. 4.3.

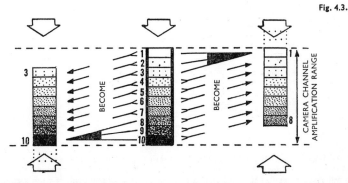

BLACK LEVEL (set-up, sit). Adjusting camera-channel black level moves the picture tones down or up the tonal scale, the effect being most noticeable in darker tones.

Left: Set down (batting down): all dark tones clipped to black, light tones down-scaled (unless compensated for by video gain increase). No lighter subject tones become visible. *Centre:* Normal set. *Right:* Set up. No dark greys or black in the picture; lightest tones may be merged, cut off by the white clipper limit. No lower subject tones become visible through set up.

Fig. 4.4.

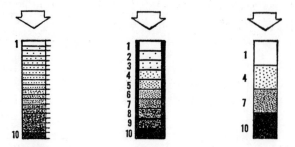

GAMMA ADJUSTMENT. This alters the subtlety or coarseness of tonal reproduction. *Left:* Low overall gamma—picture-noise increases with correction. *Centre:* Normal gamma. *Right:* High overall gamma.

Although in some situations we may set up the camera channel and leave it without further readjustment of its video controls, this will not usually produce optimum picture quality. The picture-signal unavoidably changes as cameras look around the scene:—

Subject tones of varying tone and contrast come into shot.

Pictorially-attractive lighting will rarely be of even intensity in all directions.

The brightness of most surfaces will alter with the camera's viewpoint. The brightness of smoother surfaces falls at acute shooting-angles (Fig. 5.14).

Fig. 4.5.

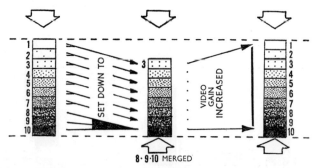

PRACTICAL PICTURE CONTROL. In practical picture control, combinations of these adjustments will be used. For example, normally a scene will be exposed (light control) to put required subject tones in the straight portion of the camera-tube's curve; video gain set to allow highlights to just clip (at white); black level set to allow shadows to just clip (at black), and gamma selected for good tonal gradation in important subject tones.

For special effects (e.g. to enhance a night shot), the scene will be properly exposed; black level then set-down to clip off the lowest tones, and video gain increased to correct the lightest tones to normal. Reduced tones (1) (2) (3) are now corrected, original (8) (9) (10) still merged to black, and tones (1-7) expanded to fill remaining tonal range.

56

Surfaces can appear to lose tonal contrast and brightness as they become defocused.

Effective surface brightness can fall as camera-distance increases.

We can adopt a fixed modulation approach to picture control and leave the cameras and their channels set for the maximum brightness and shadow they are likely to receive. Gradation in lighter tones will crush out at times; shadows and dark tones occasionally remain under-exposed.

This approach to picture control is likely, therefore, to result in variable picture quality. If the onus is placed on lighting and staging to keep tonal values within the camera's exposure range, there will still be the physical variables listed earlier.

Finally, there will be the unwanted tonal variations arising from the camera-tube's own vagaries. These can take the form of black-level and gamma changes, electronic redistribution effects, shading, etc.

The camera-tube and the picture-tube can only reproduce relatively restricted tonal ranges. There is everything to be said, therefore, both technically and artistically, for making full use of the range that *is* available, and fitting it to the subject. This leads to what might be called a peaked modulation approach to picture control. This is a method by which gradation in unimportant highlights and/or shadows is sacrificed deliberately to obtain better gradation in other tones. We expose for the most important subject tones; and that will normally mean faces.

By permitting controlled over-exposure, limited subject areas such as specular reflections from shiny surfaces, incidental papers on a desk, whitish collars, can be allowed to crush over the camera-tube's knee. This will leave the straight portion available for better gradation in faces, intermediary tones and shadows. Working too high up the curve, however, will cause even the face-tones to become over-light and unmodelled.

Without care and foresight, this control method could be disastrous. If, for instance, we allowed a light grey surface such as a khaki collar to crush out as an incidental highlight, then any lighter-toned surface coming into the shot (e.g. blonde hair; an off-white dress) would become a detailless white mass too.

If we have to reproduce good gradation in very light-toned surfaces, this can only be done in a system of restricted contrast-range, at the expense of darker picture tones. The effect nevertheless may still be quite acceptable. Light, gauzy ballet dresses, for example, would lose their nature entirely if allowed to crush out. But exposing them correctly may cause face tones to look dark, while shadows and darker tones merge.

Although allowing darkest subject tones to merge will permit better exposure for lighter tones, we must not lose important detail. In a dress display, it would obviously be foolish to allow modelling that showed form and texture of dark material to be lost. In a drama, on the other hand, dark suits may remain unmodelled masses quite readily, in order to achieve better face tones; remembering, though, that under-exposure of large dark areas can lead to distracting picture noise.

The artistic aspects of picture control

This brings us to the artistry of video operation. For, in using the system's tonal range selectively, the question of choice naturally arises.

We may decide to take incidental white peaks to the upper tonal limit. But what are these actually to be? Exactly what is to be crushed to provide better overall gradation? Is it the whitest thing in the picture? As our subject sits writing, are we willing to lose detail in the paper he writes on, his fair hair, the globe of his reading-lamp? The choice is arbitrary, and not always obvious.

Similarly, we shall be taking the lowest picture tones to the black-level. But what will they be? Do we want to see detail in his dark suit, the deep shadows ...?

We cannot expect to get subtle gradation in both the paper and the shadows, for the system will probably not be able to expose and reproduce that contrast range. Clearly, the solution will depend entirely upon the subject, the mood, and the atmosphere that we hope to engender.

Video operation has several important artistic aspects. To provide a constant flow of tonally-matched pictures, so that faces do not appear light in one shot, darker in the next; so that there is atmospheric continuity (the scene must not look like evening on one camera, morning on the next; high-key from one viewpoint, low-key from another).

Despite precise lighting, such anomalies can develop through changes in proportions of picture tones and through simultaneous contrast; causing subjects to look lighter when against a dark background, and vice versa.

Fig. 4.6.

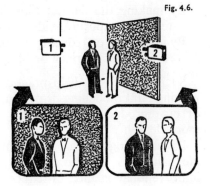

UNMATCHED BACKGROUNDS. Cutting between a predominantly low-tone shot (1) and a predominantly high-tone shot (2) can produce changes in apparent subject tones and picture brightness.

Exactly what constitutes the mood of a picture is difficult to formulate, but several tonal predominances have become associa-atively linked in most people's minds, and these cannot be ignored in lighting and picture control.

A *low-key* where grey to black areas predominate, suggests evening; cosiness; tragedy.

A *very low-key* where dark grey to black areas predominate, suggests night; mystery.

A *high-key* where white to mid-grey areas predominate, suggests cheerfulness; comedy.

A *very high-key* where white to light-grey areas predominate, suggests delicacy; ethereality.

Our impressions of quality will also be influenced strongly by the picture's tonal range. High pictorial quality will normally involve our seeing a full tonal range, from rich whites to crisp blacks. Where the picture uses only part of the complete white/black tonal range, typical observation will be that a picture without dark tones looks thin, lacking body (suitably applied, may be ethereal). A picture without lighter tones looks dull, lacks snap, vigour, sparkle (appropriate, perhaps, for foggy locales). While a picture having few half-tones looks harsh (excellent, perhaps, for highly-dramatic, sordid scenes).

Picture control can prevent such pictorial effects occurring inadvertently—especially in shots that lack a wide integral tonal contrast; but it can enhance these effects when they are required. This the video operator can do:

By making the lightest tones reproduce lighter (or darker), by greater exposure or video-gain.

By making the darkest tones reproduce darker (or lighter), by reducing exposure, video-gain; by black-level adjustment.

By increasing or reducing contrast at either end of the tonal scale (e.g. by black-stretch or white-stretch gamma adjustment).

Such control can help practically to improve *picture clarity* by making faint modelling or patterning (e.g. in textiles) more evident, or by reproducing low-contrast detail (e.g. in faded documents) more clearly.

By increasing contrast in extreme tones, we can merge indeterminate half-tones into homogeneous masses; and so we can often improve tonal balance, or reduce distracting scenic features. Wherever, for example, a black surface is intended to appear as a flat, even mass (for a caption, or in black scenic backgrounds), the darkest tones can be merged by black-level or gamma adjustment.

Sometimes a visual effect cannot be created entirely by lighting—either because that would create or intensify picture defects (through excessive contrast), or because facilities or conditions prevent more suitable light-adjustment. In these circumstances, video-operation may be able to improve matters.

A night-exterior shot, if lit with appropriate intensities, would leave the camera-tube short of light. The picture would be hopelessly under-exposed and hence noisy, patchy, with poor definition, smeary, etc. A better expedient is to use a scaled-up lighting balance of sufficient intensity for the camera-tube to function well, and then cheat the reproduced tonal values by video adjustment. It may be sufficient to lower the camera channel's black-level (i.e. to "Bat down on blacks"—U.S.: "Sit the picture down"—Brit.). Perhaps increasing the gamma and the video-gain would produce a more convincing effect. The result is a combination of lighting and video skills.

Because it is wearisome to watch a dark picture over-long, a motion picture practice is to fake the picture-density during the printing process. The scene is printed dark (perhaps over-dark) to establish the mood, and then gradually stepped up to a more comfortable level as the scene continues. The same principle can be applied in television and so help dissuade the viewer from readjusting his receiver.

5

TELEVISION LIGHTING

TELEVISION lighting is neither a mysterious ritual, nor yet an over-rated form of illumination. It has two major considerations: the mechanical and the artistic.

Mechanically we are concerned with what lighting equipment can do, and what happens when it is used in certain ways. This can be appreciated from study and experience, for the mechanics of lighting essentially follow definable principles.

The artistic aspects of lighting are harder to discuss, for they involve the question of how lighting treatment can arouse an emotional response in our audience, and this is easier to demonstrate than explain.

Unlike motion pictures, where each camera set-up can be lit for the most pleasing effect, television lighting must be a compromise. Production may be continuous, without retakes; subject positions are more arbitrary. The lighting has to satisfy the technical needs of the television camera, avoid spurious shadows, and not impede other studio operations. And while doing this for varying camera angles, however complex the production, it must continue to capture the atmosphere of the occasion, and show the scene to best advantage.

Studio lighting equipment

Most television lighting equipment consists of a combination of regular motion picture fittings, together with apparatus especially developed to meet television's particular needs.

These sources produce light of two qualities:

Soft light—coming from broad sources that are predominantly diffused and therefore shadowless.

Hard light—casting pronounced shadows, and highly direct-ional.

61

The division is not a rigid one, but soft light will generally originate from diffused or frosted lamps in an open-fronted reflector, while hard light is produced by focused spotlights.

The luminants used are already familiar to us in everyday applications: the incandescent tungsten lamp, carbon arc lamps, and fluorescent lamps.

Incandescent tungsten lamps (inkies)

The most extensively used luminant, the tungsten lamp has many advantages: it has a reasonably long life and a wide range of intensities; it can be mounted in a wide variety of housings; it is reliable and requires little attention.

Against this, tungsten lamps waste much electrical energy in heat, have proportionately large filament areas, precluding the hard, clear-cut shadows that point sources of light provide. Their light is normally deficient at the blue end of the visible spectrum, i.e. of a low colour temperature.

Incandescent soft light sources. To provide a truly soft light, a source must have a light-emitting area that is many times larger than the lit subject. We cannot obtain completely shadowless light from small-area or point sources, although a sufficient approximation may sometimes be achieved by interposing some kind of diffusing screen. Unfortunately this would cut down the available light to a fraction of its original intensity.

The more lamps included in a fitting, the more will their individually-cast shadows overlap, and the softer the light become. But the television camera is often able to discern even these faint shadows so that, despite losses, diffusion screens may still be needed.

Fig. 5.1.

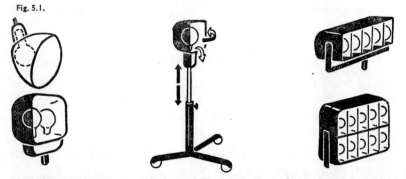

SOFT LIGHT SOURCES. *Top left:* Scoop (500 w. to $1\frac{1}{2}$ Kw.). *Centre:* Broad (broadside) in telescopic castored stand (500 w. to $1\frac{1}{2}$ Kw.). *Bottom left:* Double broad (1 to 3 Kw.). *Top right:* 5-light (striplight; border-light; trough). *Bottom right:* Bank (cluster).

Fig. 5.2.

 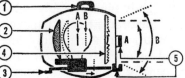

THE FRESNEL SPOTLIGHT. Available in 100 w., 500 w., 2 Kw., 5 Kw., 10 Kw., and 20 Kw. ratings. A, spot position (10° coverage). B, flood position (60° coverage). (1) hoisting loop; (2) parabolic mirror; (3) focusing handle; (4) Fresnel lens; (5) frontal brackets for lamp accessories.

Incandescent hard light sources. SPOTLIGHTS. Focusable spotlights combine a lens and parabolic mirror to control and concentrate the light from the lamp into a narrow beam of adjustable spread. While the main area of the beam is of uniform brightness, its edges are often made to fade away gradually; these soft edges enabling the borders of adjacent beams to be blended smoothly.

INTERNAL REFLECTOR (SEALED BEAM) LAMPS. By making the envelope of the lamp itself parabolic in form, and internally silvering its inner surface, a much cheaper form of light fitting can be devised that requires no housing. Although the beam of this pre-focused lamp cannot be adjusted directly, one can clip on accessories to restrict its spread. Barndoors, snoots, diffusers, all increase its potential applications.

Available in sharply-focused and semi-diffused forms, internal reflector lamps can be clipped or clamped to settings and standard lamp fittings. Supplementing larger lighting units, these lamps may be used singly or in clusters. Power ratings cover a range including 100, 150, 300, 500, 1,000 watts.

THE ELLIPSOIDAL SPOTLIGHT. By adjusting the design of the fresnel-spotlight so that the lamp's filament is now situated behind the lens's focal point and backed with an ellipsoidal reflector, a fitting with new, different features is obtained.

This ellipsoidal spotlight can provide us with several additional means of light control: metal stencils can be slipped in at its focal plane to give a variety of shadow masks (e.g. for window-shadow effects); internal shutters give hard-edged local beam cut-off, and adjust beam shape; a removable internal iris enables the lamp's brightness to be controlled.

Fig. 5.3.

INTERNAL REFLECTOR SPOTLIGHT (Sealed beam). Has a mirror-silvered internal wall coating, and a clear or frosted front.

63

OVERRUN TUNGSTEN LAMPS. When a tungsten filament lamp is fed from a slightly higher supply voltage than it is normally designed for, two important things happen. We obtain a greatly increased light output, and a markedly improved colour-temperature, the usual red-yellow light becoming noticeably bluer.

The penalty for this is a short life—if a gay one—two to nine hours against the usual thousand hours.

Where there is insufficient space for a powerful standard light fitting, or where practical lamps in a setting need to be of high intensity, the overrun tungsten lamp in conventional or sealed beam forms, can be invaluable. It may augment location interior lighting. TUNGSTEN-HALOGEN (TH) or QUARTZ-IODINE (QI). These have several advantages over conventional tungsten lamps, deriving improved performance from a halogen gas filling.

This gas, iodine or bromine, around the tungsten filament "recaptures" any vaporized metal instead of its progressively blackening inside the bulb. Light output becomes more constant, colour temperature being around 3200 K. Smaller than the normal lamp, the TH version offers designs for long life or for normal-life, high-output applications.

Carbon arcs

Although the widest-used light source in motion picture studios, the carbon arc has only limited (if important) application in television.

Arcs have the drawback that they need skilled adjustment, use trims (carbon rods) of limited burning time, and are bulky.

It provides a very concentrated point source; hence sharp, clear-cut shadows and crisper, more dynamic modelling than incandescent lamps. Where a high light-intensity is needed over a large area, the arc is without parallel, especially for single-source effects (sunlight, dance spectacles, etc.). Finally, the light's colour-temperature is high; close to that of sunlight.

Fluorescent light sources

The familiar fluorescent tubular lamp has for some years been used as a soft light source in many studios. It has many attractive features, for it gives: three times the light of a similar wattage tungsten lamp, has a long life, and an adjustable colour quality, according to the tube coating selected.

Against this, the lamp is only obtainable in low power ratings and is essentially non-directional, bulky and fragile. Original problems in dimming adjustment have been overcome by thyristor control circuits.

Fig. 5.4.

FLUORESCENT SOFT LIGHT SOURCE.
A shallow housing, containing tubular
fluorescent lamps, provides a large area
diffuse source.

It remains, therefore, a light source which in certain installations offers economy and simplicity; but has limited artistic applications.

Lamp mountings

Exactly how studio lamps are held in their chosen positions will influence the flexibility of the entire system. It must affect: the time and effort required to set up the lighting rig, its adaptability to variations in settings and production treatment, and the elaboration possible in lighting treatment.

Adjusting lamps from a step ladder is tedious in a heavily rigged production. Instead, lamps can be designed with hooked or "T" controls, to be operated by a long flexible pole.

Lamp-rigging approaches range from specifically hung lamps (for a custom-built lighting plot) to "blanket rigging", in which the entire studio is hung with dual-purpose lamps. Here the Lighting Director selects appropriately positioned lamps, which by adjustment provide soft or hard illumination as required. Such arrangements may be the inevitable economic compromise in future studios.

Floor mountings. Wheeled telescopic stands, from 6-inch dwarfs to 15 foot hydraulic giants, have considerable flexibility, for their height and direction can be adjusted quickly and easily. In motion pictures they provide the precise lamp positioning that is necessary for high-grade lighting.

But in television, where floor space is at a premium for camera and sound-boom manoeuvres, floor lighting is unpopular. In fact, where lower quality lighting is tolerated, floor mountings may be excluded altogether. Under well-organized conditions, however, floor mountings and their cables need not prove inconvenient, and a good compromise can usually be reached.

Lighting accessories and light control

When selecting a lamp for a particular purpose, there are three main things about its performance that interest us: the intensity (brightness) of its light, the quality (hardness, softness), and its coverage (spread) from the required position.

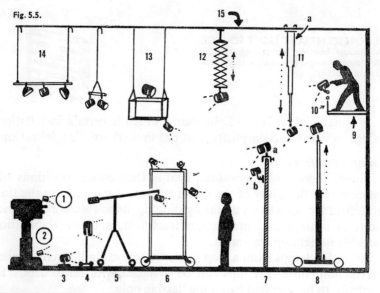

Fig. 5.5.

LAMP MOUNTINGS. (1) camera eyelight (headlamp); (2) underlighting from camera (100-500 w.); (3) turtle; (4) telescopic floor stand; (5) small boom light; (6) spot-rail (light tower), tubular scaffolding; (7) scenic flat with lamps clamped (a) to the top and (b) to its surface; (8) hydraulic extensible stand; (9) lighting gantry (catwalk); (10) lighting rail; (11) telescopic pipe (skyhook)—(a) sliding along power-rail supplying lamps; (12) pantograph (lazyboy); (13) tubular cradle, supporting heavy apparatus, arcs, etc.; (14) suspended barrels or pipes (counterweighted, or electric hoists); (15) girders, ceiling-hooks, overhead pipe-grid, or suspension bars.

Having chosen the most suitable type from those available, we shall want to adjust the light from it so that it illuminates only the area of the scene we wish it to light, at the appropriate intensity.

For this final adjustment, various lamp accessories have been developed. Most of them originated in motion picture studios, and evolved from local gadgetry to respectably standardized forms. But they still retain their colloquial names, and are often improvised as the need arises, even to-day.

Methods of light restriction. Although diffusers are effective (if uneconomic) in softening soft light even further, their primary effect upon a hard light source is to reduce its brightness.

Sliding them in front of the lamp housing, or holding them as scrims (translucent flags) in the light beam, we can adjust the effective intensity of over-bright lights, instead of employing more costly dimmer circuits.

More important, we can reduce the light falling upon localized areas. By diffusing part of the light beam, we can reveal detail in a dark dress, without over-lighting its wearer's face.

We can cut down the light falling on surfaces near the lamp, while enabling more distant parts of the setting to receive the lamp's full intensity. Without partial diffusion, we would have to use two separate lamps to give the same lighting balance.

GOBOS: Large boards placed before a lamp to prevent fresh light spilling over selected areas, or causing flares in the camera lens.

FLAGS: Small gobos (e.g. 1 sq. ft.) affixed to angle arms, for the same purpose.

TARGETS, BLADES, DOTS: Tiny flags (e.g. 6 sq. ins.) of rectangular or circular shape.

LIGHT DIFFUSION. Translucent media placed in front of lamps enable their light to be reduced and diffused. Typical materials include: Spun-glass sheeting; wire mesh (wires); gelatined wire mesh (jellies); frosted and opal glass; cello glass; oiled silk (silks); each causing its own characteristic light change.

REFLECTORS. Light bounced from reflector boards is sometimes used to augment direct lighting. Its colour and hardness will depend upon the reflector's surface. Clumsy and uneconomical, reflectors are chiefly used as boosters to fill in deep shadows and reduce lighting contrast, when working in sunny locations.

Within the television studio, reflectors can be used to simulate glare in snow scenes, and to produce certain ethereal effects.

Fig. 5.6.

RESTRICTING THE LIGHT. *Left:* Barndoor. Independently adjustable flaps on a rotatable frame, to cut off light-beam selectively. *Centre Left:* Spill Rings. A series of shallow concentric cylinders, restricting edge-spread of light-beam. *Centre Right:* Snoot and Funnel. These restrict the spread of spotlights to a localized beam. *Right:* Cookies. Inserted into lamp spigot. Made from opaque or translucent materials, they transform hard light-beams into dappling. Stencil sheeting can be slung or frame-clamped before lamps. Similar effects can be created with metal-foil stencils in projector spotlights.

Methods of controlling light intensity. To achieve a pleasing lighting balance, we must be able to control the relative brightness of the lamps lighting the subject.

This can be done in several ways: by changing the lamp/subject distance; by the type and power of lamp we choose; by introducing

Fig. 5.7.

MECHANICAL CONTROL. *Left:* Venetian Shutter. A set of adjustable louvres (hand or motor operated) controlling the light-flow. *Right:* Adjustable Iris. By adjusting the aperture, a spotlight's intensity can be varied.

diffusers; by adjusting the amount of concentration (i.e. by flooding or spotting) in the case of spotlights.

Each of these solutions has its limitations.

For a fixed lighting set-up, such methods of intensity control are satisfactory enough. But where we want atmospheric and mood changes, for dramatic and scenic effects, and where our aim is a sensitive lighting balance, we must be able to control the brightness of selected lamps from maximum to black-out.

Both mechanical and electrical methods enable us to do this.

ELECTRICAL CONTROL. For tungsten-filament lamps, dimmers provide an extremely flexible control of light intensity.

Several types are available, as we shall see, each with its own merits and shortcomings.

The elaboration of dimmer facilities and their associated switch-gear will vary from one studio centre to the next.

For less ambitious productions, mobile dimmer-boards, handling 10-25 kw., will suffice.

More complex installations may run to permanent lighting panels handling up to 200 kw., and able to be coupled for straight-forward group-cueing and effects operations. But the most comprehensive facilities are to be found in the complete electronic lighting consoles used by several major networks, handling 300 kw., or more.

These possess enormous flexibility, with switching and intensity adjustment of single and group lamps, at variable speeds; automatically memorized group selection; automatically timed fades; etc., etc. Operations can be stored, too, on cards or tape.

How great the need for such elaboration really is—and how reliable—is conjectural. Ultimately, we could end up with lighting being switched to suit each new shot, in filmic style. But we might equally well find elaborate mechanization substituting automation for art.

Basic lighting principles

These, then, are the tools available to the lighting director. Now to the ways in which he uses them, and the principles underlying his art.

TABLE 5.I.

METHODS OF ELECTRICAL CONTROL

Method of dimming	Principle	Advantages	Disadvantages
Resistance Dimmer	Lamp current varied by adjusting amount of resistance wire in series with lamp.	Simple. Medium Cost. Reliable. Suitable for A.C. and D.C.	Power dissipated as heat in resistance wire. Fairly bulky. For even dimming, small load must not be fed from high load dimmer.
Auto-Transformer Dimmer	Iron-cored coil across the power supply; variable voltage can be tapped off to feed lamp.	Cool working; compact; reliable; smooth control with varying load. Excellent power economy.	Suitable only for A.C. Relatively expensive. Bulky.
Saturable Reactor. (Reactance Dimmer) (Choke Control)	An iron-cored coil in series with an A.C. fed lamp will cut down its current; and hence lamp brightness. By adjusting a D.C. control-current flowing through an overwound control-coil, the impedance can be varied.	A small D.C. control supply can be used remotely to adjust large loads. Fairly low cost. Occupies small space.	Suitable only for A.C. Heavy. Dimming action can vary considerably with loading.
Magnetic Amplifier (low current system)	A derivation of the saturable reactor. An auxiliary D.C. amplifier adjusts control-coil current, using feedback developed from the load circuit.	Smooth control for various loads.	Suitable only for A.C. Small dependence on load.
Silicon-Controlled Rectifier (Thyristor)	A semi-conductor device which controls lamp current according to the timing of a stream of electrical "gating" pulses applied to it.	Lightweight; smooth control with varying load; compact; high efficiency. Medium to low cost.	Can generate interference: (a) acoustical, as lamp filaments vibrate at audio frequencies; (b) electrical noise induced into nearby mic. cables. Filter circuits and special mic. cables reduce.

Light has several aspects that we need to consider here:

Its direction—the angle at which it strikes the subject relative to our viewpoint;

Its coverage or distribution—the area lit by the source;

Its intensity, and

Its quality—how hard or soft; whether white or coloured.

Light direction will determine how the contours of a subject are revealed. Surfaces face-on to the beam will be brightest, while planes at an angle to it, or hidden from it, will be proportionately darker.

This results in tonal gradation, which the eye interprets from experience as due to surface contours.

A subject's appearance will depend largely upon which facets we choose to reveal and which to conceal; which we emphasize, which we subdue.

This, in essence, is the Art of Lighting.

Light and shade alter our assessment of size, shape and distance. And, as we shall see, the illusions they create have widespread applications, not only in lighting, but in the design of settings, make-up and costume.

Hard light produces shadows, which may be cast across adjacent planes; revealing surface undulations; or revealing its texture.

Hard light can be restricted to localized areas, but its clear-cut, vigorous qualities may prove too harsh, and more strongly contrasted than we require.

Soft, diffused light is less readily localized, but has numerous practical features:

It avoids distracting multiple shadows that numerous hard sources would give.

It can lighten hard light shadows, without creating further shadows.

It can, therefore, reduce tonal contrast, or excessive modelling in the subject.

It can suppress unwanted subject-modelling, preventing wrinkles and bumps from becoming prominent, whether in skycloths or faces.

It can provide the shadowless base-light (foundation light) with which to flood the studio scene.

Soft light has a plastic nature, providing subtle half-tones, delicate gradation. Excessive soft light can produce flat, muddy pictorial quality, however, and a careful blending of hard and soft will generally be needed.

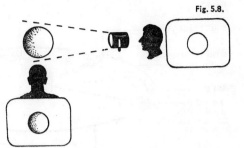

Fig. 5.8.

RELATIVE VIEWPOINT AND LAMP POSITION. A ball lit by one lamp and viewed from different angles appears unlit (a black disc), half-lit (a solid round object) and flatly-lit (an unmodelled white disc) depending on the viewpoint. If we keep our viewpoint still and move the lamp, we get similar results. The final effect depends, therefore, upon both viewpoint and lamp position; changing either alters the effect.

Direction of the light

Here, then, are the three extreme positions from which the subject may be lit (light immediately above or below the subject behaves similarly to side-lighting). Any light directions we choose must have something of these characteristics.

FRONTAL LIGHT. Lighting along the axis of the lens, shadows will be cast directly behind the subject and obscured. Texture is reduced to a minimum, the subject looks flat. Dead frontal lighting might be used, therefore, to disguise wrinkles, to avoid confusing shadows (e.g. on multiplane captions—Chapter 19).

Where this frontal light source is very small compared with the subject, there will be edge-shading as light falls off at its periphery—especially for light-toned, smooth, curved surfaces. The frontally-lit ball of Fig. 5.8 would, in such circumstances, have a distinctly shaded border if a tiny point-source had been used.

EDGE-LIGHTING. Edge or side-lighting reveals maximum texture, maximum surface-contour. Invaluable, then, for displaying shallow relief (e.g. to show embossing, wood-grain, stonework, fabric, etc.) to best advantage.

Frontal lighting here would be quite ineffective. Equally clearly, then, we must avoid edge-lighting when lighting skin (unless we want to emphasize irregularities), or when lighting surfaces that are supposed to look flat (e.g. skycloths).

BACKLIGHT. Direct backlight would, of course, be hidden by the subject itself, and hence largely ineffective. But as it moves off-centre, backlight will illuminate the edge(s) of the subject, emphasizing its outline at that point. This helps us to pull the subject away from its background; to create a pseudo-stereoscopic effect.

71

Fig. 5.9.

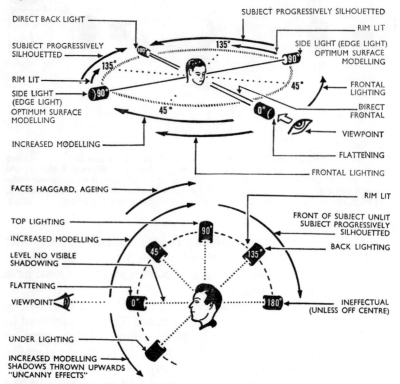

LIGHTING ANGLE. The lighting angle we choose depends on which particular features we want to display, e.g. roundness, surface-texture, relief.

Lighting the human head

The fundamental effect of each of these light directions will remain substantially similar, whatever the subject.

What will vary from one subject to the next, under identical lighting, is exactly which surface contours are revealed or concealed, for these will alter with our subjects' shape and position. We edge-light, for example, not just to show modelling, but to show particular modelling.

Over or under emphasis—except for special effect—tends to make our subject look unnatural, strange, even unrecognizable. One's preconceived ideas about what that subject should, or could, look like, have been violated.

Faces vary considerably in the size and proportions of their parts. The more prominent the features, bone-structure, skin irregularities, the more pronounced the modelling.

The fundamental effect of any light direction will be similar for everyone. The extent of its effect will depend upon the individual subject.

From the effect of a single lamp upon the subject from various positions, let us turn to a practical situation.

Only occasionally will one lamp be sufficient to light a person. We shall need a combination of lamps. Basically, we have to reveal the subject; reduce excessive contrast in shadows thrown by that keylight; pull the subject away from its background.

Fig. 5.10.

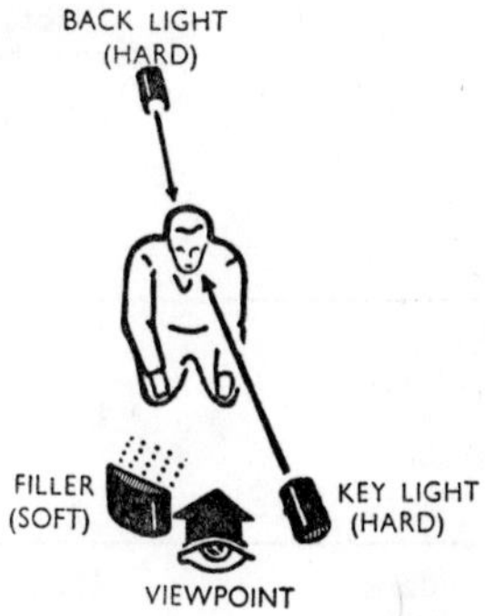

THREE POINT LIGHTING. The key light illuminates and models the subject, and the filler (fill-light) reduces the harshness of shadows cast by the key, revealing detail. The backlight outlines the subject in light, separating it tonally from its background.

BACK LIGHT
(HARD)

FILLER
(SOFT)

KEY LIGHT
(HARD)

VIEWPOINT

Fig. 5.11.

PRODUCTIONAL USE OF LIGHTS. (1) Backlight: light from anywhere behind the subject; typical elevation 45° to 75°; (2) Hair Light: localised light revealing hair detail; (3) Base Light, Foundation Light: diffuse light uniformly flooding the whole setting, preventing light intensity falling below the minimum acceptable. In general, television practice establishes a flat base light and builds up other lights from this; motion picture practice establishes a suitable key light level and reduces contrast by adding filler; (4) Key Light: predominant direction of light, usually frontal for full-face and side light for profile; (5) Clothes Light: hard light revealing texture and form of clothing; (6) Eye Light: a low-power lamp providing eye catch-light without modifying exposure; (7) Filler, Fill-in, Fill Light: soft light used to lighten shadows cast by spot-lights. The balance between key and fill lights determines lighting contrast; (8) Underlighting: light from below camera lens axis, usually at floor level; (9) Set Light, Background Light: any light that illuminates the background alone.

We shall also encounter the terms—Frontal: any light from around camera viewpoint; Modelling Light, Accent Light: any hard light revealing texture and form; Cross Light: frontal spotlight of any height placed about 45° laterally to the subject; Kicker, Cross Backlight: back light of any height placed about 135° to the subject.

73

TABLE 5.II

LAMP POSITIONS AND PORTRAIT LIGHTING

Position	Back Light	Frontal Light
Lamp's Vertical angle too steep	Nose becomes lit while face in shadow ("white nose"). Effect most marked when head tilted back. "Hot top" to forehead and shoulders.	Harsh modelling. Haggard appearance. Ageing. Black eyes; black neck. Long vertical nose shadow. *Emphasizes:*—forehead size, baldness, deep eyes. Figure appears busty. Forehead shadows from hair. Hat brims and spectacles shadow badly.
Lamp's vertical angle too shallow	Lens flares, or lamp actually in shot.	Picture flat; subject lacks modelling. Shadows on background, behind subject.
Under-lighting	Largely ineffectual. In women, under back hair lit. Shadows from ears and shoulders cast on face.	Inverted facial modelling. Shadows of movement beneath head level (e.g. of hands) appear on faces. Shadows cast up over background. "Mysterious" atmosphere when under-lighting used alone. Useful to soften harsh modelling from steep lighting. Reduces age-lines in face and neck.
Light too far off camera axis	Ear and hair shadows on cheek. One side of nose "hot". Eye on same side as backlight appears black—being left in shadow while temple is lit.	In full face, long nose shadow across opposite cheek. An asymmetrical face can be further unbalanced if lit on wider side. One ear lighter than the other.

Note: All lamp positions are relative to camera viewpoint.

TABLE 5.III

LIGHTING BALANCE

Light direction	Too bright	Too dim
Frontal Light	Backlight less effective.	Backlight predominates, often becomes excessive.
	Skin-tones high.	
	Facial modelling lost.	
	Lightest tones tend to be over-exposed.	Darker tones underexposed. Can lead to muddy, lifeless pictorial effect.
	Harsh pictorial effect.	
Backlight	Excessive rim-light.	Two-dimensional picture; lacks solidity.
	Hot shoulders and tops of heads.	
		Subject and background tend to merge.
	Exposing for areas lit by excess backlight causes frontal light to appear inadequate.	Picture appears undynamic, in-definite.
Filler	Modelling from key-light reduced and flattened.	Excessive contrast.
		Subject harshly modelled.

We are discussing the human head, the most familiar subject that we have to light. But a careful study of the problems of portrait lighting will lead us to the principles of lighting that apply to all solid subjects. For these other subjects, the aspects we want to emphasize will be quite different, but our methods identical.

Surface tones and their visual effect

When we look at a picture, our impressions of depth and solidity come from two general groups of visual clues: *Compositional line*—the clues due to linear perspective, overlapping planes, etc. (Fig. 8.15); *tonal values*—the relative brightness of surface areas and tonal gradation. Apart from any effect that cast shadows may have upon compositional line, lighting is chiefly concerned with exploiting tonal values.

Here we meet certain well-established maxims. Concerning *tonal areas* we find that a light-toned area appears to the eye to be larger and more distant than it really is. Conversely, a similar dark-toned area looks smaller and nearer. *Tonal gradation* (shading) will cause the darkened part of the subject to appear to recede, while highlighting makes that area appear to protrude.

These effects, which are the basis of lighting balance, we can see more readily in a series of simple diagrams.

Fig. 5.12. Pt.1.

TONAL PLANES AND TONAL GRADATION. *Left:* A light-toned object or plane tends to look larger and more distant than a similar dark-toned area. *Right:* Compare the effect of distance where our vision stops at the black openings and where the white openings suggest further planes beyond the walls.

Fig. 5.12. Pt.2.

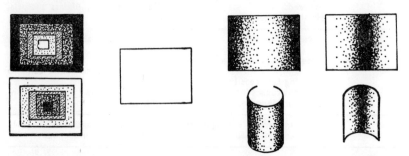

Left: By lighting progressive planes we get an illusion of recession, whereas darkening more distant planes destroys all sense of depth. Now we have the principle of tonal gradation (shading), which will affect our impressions of form. *Right:* The flat, plain area, when shaded one way suggests one thing but shaded differently, suggests something else.

Fig. 5.12. Pt.3.

Perspective alone gives us a partial impression of space (*left*). Tonal change, suitably graded, gives us an impression of both depth and solidity (*centre*), while unsuitable gradation will start to destroy the illusion already provided by perspective (*right*).

Fig. 5.13.

SURFACE BRIGHTNESS. The apparent brightness of a surface depends upon its actual *surface tone* and the amount of light directed upon it. Each of these examples would appear equally bright.

Fig. 5.14.

LIGHT REFLECTION. Brightness changes also with the texture and finish of the surface, and the angle at which we view it. *Left:* Complete absorption indicates little or no reflection (e.g. black velvet), and the surface appears dark from all viewpoints. *Centre left:* With diffuse reflection light scatters in all directions (rough, irregular surfaces) and the surface is fairly bright from all viewpoints. *Centre right:* Spread reflection occurs on glossy surfaces, and their appearance is fairly dark from viewpoint (1), bright from viewpoint (2) and fairly bright from viewpoint (3). *Right:* Specular reflection is uni-directional, and comes from polished metal, mirrors, etc. From viewpoints (1) and (2) the surface seems dark, from viewpoint (3) it appears bright.

Fig. 5.15. Pt. 1.

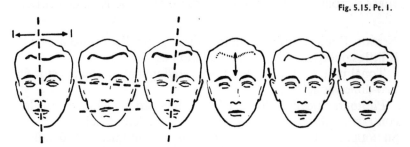

CORRECTIVE LIGHTING. The human face is rarely symmetrical. Most faces have characteristic unbalance (*left*) or disproportion (*right*).

Fig. 5.15. Pt. 2.

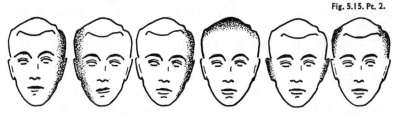

On camera, such irregularities may be emphasized, but can be disguised with varying success by make-up and, where the subject is stationary, by lighting.

Fig. 5.15. Pt. 3.

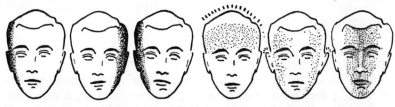

Where the subject is mobile, corrective lighting is impracticable, and irregularities may even be exaggerated if incorrectly lit. For example, by the key light on the wrong side (*left*), by top light, combined top light and backlight, and double-rim lighting (*right*).

77

Lighting techniques

Styles of pictorial treatment

Roughly speaking, there are three broad styles of pictorial treatment through which we can depict the three-dimensional world on a flat screen: Notan—Silhouette—Chiaroscuro.

Fig. 5.16.

THREE PICTORIAL STYLES. *Left:* Notan. *Centre:* Silhouette. *Right:* Chiaroscuro.

NOTAN: Depicts surface detail; outline; generalized tonal areas; but is little concerned with tonal gradation as such.

Tonal pattern, rather than form, predominates. The effect is flat, two-dimensional.

Photographically, Notan effect comes from high-key (i.e. low-contrast), reduced backlight, absence of modelling light.

SILHOUETTE: Concentrates upon subject outline. All details of texture, tone and surface modelling are suppressed.

Silhouettes we shall find photographically in against-the-sun shots (contre-jour), and for highly dramatic and ornamental shadow effects.

CHIAROSCURO: Here we use the principles of shading and tonal separation discussed earlier, to build up an impression of solidity and depth.

Photographically, this is the most familiar pictorial style. In fact, through its use in motion pictures, we are liable to accept it so readily that we forget it is actually a contrived style.

The Chiaroscuro style, when well handled, possesses arresting, vital qualities, as the Dutch Master Painters found, and depicted so convincingly in their interiors.

Atmospheric lighting

Man's constant reactions to the lighting of his surroundings tend to be of a very fundamental nature. We can often see a direct link between how he feels and the sort of conditions he normally associates with a particular kind of lighting: the suspense of a shadowed forest, the exhilaration of sunshine, or the tranquillity of dusk.

78

By imitating the essentials of certain environments in the way we arrange the lighting treatment, it becomes possible to recapture something of their moods. "Moonlight and Romance" is a heavily-worked example of this idea.

When we light for a particular mood, this is basically what we are doing; reminding the viewers (subconsciously) of corresponding conditions, and the kinds of experiences that go with them—whether by natural or stylized convention.

Darkness can be concealing, mysterious, dangerous, treacherous, shielding. Sunlight can be revealing, invigorating, exhausting.

Lighting can remind the viewer of these emotions, and affect his feelings about the picture he is seeing.

By suitable lighting we can change the character of a setting completely.

Some people imagine that realism is the lighting director's ultimate aim. Nothing could be further from the truth. Occasionally an atmosphere may be suggested by directly substituting studio lamps for the original light sources, but this is the exception. If we scrutinize our normal surroundings closely, we shall see how unwise direct imitation would usually be. The half-lit faces, black eyes, "hot" tops, ugly shadows of the real world, would look pretty poor on the screen.

Properly-lit simulation of Nature will, by comparison, be cleaner, more dynamic, more convincing, in multi-camera production.

One of the most difficult effects to reproduce is, paradoxically enough, complete darkness. Taken literally, the result would be artistically and technically ineffectual. But action is quite often set in darkened rooms, night exteriors, tunnels, etc., and some solution must be found.

Wherever possible, some sort of illumination is insinuated into the scene. Otherwise, one has to rely on the acceptable convention of using restricted backlight to reveal the general shape and position of the subject.

Atmospheric continuity

Because the effect of lighting changes with our viewing position, the varying viewpoints of continuous multi-camera production present many problems.

What looks like moonlight on one camera can resemble sunlight on a reverse-angle shot.

Once an environment has been established, the lighting continuity will rest largely upon the average brightness of backgrounds in successive shots. Widely contrasting background tones will make picture-matching difficult. In extreme cases, cutting from dark to light can even make it seem as if we had cut from night to day.

TABLE 5.IV

APPROACHES TO PICTORIAL TREATMENT

Type	Approach	Method
Ilumination	Overall (flat)	Main consideration is visibility. Lighting is almost entirely frontal, and flat. Tends to NOTAN. At its best, subjects distinguishable from each other, and from their background. At worst: flat, characterless pictures; ambiguously merging planes; low pictorial appeal.
	Solid	By a careful balance of frontal and back-light, subjects are made to appear solid. This and clarity are the principal considerations of this chiaroscuro approach. The lighting suggests no particular atmosphere.
Realism	Direct Imitation	*Direct* imitation of effect seen in real life. e.g. sunlight through a window, imitated by a lamp shining through it at a similar angle.
	Indirect Imitation	Imitation of a natural effect; but achieved by a *contrived* method. e.g. simulating sunlight by lighting a backing beyond a window and projecting a window shadow on to an adjacent wall, from a more convenient position.
	Simulated Realism	An imitation of a natural effect, where there is no direct justification for it from within the visible scene. e.g. a window shadow on a far wall, that comes from an unseen (probably non-existent) window. (This may be a projected slide, cast from a cut-out stencil, or a real off-stage window.)
Atmospheric	"Natural"	A lighting treatment in which natural effects are not accurately reproduced, but *suggested* by discreet lighting. A pattern of light that highlights and suppresses pictorial detail selectively, to create an appealing pictorial effect. Suggesting realism in most instances, but seldom permitting dogmatic justification for each light source and direction. A typical motion picture approach.
	Decorative	Associative light patterns. e.g. of leafy branches on a plain background.
	Abstract	Light patterns that have no direct imitative associations, but create a visual appeal. e.g. a silhouetted unknown person, lit only by a rectangular slit of light across his eyes. A flicker-wheel pattern cast over an exciting dance sequence.

By lightening such dark surfaces, or shading light ones, and by judicious picture control, matters can be improved somewhat. But when cutting over wide angles, atmospheric effects in continuous production must normally be a compromise.

Lighting a mobile subject

The more static the action, the greater will be the opportunity for perfection in lighting. (And for the viewer to inspect and criticize the results!)

The greater the amount of movement, the more arbitrary must the lighting be, but with less opportunity for the viewer to detect any imperfections.

If we examine television and film lighting techniques closely, we shall find them falling into variations on four basic themes. (See Fig. 5.17.)

(i) Lighting individual shots would provide optimum lighting treatment for each camera. But for a show of any complexity, such a procedure would prove extremely cumbersome. Each position would have to be planned accurately, and corresponding lighting arranged.

We can, however, provide tailor-made lighting for the most important key-shots in a sequence, allowing intermediary action to reach a slightly lower standard. This is frequently done in television lighting.

(ii) Lighting localized groups. By subdividing the complete setting into a series of localized acting regions, each area can be lit to suit the action within it. How successfully, will depend upon how varied this action proves to be.

In many scenes, action is centred around fixed points, such as tables, settees, etc., and one lighting set-up will suit several shots in their neighbourhood. If need be, slight changes can be made between-whiles.

Where successive shots are reasonably similar in the same area, results can approximate to those of Method (i). But where subject and camera positions vary considerably, the compromise may be inadequate.

(iii) Systematic overall treatment of the whole acting area. By this means we can cope with situations where action details are vague, action is widespread, or time and facilities are limited.

Using the basic three-point scheme of lighting, this is essentially a mechanical approach to the lighting problem, but with care the results can be almost as pleasing as more precise methods.

The hard frontal key-light is positioned to suit the sound boom rather than the inherent needs of the subject itself.

(iv) Soft frontal light with modelling rearlight. The television camera invariably needs a soft overall base (foundation) light to flood the entire studio scene. This reduces lighting contrast and prevents under-exposure of shadow areas.

By making this base light sufficiently strong, we shall have enough illumination to see our subject quite clearly, and no hard frontal will be required for that purpose.

However, excessive soft light is not conducive to dynamic results, and some form of hard modelling light is also essential.

There are two broad approaches here. One is akin to Method (iii); the other to Method (ii).

(iv/a) Lighting the setting as a whole, soft frontal lighting is augmented by three-quarter back spotlights. Suitable for wide-area movement, this scheme is satisfactory enough when lamp positions are carefully chosen. At worst, results can be mediocre and somewhat arbitrary.

(iv/b) Using a more subtle technique, localized hard light provides backlight and modelling light for specific positions. Results akin to Method (ii).

In multi-camera production, cameras tend to shoot obliquely to the setting, so frontal soft light for the set becomes off-centre filler as far as cameras are concerned, and lamps along the side walls become frontal to the lens.

As well as avoiding boom shadow problems, this technique can provide lighting of an extremely high quality. But an over-abundance of soft light must be prevented from flattening modelling, or over-lighting walls.

Follow shots

Where a character in a play is to move about carrying his main source of illumination with him, we meet a special type of problem. Whether he walks down an unlit spiral stairway with a lighted candle, or searches through a wood at night by the light of a hurricane lamp, each situation presents its own particular problems.

Firstly, his carried light source must not be over-bright, causing spurious electronic effects. Nor must it appear disproportionately dim on-camera. If the light is enough to light its surroundings, it will usually be excessively bright for the camera-tube.

We must, therefore, simulate the overall effect rather than try to reproduce it directly.

Fig. 5.17. Pt. I.

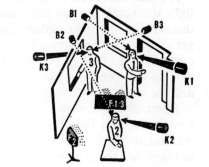

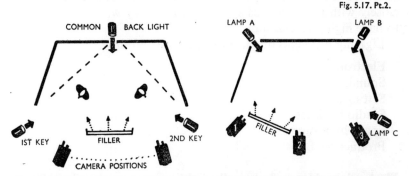

KEY

BACK LIGHT

FILLER

LIGHTING A MOBILE SUBJECT. *Left:* Using three-point lighting principles, we can light for individual positions (Method 1) or localized areas (Method 2). *Right:* Actor (1) is lit by key K1, filler F1, and backlight B1; actor (2) is lit by K2, F2 and B2; and actor (3) by K3, F3 and B3.

Fig. 5.17. Pt.2.

COMMON BACK LIGHT

LAMP A LAMP B

IST KEY FILLER 2ND KEY

CAMERA POSITIONS

FILLER LAMP C

Left: On the same principle, we can systematically light the whole acting area (Method 3). With this basic set-up we get key, backlight and filler of some sort for most camera positions. *Right:* For camera (1) key light is A, backlight B and C; for camera (2) key light is C, backlight B and A; while for camera (3) key light may be either C with B and A as backlight, or with A as backlight and C as side light. We see, too, how a key light for one camera can act as backlight for another at the same time.

Fig. 5.17. Pt.3.

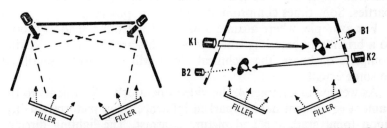

FILLER FILLER

K1 B1

B2 K2

FILLER FILLER

Left: Method 4a lights the setting as a whole, using only soft frontal light. *Right:* Method 4b relies on soft frontal for universal filler, with hard key and backlights along the walls of the set, lighting individual positions.

83

A studio lamp can be manipulated by an operator out of shot, to light the actor and his surroundings as he moves. However, the direction of the light seldom agrees with the visible source's. We get shadows of the carried light, on the person's body. The appearance of the set does not change during the "follow-shot". Anything coming between the studio lamp and the actor becomes brightly lit, and the actor is shadowed.

Alternatively, a series of lamps can be set up along the actor's route. We then fade them up successively as he walks. But this method requires several lamps, precise operation, and may still not simulate the correct light direction.

Television lighting problems

Studio lighting both affects and is affected by many other studio activities. Consequently, the lighting director needs to co-operate closely with these various specialists, and they with him, if we are to have high pictorial standards.

Lighting problems divide loosely into six groups:

Electronic.
Sound.
Settings.
Studio facilities.
Space limitations.
Productional.

Electronic problems. Technically, lighting intensity must be sufficient for the camera-tube. Brightness must be reasonably even over each set. Contrast must not be excessive for the system.

Technically these things are not difficult to achieve. But, unfortunately, these requirements are not always compatible with our pictorial aims, and one tries to find a compromise suitable to all parties. Sometimes compromise is almost impossible.

For example, where a camera has to pass from a shadowy room to a brightly-lit exterior; where room lights are turned out, leaving complete darkness, the technical ideals will have to be sacrificed to some extent.

As we saw in Chapter 4, the video operator's adjustments to the camera equipment have a marked influence upon pictorial quality, upon tonal gradation and picture contrast. The lighting director relies, then, upon the artistic sensitivity of those operating the camera equipment for a faithful interpretation upon the screen of his own lighting treatment. The programme director, designer, and make-up artist, are in the hands of both parties!

Sound. The telescopic arm of the sound boom is continually swung, to keep the microphone within a short distance of the performers. Shadows from the boom-arm may therefore fall across people or backgrounds as the microphone moves and shots change.

Because boom-operation can influence lighting treatment even more than most other aspects of studio operations, we might usefully look at practical ways out of this dilemma.

Fig. 5.18.

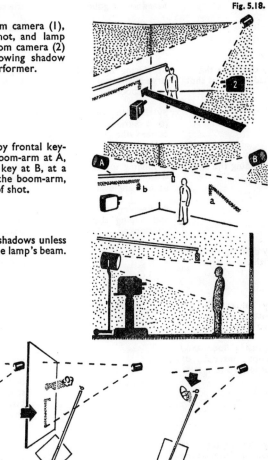

BOOM SHADOWS.—From camera (1), mic. shadow is out of shot, and lamp provides backlight, but from camera (2) lamp is now frontal, throwing shadow on background behind performer.

Mic. shadow **a** is caused by frontal key-light being in line with boom-arm at A, whilst by placing frontal key at B, at a wide horizontal angle to the boom-arm, shadow **b** is thrown out of shot.

Floor lighting avoids mic. shadows unless the boom-arm dips into the lamp's beam.

Some common circumstances in which boom shadows occur are (*left*) when the boom-arm is parallel with a wall, (*centre*) when the boom operator cannot see his boom shadows, and (*right*) when the performer is playing away from the boom. Also, where two booms pick up near and far action, the latter can shadow the closer source.

TABLE 5.V

METHODS OF ELIMINATING BOOM SHADOWS

Approach	Method	Result
By removing the shadow altogether	Switching off the offending light.	This interferes with lighting treatment.
	Shading off that area with a barndoor or gobo.	Normally satisfactory, providing the subject remains lit.
		Shading walls above shoulder height is customary, to provide better prominence to subject.
By throwing the shadow out of shot	By placing the keylight at a large horizontal angle relative to the boom-arm.	A good working principle in all set lighting.
	By throwing the shadow on to a surface not seen on camera when the microphone is in position.	The normal lighting procedure.
By hiding the shadow	Arranging for it to coincide with a dark, broken-up, or unseen angle of background.	Effective where possible, to augment other methods.
	By keeping the shadow still, and hoping that it will be overlooked.	Only suitable when inconspicuous, and when sound source is static.
By diluting the shadow with more light		Liable to over-light the surface, reduce surface modelling, or create multi-shadows. Inadvisable.
By using soft light instead of hard in that area of setting		Occasionally successful, but liable to lead to flat, characterless lighting.
By using floor-stand lamps instead of suspended lamps		Light creeps under the boom-arm, avoiding shadows.
		But: floor-lamps occupy floor-space; can impede camera and boom movements; can cast shadows of performers flat-on to walls.
By altering the position of the sound boom relative to the lighting treatment		May interfere with continuous sound pick-up or impede camera moves.
By changing method of sound pick-up	By using slung, concealed, personal mics., pre-recording, etc.	These solutions may result in less flexible sound pick-up.

Settings. The layout, shape and finish of sets necessarily affect how we can light them. In Chapter 8 we shall be discussing design techniques. Meanwhile, let us see how design arrangements influence studio lighting.

SIZE. The larger the setting, the broader will production treatment usually be. But size as such does not signify. Difficulties only begin to arise when precise lighting is needed for many positions in a large area (e.g. for accurately-lit close-ups).

HEIGHT OF SETTINGS. The tall sets that low-angle and vista shots require, force back and side lighting to become steep, especially where the setting is relatively small (e.g. 12 ft. × 20 ft.), and action moves in the up-stage area.

OVERHANGS. Overhangs (e.g. chandeliers, ceilings, canopies, beams, branches, etc.) of any kind can create shadowing problems. Supplementary lamps may light out the shadows, but create their own new snags.

SHAPE OF SETTINGS. Deep and narrow sets (e.g. corridors) tend to make back and side lighting undesirably steep.

Wide and shallow sets may preclude a single frontal-key, but are even more troublesome camerawise (cameras easily shoot past the edges of the set).

SURFACE TONES AND FINISH. Over-dark walls and furnishings tend to lack modelling, causing faces to appear unduly light; creating high tonal contrasts.

Extra lighting on them may improve their tonal value, but is not always practicable. Similarly, lightening by spraying or painting is liable to destroy surface detail.

Over-light surfaces appear "hot" and glaring on-camera, and make neighbouring faces look unduly dark. Although one can try to keep direct light off them, "spill" or "leak-light" from adjacent soft lights can still leave them over-bright. Surface spraying may unwittingly cause ageing or dirtying, if used to darken them down.

The illusion of depth relies largely upon tonal separation of picture planes. Tonal differences of around $1\frac{1}{2}$: 1 to 2 : 1 are typical between faces and backgrounds. 3 : 1 is seldom reached, except for dramatic or special effects. Backgrounds rarely exceed $1\frac{1}{2}$ times the face brightness. As the reflectance of white skin is 30-40 per cent, this gives us an approximation for preferred tonal values in the set walls after lighting.

Overbright surfaces are a regular television problem. Newspapers, scripts, costume, table-tops, are always having to be sprayed down, dipped, or dulled down, to prevent excessive light bounce. Over-accentuated backlight can aggravate the situation. But where a surface is relatively smooth, even reduced light can still give rise to annoyingly bright specular reflection and glare.

To ban all light or shiny surfaces outright is no practical solution, and each difficulty will usually have to be treated as it arises.

Studio facilities. The elaboration of lighting treatment must be conditioned by the studio lighting facilities themselves. But lavish equipment does not necessarily make for better lighting. Many of television's lighting limitations lie in the problems of continuous live action, unco-ordinated camera treatment, and sheer mechanical restrictions.

Lamp-rigging problems, power limits and, above all, time, have a direct bearing upon the finesse of studio lighting.

The greater the number of sets, the more acute do these considerations become.

Flexible equipment, co-operative planning and sufficient time, can ease many lighting difficulties, though many must remain inherent in the medium.

Space limitations. Almost irrespective of studio size, space remains at a premium. Even in large studios, each set must be kept reasonably compact, or it would seem unduly large on the screen. So cameras, sound boom and lamps will still vie for floor space in front of each set.

In settings that are unavoidably small, lamp positions may be very much of a compromise.

Backings outside windows and doors are usually set quite close to them, to avoid wasting studio space. But in doing so, it becomes hard to light the backing properly. Naturalistic window shadows from an off-stage "sun" are impracticable or poorly simulated.

Multi-camera production. To achieve good lighting balance demands that we adjust the relative intensities of hard key light and soft filler with care. These proportions will vary with the mood and atmosphere we are seeking.

Theoretically they should vary too, to match the length of shot. A close-up requires proportionately more filler than a long-shot, generally speaking. Otherwise, with too little the close-up will look severe and dramatic, while an over-filled long-shot will lose its dynamic value.

Proper adjustment for these varying shots is not easily achieved in television lighting, for two reasons:

(i) The director can cut from long to close shot at any time, simply by using different lens-angles on the respective cameras.

(ii) The soft light used as a filler to the spotlights has an unfortunate physical property. Its intensity falls off rapidly as we increase its distance from the subject. (The inverse-square law causes the light two yards away from it to be only $\frac{1}{2}$ of that one yard from the lamp. At 4 yards it has fallen to around 1/16; while at 8 yards it is only 1/64.) Hard light from spotlights does not fall off so rapidly with distance.

The effect of such soft light will change considerably, therefore, between the downstage and upstage regions of the set. And, as we have to position the lamps themselves at the front of the set (if we want to avoid patchy, toppy, soft lighting), results yet again are a distinct compromise.

Lighting interpretation

In this chapter we have discussed briefly the principal mechanics and techniques of television studio lighting.

We have not included how-to-do-it plots of suitable lighting set-ups for typical situations. This is no accident. For, with an understanding of fundamentals, we can develop our own techniques. We can express ourselves. But formulae breed slavish imitation.

Certain techniques will be found in the work of all good lighting directors. And we cannot deviate from these and still get attractive pictures, for these are the genetic truths of the art.

This does not mean, though, that the individual is bound to come around finally to a few rule-of-thumb approaches that must limit

Fig. 5.19.

THE PERFORMER. *Left:* Mutual shadowing is not readily avoided in a mobile production, where people stand close together. Over a wide angle, lights behind will cause him to shadow B. The results, especially in close-up, can be unattractive. Re-arranging the lighting may modify the quality of other shots, and it is often preferable to readjust the performers, having them play to the light.

Right: Where performers work close to a background, good portrait lighting may be impracticable, as backlight tends to be too "toppy" or "edgy" in full-face shots, causing harsh facial modelling.

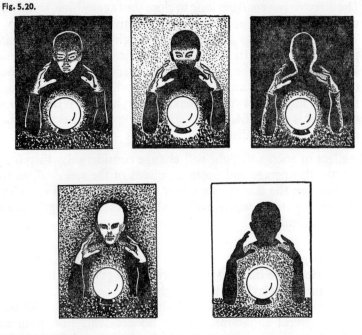

LIGHTING INTERPRETATION. *Top left:* Underlighting. *Top centre:* Localized eyes and hands. *Top right:* Rim-lighting only. *Bottom left:* Stark top lighting. *Bottom right:* Silhouette.

his self-expression. And, in a small way, this has been illustrated in Fig. 5.20.

Here we see a well-worn theme, the fortune teller gazing into his crystal ball. In only five sketches we can see some of the personal lighting interpretations that can be applied to this single hackneyed subject.

Techniques arise from personal selection; from deciding exactly which features in a subject or a scene are to be emphasized or suppressed. And here we enter the realm of the artist.

Fig. 5.21.

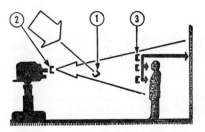

LIGHT MEASUREMENT. Light measurement facilitates consistent, accurate, lighting methods. Lighting treatment involves sufficient light intensity for the camera-tube at the chosen lens aperture; reasonably constant intensity from varying camera positions, and suitable contrast range for the system.

Basic methods (1) incident light; (2) average reflected light; (3) surface brightness.

TABLE 5.VI

METHODS OF LIGHT MEASUREMENT

	1. *Incident light method*	2. *Reflected light method*	3. *Surface brightness method*
Meter position	Beside the subject, pointing at light sources.	Beside the camera, pointing at subject.[1]	Beside the camera, pointing at the subject.
Measuring	Light intensity falling upon subject from each lamp direction in turn.	Average amount of light reflected from scene and received at camera lens.	Brightness of surface at which the instrument is directed.
Providing	For "average" subjects of fairly restricted tonal range, typical incident light intensities and balance suitable to camera can be assessed. Base-light, key, filler, and backlight measured in turn.	Average reflected light levels suitable to camera's sensitivity, can be checked.	(i) By measuring surfaces of known reflectance (skin, "standard white", and black) one can deduce the suitability of light intensities falling upon them. *Also* allows scenic tonal contrasts to be measured to prevent over-contrasty lighting, over-lit highlights, under-lit shadows.
Relative simplicity	Simple, consistent. Does not require experienced interpolation. Widely used in motion picture lighting.	Readings vary with meter angling, and experience is needed to make allowance for subject tones, and contrast. Large dark areas cause readings to be falsely low,—encouraging over-exposure of highlights. Large light areas give high readings; shadows may be underexposed if this measurement is accepted.	Method requires some experience in judging the *importance* of individual surfaces' brightness relative to over-all exposure.

	1. Incident light method	2. Reflected light method	3. Surface brightness method
Advantages	When a show is to be repeated original levels can be duplicated readily. Method facilitates even lighting. Balance between various light directions readily checked.	Method provides a quick rough check of average light levels. Can facilitate evenness of lighting.	Method is capable of assessing surface brightness and contrast very accurately.
Shortcomings	Arbitrary allowance has to be made for subject tones.	Meter readings are only of an "average" nature; which varies considerably with tonal values and proportions.	Several separate readings are necessary to check evenness of lighting, and contrast.
	The amount of light required depends upon the subject,—which this method cannot assess.	Method does not indicate contrast range of subject or lighting.	Method measures scenic tones, but does not distinguish their relative importance; and hence the desired exposure.
	Method only directly useful for "average" subject-tones.	Meter's "angle-of-view" seldom identical with the camera-lens'.	Tonal contrast measurements may not signify:—
	Does not take into account tonal values, proportion of tones, tonal contrast.	Where a single surface (e.g. a face), is to be equally exposed in a variety of settings, measured exposure *should* be constant; but will vary as adjacent tones change.	If the tones measured do not appear together in picture; if their proportions are small and unimportant; if they *may* be acceptably "crushed out", without injuring pictorial quality.

Note: Where meter is held close to subject, measuring individual surface brightness, method becomes as 3.

6

TELEVISION SOUND

To appreciate the sound engineer's problems and techniques, we can best begin by taking a look at the nature of sound itself.

Fig. 6.1.

THE SOUND WAVE.   *Left:* Sound waves are compressions and rarefactions in the air, emanating from the sound source, diminishing in strength with distance. *Centre:* The air-pressure fluctuations at any fixed point over a period of time are in many instances regular and rhythmical (periodic). *Right:* They can be plotted graphically.

We shall hear several terms regularly used to describe various features of this sound. There is the *pitch* or frequency. This is simply the number of complete vibrations (cycles) it makes each second. (Measured in cycles-per-second. The newer term "hertz" is synonymous.)

The sound's *loudness* is the strength of these vibrations, i.e. how violently the air has been disturbed, or the vibrating-body displaced. We measure its *amplitude* (or intensity) in decibels (dbs.) or sometimes phons. Because our ears are so constructed, we do not hear volume changes in the way we might expect. If we double the true intensity of a sound, we shall not hear anything like a twofold increase in volume, for our ears' sensitivity is logarithmic, not linear. A sound's volume has to change some 25% before we detect any volume difference at all. It is thanks to this non-linear behaviour of the ear that we can accommodate such a wide volume range.

Wavelength is the distance a sound-wave travels through a particular medium before another identical cycle of movement begins.

93

(More technically, the distance between points of similar amplitude and phase, in successive cycles.) This distance is measured in feet or metres, and for any frequency depends upon the material the sound is travelling through. (Sound moves through some more quickly than others.) Wavelength is always inversely proportional to frequency.

Phase is a term used, when comparing how far one vibration has gone towards completing a whole oscillation, with another similar oscillation. We measure phase in degrees or radians.

Even if they are present, we cannot hear sounds below about 20 cycles-per-second, or above some 15,000 cycles-per-second. Below 20 c/s, we cease to hear vibrations as sound, but begin to feel them instead. They are subsonic. Above our upper frequency limit, the supersonic vibrations are too high for us to appreciate, although certain animals may still hear them quite clearly.

Nor do we hear equally well over our 20-15,000 cycles audible-range. Best around 3,500 cycles, our hearing sensitivity falls off for sounds of higher and lower pitch. Exactly how it varies depends on the sound's intensity, and, of course, upon the individual. (As we age, our hearing at the highest frequencies progressively worsens.)

For quiet sounds, the fall-off of the upper and lower ends of our audibility-curve is considerable, while at greater volumes the ear's response is more even. So this accounts for the differences we notice between the quality or "colour" of the same sound heard distantly and close-to.

Fig. 6.2.

VOLUME OR AMPLITUDE

· ONE CYCLE ·

TIME

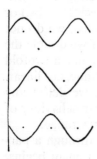

DEFINING THE SOUND WAVE. *Top Left:* The number of cycles per second is the sound's frequency. *Centre:* Loudness: soft sounds produce slight fluctuations, loud sounds strong fluctuations.

Right: Pitch: upper waveform illustrates high pitched sounds (high frequency), lower waveform illustrates low pitched sounds (low frequency).

Bottom Left: Phase: the phase difference between the middle and top waveforms is 90°, and between the bottom and top waveforms 180°.

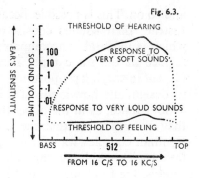

Fig. 6.3.

AUDIBILITY CURVE. For soft sounds the ear is relatively insensitive to low and high frequencies, but for loud sounds its response is more even.

EAR'S SENSITIVITY

SOUND VOLUME

THRESHOLD OF HEARING

RESPONSE TO VERY SOFT SOUNDS

100
10
1
.1
.01

RESPONSE TO VERY LOUD SOUNDS

THRESHOLD OF FEELING

BASS   512   TOP

FROM 16 C/S TO 16 KC/S

Sound quality

How individual sounds add together to form a combined complex wave is not too hard to appreciate. But how we manage to disentangle the component sources on hearing it, is a speculative field for thought. By electrical or mathematical analysis, one can discover the proportions of component pure tones that comprise any complex waveforms, and this helps us to see how a particular sound colour is derived. But exactly how the ear and brain pick out one voice from a crowd, or one instrument from an orchestra, is another matter.

Quality depends upon harmonics

When various musical instruments play the same note, we can detect a characteristic tonal colouration in each, from which we identify one as being an oboe, another a bassoon, another a 'cello,

Fig. 6.4.

TUNING FORKS

ELECTRONIC OSCILLATORS

FLUTES

SOUND QUALITY. The simplest possible sound is a regular, uncomplicated vibration, called a sine-wave. *Left:* Certain sources produce such pure sounds. *Centre:* A pure tone will vibrate the air in a sinusoidal pattern. *Right:* Most sound sources produce more complex, rich sounds, but if analysed these can be seen to comprise proportions of component pure tones.

95

and so on. This is possible, because few sources produce pure notes. Instead, their sounds consist of a main note (the fundamental), which is inexorably accompanied by a number of spurious notes, formed at the same time. These overtones or harmonics will be present in varying proportions, according to the type of instrument, its design, and how loudly it is being played.

Sometimes, as in the oboe, the harmonics may be so strong that they are even louder than the fundamental. For transient sounds, the resultant complex waveform can contain a random mixture of any frequencies within the audible spectrum. But for steady musical sounds, the fundamental will be accompanied by related multiples of its own fundamental frequency.

It is interesting to find that even complicated natural sounds can be simulated artificially by adding pure sine-waves in the same proportions as in the original. The result can be indistinguishable from the real thing. Conversely, taking a complex sound, we can by audio-filters weaken or strengthen selected harmonics, and change its colouration completely. A trumpet or bassoon can be made to sound like a flute. These are admittedly laboratory stunts, but they demonstrate the nature of sound quality, how it originates, how it can become changed through inaccurate reproduction.

For completely faithful reproduction, the complex sound-wave would need to be followed extremely accurately by all parts of the audio chain, from microphone to loudspeaker. Save for the simplest sounds, this is not readily achieved. Fortunately for us, however, our brain is astonishingly tolerant. It accepts wild travesties as

Fig. 6.5.

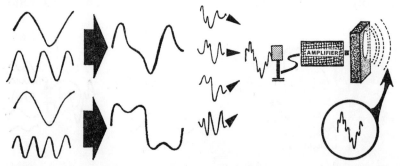

FORMING THE COMPLEX WAVEFORM. *Left:* The wave shape varies with the frequencies involved, their volume and phase relationships. The more complicated these, the more complex the combined waveform. *Right:* This combined effect will be that traced by the vibrating microphone diaphragm, the resultant electrical signal, and the loudspeaker-cone in a sound system.

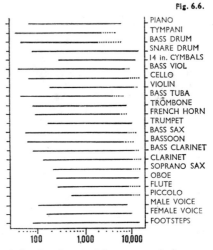

Fig. 6.6.

AUDIBLE FREQUENCY RANGES
of musical instruments and natural
sounds. In cycles per second.

PIANO
TYMPANI
BASS DRUM
SNARE DRUM
14 in. CYMBALS
BASS VIOL
CELLO
VIOLIN
BASS TUBA
TROMBONE
FRENCH HORN
TRUMPET
BASS SAX
BASSOON
BASS CLARINET
CLARINET
SOPRANO SAX
OBOE
FLUTE
PICCOLO
MALE VOICE
FEMALE VOICE
FOOTSTEPS

100   1,000   10,000

reasonable substitutes for the original, often without our being too conscious of the fact; as listeners to miniature radios must have appreciated.

The characteristics of reproduced sound

Reproduced sound can differ in a number of important respects, from what we would ourselves hear in the vicinity of the source. Realizing these differences can help us to appreciate the problems of sound control.

The limitations of reproduced sound

It is *monaural* (one-eared). Sound pick-up and reproduction are through a single channel. Consequently the direction of the sound cannot be indicated; only its distance suggested. Sounds cannot therefore be spatially separated; only segregated by volume, pitch and quality differences. (Television sound not being stereophonic.)

The amount of reverberation accompanying a monaurally reproduced sound appears greater than we would have noticed if listening binaurally (i.e. stereophonically).

The ear demands a wider frequency range from monaural reproduction to equal the same apparent sound fidelity under binaural conditions.

The microphone itself is not discriminating. It gives equal prominence to any sounds in its pick-up range, whether we want them or not. Skilled microphone positioning is often necessary, therefore, to avoid extraneous masking sounds.

The volume-range (dynamic range) acceptable by the sound system is technically restricted, if we want to avoid background noise or distortion. The volume of the reproduced sound may not be comparable, therefore, to that of the original. This leads to scale-distortion and to volume-compression.

Moreover, the sound-channel adds various types of distortion to the original sound; principally in the reproducer. So the character and relationships of sounds can become blurred or confused as a result.

For recorded sounds, the pitch of the original may be reproduced inaccurately.

Adjustments possible with reproduced sound

As we shall see in later sections, we can control the characteristics of reproduced sound in several ways.

Its volume, by adjusting audio amplification, or the microphone's distance.

By microphone placement and type we can adjust the proportion of direct sound pick-up to acoustical reflections.

The extent and quality of these reflections is adjustable (by acoustical treatment).

The tonal quality of the sound may be changed by deliberately increasing and/or decreasing parts of the audio spectrum or by deliberately adding distortion, for special effects.

In recorded sound we can also deliberately alter its pitch and duration; reverse the sound; create synthetic noises; continuously repeat a selection of sounds.

Sound in the television studio

Studio acoustics

Acoustics concern us from two points of view. Firstly, there are the acoustics of the studio and the settings, and how the studio sound becomes coloured or reverberated by them. This will principally be a matter for studio design. But the shape and materials of our settings must also affect the quality of the sound pick-up within them.

Secondly, wherever the studio settings aim at suggesting realism, we shall need to take into account just what kind of sound one ordinarily hears in the places depicted. If we allowed a picture of an open seascape to be accompanied by echoing sound, the overall effect would hardly be convincing.

In radio broadcasting studios one has considerable control over the acoustic surroundings. In television the situation is less flexible.

Most of the acoustical principles we need to know are familiar to us. Except in free space, any sound we hear will consist of direct waves accompanied by others reflected from the surrounding surfaces.

Most surfaces absorb sound to some extent. Materials such as drapes, cushions, carpets, are strongly absorptive.

Hard surfaces—tiles, brick, wood—will reflect most of the sound falling upon them. Materials do not behave similarly for all frequencies, however. A surface may absorb the highest notes, while reflecting lower notes. The reflected version of the sound would then lack top compared with the original. Where surfaces have been frequency-selective in this way, the reflections are said to be coloured.

The kind of reflection we get from a surface will depend, too, upon:

Its finish—whether shiny, rough, cellular, etc.

Its shape—whether flat, curved, corrugated, etc.

Its structure—whether it is solidly fixed, freely hung, clamped, thick, thin, etc.

Where the construction is slight or unsupported, the material is liable to resonate, emphasizing certain notes as it vibrates in sympathy with them.

Where the surroundings are highly absorptive, the studio or room will seem dead; lacking reverberation. We shall hear only the direct sound.

If there is little absorption, strong reflections will reinforce the original sound, making it appear louder. Where these reflections mingle with the direct sound, we talk of the sound being reverberant or live. But if these reflections become too strong, they are liable to mask the original sound, making it less intelligible. Furthermore, the reflection's quality may become changed (through frequency-selective absorption), giving the result a hollow, hard or "woofy" colouration.

Where hard reflective surfaces are sufficiently far away to allow a lapse of 1/18th second or more between our receiving the direct and reflected versions (i.e. over 60 ft.), we shall hear an echo—a distinct repeat following the original. Too much echo can cause a confusion of repeated sounds that make the overall effect unintelligible.

From these acoustical principles, we can see that it should be possible to construct a studio with whatever tonal and reverberation characteristics we choose. The natural hangover (decay) time can be reduced by affixing sound-absorbent materials to walls and ceiling. Fibre panels, quilting of flock, seaweed, glass wool, etc., are commonly used. By having vari-faced panels of movable hard and soft surfaces, the acoustics can be adjusted within limits to suit the occasion.

The live (reverberant) studio. Although excessive echo is undesirable, many kinds of sound strike us as more pleasing when heard in fairly reverberant surroundings. Their tone becomes enriched and strengthened.

Musical items particularly are improved in this way. Reverberation conveys an impression of spaciousness and vitality.

On the other hand, reverberant surroundings bring problems. Extraneous studio-noise from cable-drag and camera manoeuvres will carry more readily than in a dead studio. We shall hear footsteps, scenery changes, ventilation and camera cooler-fan noise.

Anyway, for many occasions reverberant sound would be out of place, in outdoor scenes, small rooms, the interior of vehicles, for example. Even close microphones cannot completely exclude reverberation.

There is much in favour, therefore, on purely practical grounds, of having a non-reverberant studio and simulating reverberation, which is mixed with the direct sound as required. Although we can synthetically add reverberation to a dead sound, we cannot deaden a reverberant sound.

The dead (non-reverberant) studio. As sound does not carry well in dead surroundings, extraneous noises are no longer troublesome. Exterior scenes have the naturally non-reverberant acoustics they need.

On the debit side, the acoustics of a very dead studio are psychologically depressing to work in. Because sounds do not carry, performers may need to speak or play more loudly. Musicians may have difficulty in hearing other sections of an orchestra or remote singers.

Because closer microphone-placing is necessary, the sound-coverage problems and the likelihood of boom-shadows will increase. Sensitive, highly-directional microphones may help here.

As always, the answer is a compromise. Adjustable acoustics in the television studio are more readily suggested than accomplished, especially in multi-set production. Typical reverberation times found to suit critical listeners range from $\frac{2}{3}$ sec. (for speech) to 1 sec. (for choir) in small halls and from 2 to 3 secs. in large halls.

The microphone

When we come to use a microphone, the main things we want to know about it are:

Its physical features—i.e. its size, robustness, freedom from handling-noise, stability, general reliability.
The kind of audio quality it provides.
Its directional properties.

Other factors, such as its sensitivity, and design features, are more the concern of the audio engineer.

Physical features
For the most part, size only becomes a consideration when we are trying to hide the microphone, make it unobtrusive, or when we have to carry it around. Miniaturizing can result in reduced performance.

Many high-performance microphones cannot withstand the rough handling they might encounter during a remote telecast in the field. We should have to use a more rugged, if lower-performance, microphone.

Fig. 6.7.

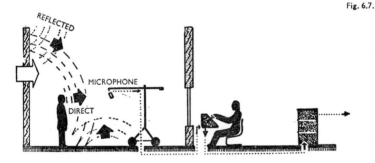

STUDIO ACOUSTICS cause reflection and absorption of sound waves. High frequency sound waves follow straight paths, they are readily shielded off, and reflect strongly from hard surfaces. Low frequency waves (below 100 cycles) spread, and consequently are more easily heard round corners. They require larger surface areas for efficient reflection, and are not so readily absorbed.

Direct and modified reflected sounds reach the studio microphone, generating a corresponding electrical signal. This audio signal is taken to the audio mixing console in the Sound Control Room, adjusted and blended by the sound mixer as necessary, amplified, and routed via Master Control to the sound transmitter.

101

Various high-grade microphones cannot be handled, or used in windy surroundings, without causing extraneous noises. Attached wind-shield bowls of silk or chiffon cut down wind-rumble in many microphones to acceptable proportions.

There are microphones whose performance may vary with the temperature of the surroundings. Others are versatile—but temperamental.

The perfect all-round microphone is still a matter for personal choice—if it exists at all.

Audio quality

The ideal microphone should respond equally well throughout the whole audible spectrum. It would have a smooth, even response without dips, peaks, resonances, or fall-off at the extremes. We would then have no colouration of the sound pick-up.

The microphone should respond immediately to sharp, transient sounds, and have no hangover, ringing, cross-modulation, or other undesirable after-effects. The sound would then be more cleanly reproduced, without a trace of buzz or general muzziness to the string-tone of a full orchestra, for example.

The microphone should not overload and cause distortion when placed near loud sound sources.

Its frequency-response should be constant over the pick-up area.

As we shall find in practice, some microphones are nearer to these ideals than others. One may have a good-looking frequency-response curve, but may sound just that bit unrealistic when used to balance a piano recital, compared with another type of microphone of more restricted range, but with better transient characteristics. The experienced audio man finds himself able to recognize the sound quality from different microphones, especially when they are directly compared.

Microphone types

There are four basic types of high-grade microphones:

Dynamic (moving coil).
Ribbon (velocity ribbon; pressure-gradient ribbon).
Crystal.
Electrostatic (condenser).

Although we are primarily concerned with their operation, and not their internals, a rudimentary idea of how these microphones work can give us a fuller appreciation of how they handle mechanically. Moreover, each type has features that affect the usage.

Fig. 6.8. Pt.I.

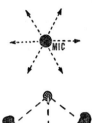
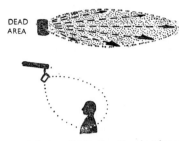

DIRECTIONAL PROPERTIES. Microphone type and design affects directional pick-up. *Left:* Omni-directional has equal sensitivity all round (dynamic, crystal and electrostatic microphones). Most useful for all-round pick-up from a group, but extraneous sound is also picked up. This necessitates closer microphone positioning. *Right:* Uni-directional has only one-sided sensitivity (electrostatic or special ribbon). Its uses include beamed single-source pick-up, balancing soft sources against louder ones, and more distant microphone positioning. It has good source isolation but sensitivity may be too beamed for close pick-up of broad sources such as spaced 2-shots.

Fig. 6.8. Pt.2.

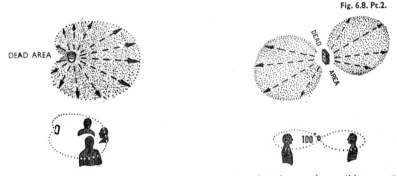

Left: Cardioid has a heart-shaped pick-up field (combined crystal or ribbon, and dynamic, assembly; special ribbon or special electrostatic microphones). Broad single-sided general purpose pick-up, readily accepting spaced sources. May be used for balancing soft against louder sources, and where fairly distant microphone positioning is desirable. Unfortunately, restricted pick-up is possible from the rear side. *Right:* Bi-directional has two 100°-plus pick-up areas in a figure of eight (ribbon or special electrostatic types). Useful for two-sided pick-up, i.e. facing speakers, soloist and orchestra, etc., and for balancing soft sources against louder ones.

Fig.6.8. Pt.3.

Highly directional microphone employs a parabolic reflector (3-6 ft. dia.) or acoustic tube or line type attachment (machine-gun; rifle) 3-8 ft. long. Pick-up angle is 10°–40°, which is useful for clear pick-up of distant sources, and for isolating sources in noisy surroundings, hence it is mostly used for remote telecasts. It is, however, cumbersome, and less sharply directional at middle and lower frequencies.

103

Fig. 6.9.

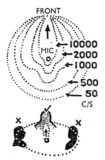

POLAR DIAGRAM. Although all micro-phones have one or more of these pick-up (polar) diagrams, any given micro-phone's response may vary with the sound frequency. For instance, because of fall-off at higher frequencies, sources in line with the microphone may be re-produced well, while others off the microphone axis lack definition.

Dynamic. Pressure variations in the sound-wave vibrate a diaphragm to which a small coil of wire is attached. This coil moves within a magnetic field, so generating an electric current. This current's waveform corresponds with the original sounds.

Above all else, the dynamic microphone is robust.

It tends, however, to be strongly uni-directional to high fre-quencies, while remaining non-directional to low frequencies. This weakness can be utilized discreetly, by turning the microphone slightly off-axis to very shrill, toppy, or sibilant sound sources.

Ribbon. Here a thin corrugated-foil strip is held between the poles of a magnet. As the sound-wave impinges upon this ribbon, differ-ences in the air pressure on either face cause it to move. The ribbon's vibrations generate an electric current in it.

(The pressure-difference—hence the ribbon motion—corresponds to the particle-velocity in the sound-wave—thus the term velocity-ribbon microphone.)

The ribbon is not as robust as the dynamic microphone; is gen-erally larger and heavier, and is most susceptible to wind-rumble.

Respecting sound quality, the ribbon generally has an even frequency-response over its entire pick-up field.

For very close sound sources, however, bass notes are dispro-portionately emphasized; although this is again a phenomenon that can be turned to advantage. Singers and speakers with "thin" voice-tone can enjoy bass benefits.

Where sound strikes the ribbon nearly end-on, the pick-up can suffer a noticeable top-loss. So the microphone would normally be tilted towards sources to avoid this effect.

Crystal. Certain types of crystal produce voltages between their faces when bent or twisted. In this microphone the air-wave impinges upon a diaphragm, which is connected to the small crystal slab. The voltage produced is the audio-signal.

Its small size, together with high sensitivity, makes the crystal ideal for lapel button microphones, although it remains a fairly fragile unit.

Electrostatic. Here a thin, tightly-stretched metal diaphragm is held close to a flat metal plate. The sound-wave's pressures cause the intervening space to vary, and hence the intercapacity of the assembly. Unlike all other microphones, the electrostatic requires a D.C. voltage supply and an adjacent pre-amplifier—features that can add to its bulk and complexity. It does, however, lend itself to remote electrical adjustment, to provide variable directional properties.

Microphone mountings

How the microphone is supported, will determine the range of action we can cover with it. The most versatile device is unquestionably the *sound boom*. Additionally, we shall find widely-used mountings including:

The small boom.
The hand boom (fishing rod).
Floor stands.
Desk fittings.
Slung microphones.
Personal microphones.

Fig. 6.10.

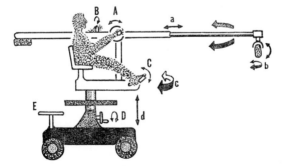

THE SOUND BOOM. A, Boom-arm extension handle (motion a); B, microphone tilt and rotate (motion b); C, foot pedal to rotate boom platform (motion c); D, boom platform raise/lower (motion d); E, steering wheel.

Fig. 6.11. Pt.1.

OPERATION OF THE SOUND BOOM. *Left:* Boom operation requires careful timing and co-ordination of movement. As we saw in Chapter 5, a constant watch must be kept for shadows on settings and performers, always remaining out-of-shot while maintaining good sound balance and aural perspective. Microphone height and distance must be constantly adjusted to remain outside the camera's view, as in a cut from close-up to mid-shot. *Centre:* A two-shot may be treated with a central, static microphone—a compromise, but with both speakers equally audible. *Right:* Alternatively the microphone may be beamed to each speaker individually. This produces optimum quality sound, but may leave a speaker off-microphone if timing or sequence are not exact.

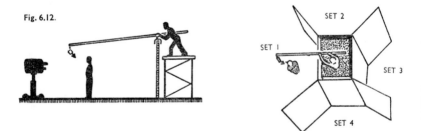

Fig. 6.11. Pt.2.

Left: The simplest subject movement can entail complicated boom operation as, for instance, if the microphone twists to follow the turn of a head, or if the boom-arm has to be rapidly extended and the microphone twisted to keep a speaker on microphone as he turns round. *Right:* As a speaker walks upstairs, the microphone boom has several concurrent movements: the boom-arm swings, is raised, is retracted, while the microphone is swivelled and tilted.

Fig. 6.12.

UNORTHODOX BOOM POSITION. Boom operationf rom above the setting isf easible, but not very practicable. Removing the boom-arm from its pram, and clamping it behind the set may serve one or more settings in a very confined area, leaving the floor free for cameras; but against this advantage, it becomes difficult for the operator to judge microphone height and to see microphone shadows.

106

THE SMALL BOOM is used for semi-permanent microphone locations, supplementing the larger sound boom in situations with little movement. A 3-section telescopic boom is preadjusted to the required length, but it cannot follow movement and must cover any action by boom swings alone. (1) Twist grip: microphone tilt and pan control; (2) adjustable counter-weight.

THE HAND BOOM (Fishing rod). The microphone is simply attached to a pole, and tucked under one arm. An extremely manoeuvrable—if tiring—arrangement, which can be used for sound pick-up under otherwise impossible conditions.

FLOOR STAND. (Stand microphone). A metal tube of adjustable height, on a weighted base, used for general static microphone treatment such as orchestral section pick-up, confined action, spot effects, incidental sound.

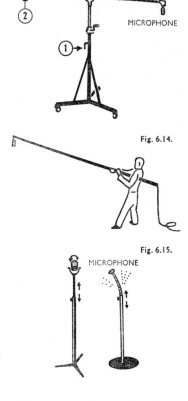

Fig. 6.13.

MICROPHONE

Fig. 6.14.

Fig. 6.15.

MICROPHONE

Slung microphones (hanging mics.). Where space is restricted, or orthodox mountings cannot be used, the microphone may be suspended over the pick-up area. Hallways, and sets with overhanging or masking scenery, are typical situations where local slung microphones can be invaluable.

Because the slung-mic. is static and uncontrolled, it has drawbacks. Its field of pick-up will depend upon its directional characteristics. If non-directional, it is liable to receive extraneous noise and acoustical reflections. If directional, people will have to play to the microphone to avoid moving out of its range. In any event, we cannot adjust the sound-perspective to suit the camera's shot. In long shots we shall have to raise it, have it disguised (e.g. as a domestic light fitting), or simply accept it for what it is.

Desk fittings. Table microphones are standard units held in clamps or low stands, in full view of the audience. Their use tends to give an authoritative air to certain types of programmes, e.g. newscasts, while appearing distracting for others. The smallest microphones remain fairly unobtrusive in any case.

Microphones are occasionally hidden where sound pick-up is difficult. Typical hiding places include faked-up furniture, ink-wells, books, floral decoration. The drawbacks of this method are similar to those with slung microphones.

Personal microphones. Microphones which are carried around by the speaker, instead of remaining under the control of a sound man, are mostly used in remote telecasts. With the personal microphone, wherever the speaker can go, the microphone goes too.

We shall meet four forms of personal microphones in television broadcasting:

Hand-held microphones.
Lapel microphones.
Lip microphones.
Neck microphones.

These may have attached trailing cables, or be connected to small personally-carried transmitters which are picked up by special local receivers and fed into the audio system.

HAND-HELD MICROPHONES. The hand-held microphone has long become a familiar appendage for commentators and interviewers. Stick or pencil microphones are commonest, although small units attached to a short, light stem are found. The sound mixer is largely in the hands of the person holding the microphone. Unless the user is reasonably careful to direct it as needed—especially during interviews—pick-up can be haphazard.

LAPEL MICROPHONES. The tiny, light-weight lapel microphone (usually crystal type), is attached to the speaker's clothing. This gives him considerable freedom of movement, particularly when a midget pocket-pack transmitter is used with it. Provided too with a pocket radio-receiver which feeds a deaf-aid earpiece, he will be able to hear producer's talkback from a special short-range transmitter in the control room—to receive instructions and link with other programme sources. The opportunities this scheme offers are immense.

The lapel microphone method of pick-up has important snags, though. It is very susceptible to extraneous noise from clothing, jewellery, etc., and its environment. Volume and quality are likely to change as the speaker's head is turned. During interviews, particularly when both parties wear lapel microphones, sound balance problems can be acute.

LIP MICROPHONES. The lip microphone is, perhaps, the most ingenious personal microphone. With a shaped mouthpiece, the microphone is held against the speaker's face. This both shields off surrounding sounds and enables his voice to be heard clearly, even in extremely noisy locations. Amidst loud noises, large crowds, machinery, etc., this is certainly the most effective means of sound pick-up.

NECK MICROPHONES (Lavalier microphones). Slung on the chest by a cord, the small neck microphone has some popularity. Worn about six inches below the chin, it is often concealed behind the necktie. Quality is adequate, but varies with the wearer's head-position.

Sound control

Located before his console in the sound control room, the sound mixer selects, controls and blends the programme sound sources. His attention is divided variously between:

The talkback loudspeaker relaying the programme and technical director's instructions.

His high-grade loudspeaker reproducing the audio programme he has helped to contrive.

The picture monitors showing the transmitted and preview shots

STUDIO AUDIO CONTROL CONSOLE (Sound Mixing Desk).

Fig. 6.16

Reading from top to bottom:

1. Faders pre-set each channel's amplification.
2. Channel faders control and fade-out each.
3. Push buttons allow pre-fader monitoring.
4. Group fader controls channel combinations (e.g. choir group and orchestra group.)
5. Master fader controls entire board.
6. Selecting sources on main effects switch permits cuts between groups for changes in quality, perspective, balance (e.g. intercutting a telephone conversation).
7. Individual sources can have reverberation added by mixing studio and echo room outputs by cross-faders. (Enlarged detail *bottom right*.)

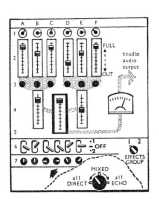

Instructions he is giving over his own private-wire talkback to the boom operators, gram-operator, spot-effects men, etc. In some set-ups they can reply over their reverse-talkback microphones.

We can consider the sound mixer's job under several headings:

Dynamic range control.
Balance.
Sound perspective.
Sound quality.
Acoustics.

Dynamic range control

When the audio system is overloaded by too loud a sound, distortion results.

If the system's amplification has been reduced to cope with very loud sounds, quiet ones will be inaudible. Adjusted so that quiet sounds are easily heard, loud sounds cannot be handled. So clearly we have a volume-range for the system, within which sounds must be fitted if they are to be reproduced well. This is the *dynamic range* of the system—usually around 20 to 30 dbs.

(The dynamic range is simply the volume ratio of the loudest to the softest sounds. Where volume extremes are great, the source is said to have a large dynamic range.)

By sensory-controls our ears tense themselves to withstand loud sounds, relaxing to receive quiet ones. They can, by this self-adjustment, hear over a wide range of volumes, although not at one and the same time. Electronic equipment does not possess this innate feature. Some form of amplification adjustment has to be introduced, to keep the volume-levels within the capacity of the audio system; to hold back loud sounds and make the softest sounds louder.

How much control is needed will depend upon the sounds that are involved. A talk may require negligible adjustment; a full orchestra can require continual control.

Needless or clumsy compression of a source's volume range is indefensible. But some degree of control is both technically necessary and artistically desirable. And, in any case, the fully reproduced dynamic range of many sound-subjects would be overpowering in our homes.

Dynamic control is common in all audio systems. Ideally, only the sound volume likely to go beyond the system's accepted limits will need adjustment. But any control should be anticipatory, and gradually introduced and relaxed. So, by slowly reducing audio-gain before a loud passage, the ear is unaware of falsification, and the crescendos and diminuendos of the original sound are apparently preserved. Excessive or erratic compression cause the sound to be reproduced as strangled and lifeless.

The tonal quality of many sounds alters with the loudness with which they are made. A shout does not seem like a whisper when reproduced quietly; nor vice-versa. Consequently, a fair amount of compression can be tolerated without destroying the aural illusion.

The most satisfactory method of controlling the reproduced dynamic range is undoubtedly by manual adjustment of the audio-gain. Here the sound mixer will adjust the amplification by the *channel fader* for the source concerned, or the overall sound volume by the *master fader* (Fig. 6.16). His *volume indicator* meter will show him the system's limits.

Electronic devices can supplement, rather than replace, manual control. The *limiter* will automatically prevent audio peaks from exceeding the maximum limit and the *compressor* will automatically make quietest sounds louder while limiting loud peaks.

Although useful technical safeguards in case the operator is caught out, they are non-anticipatory and can lead to volume inflections being ironed out, or to irritating breathing effects on strong transient sounds.

Sound balance
Position of the microphone. What is good balance? We might define it loosely as sound pick-up that enables individual sources to be heard clearly within a group of sounds: reproduction in proportions, tonal quality, and perspective apparently similar to the original. Clarity and naturalistic tone are probably the generally accepted hallmarks of good balance.

Listening with two ears not only helps us to judge the distance and direction of sound sources, but causes these sources to stand out distinctly from each other. So, unless other sounds mask the particular one we are listening to, we can concentrate upon it and be almost oblivious of our surroundings.

Most sound reproduction is still, however, one-eared, and the single sound-channel does not have these discriminating powers. Nor can the listener himself compensate for this difference, for the reproduction is necessarily a localized, homogeneous outpouring.

The most he can do is to distinguish between sources through their differences in pitch and volume. But in well-balanced monaural reproduction this will still be sufficient for him to be able to pick out instrumentalists in an orchestra, or one voice from a gathering.

It is not enough to place a microphone in a good listening position near the sound source, for our two ears have unconscious selectivity, that the microphone has not. Instead, we must position the microphone to achieve the aural effect we are seeking to convey, and that is not necessarily the same thing.

As we bring a microphone nearer our subject, we shall pick up more direct sound and less acoustical reflection. And the reproduced quality will be more realistic, for reflections have colouration.

But too close positioning results in a coarse, closed-in rendering of the sound, that is unnatural and out of perspective.

With distant microphone positions, random reflections will predominate over direct sound, and we are liable to get a highly-coloured rendering of the original. (For, remember, surfaces reflecting the sound will be frequency-selective, and will be absorbing and reflecting various parts of the whole audible spectrum, according to their material and structure.)

Some indirect sound pick-up is desirable, though, if reproduction is to have depth and vigour. So an optimum position has to be found, where direct and indirect sounds (room tone) are blended in satisfying proportions. This will vary with the subject, its surroundings, and one's taste.

The directional properties of the microphone used will influence its position. The less directional it is, the closer must it be to avoid undue reflections. It may even need acoustical screens to help localize pick-up.

The nature of the sound, too, will affect the optimum position. Loud, transient noises, for instance, can become blurred and confused when accompanied by excessive reverberation. Sustained sounds may be tonally enriched by a reverberant background.

There are two fundamental approaches to the problem of achieving good sound balance with multiple sound sources such as an orchestra. By using a single microphone at an optimum overall pick-up point and by using several microphones, each in the most favourable position for a part of the pick-up area, and mixing their combined output.

SINGLE MICROPHONE TECHNIQUES are possible, using a highly sensitive (usually non-directional) microphone. They have this artistic advantage, it is argued: that one creative artist (e.g. the orchestral conductor) can more readily select and control the sound balance than is feasible when a technician blends multiple pick-up points. Technically, the method is said to avoid the blurring of quality that can arise when one sound arrives at several microphones in turn, instead of making one clear-cut impression. (Phase distortion.) It avoids, too, the false changes in sound-perspective that occur when we alter the balance during performance, taking prominent instruments on solo microphones, and backgrounding others.

THE MULTI-MICROPHONE school of thought maintains that our natural impression of a source's characteristic quality comes largely from two-eared pick-up, and that the slight time-difference between a sound's reaching both ears contributes substantially to our assessment. This distinctive quality is not achievable, they say, with a single microphone pick-up where no phase-comparison exists.

They contend, too, that it is seldom possible to get one good microphone position for a variety of instruments having different quality and attack, owing to the irregularities of layout and acoustics. Distant pick-up of soloists results in thin quality. Without excellent acoustics, optimum layout, and sufficient rehearsal time, the multi-microphone approach is a better compromise, it is held, than a single microphone will permit.

It would be unwise to support either extreme dogmatically. Circumstances, the type of sound, and personal opinion must affect one's judgment. When intimate presence is required, for a small band, perhaps, where sectional close-up pictures are featured in turn, several microphones may be the answer. A full symphony orchestra treated identically, however, could lack expanse and tonal-character, so that a single microphone—backed at most by only slight fill-in from selected auxiliary microphones—may suffice.

Sound perspective

Superficially, we would assume that the principle of matching sound perspective to the picture scale was a foregone conclusion. A close-shot seems to demand close-up sound. A long-shot requires distant sound. In fact, though, it can be preferable to cheat the perspective to suit the occasion.

For example: if we were to intercut between an interior shot of a bedroom where an old lady is dying, and a street scene where her long-lost son is returning with his regiment from overseas, the cuts from the quiet room to the blaring military band would be disturbing. Rapid volume-jumps, especially if frequent, could be avoided,

113

either by making the change less pronounced, or by spreading the same sound over both scenes.

Again, if a singer is moving around a spacious set, should the voice-level remain constant as the distance varies? There are the naturalistic and the cheating schools of opinion.

Strict naturalism would prevent our adjusting the relative prominence of sounds to achieve dramatic effect when there was no realistic motivation.

Cheating can be equally overdone, though, as when the singing cowboy gallops away to the distant horizon, remaining in close-up sound throughout.

Sound perspective can be achieved by altering the microphone-distance to suit the length of shot (a hazardous procedure when cutting from close to long shots); occasionally, by mixing from a close microphone to a more distant one; by using a dialogue-equalizer. Because close-up sound is louder, and has generally stronger bass and top than more distant pick-up, an audio-filter has been used by some organizations to simulate the difference between close and remote sound quality.

Sound quality

Although high fidelity is our normal aim, there are several good arguments for "cooking" the sound quality. Purists might not agree.

Technically, cutting the lowest bass can help to keep down boom-rumble and hum, improve speech clarity and cut down the hollow or boomy quality of pick-up in boxy settings, etc. Cutting the highest part of the audio spectrum may reduce sibilant speech (S-blasting), quieten studio noise, hide noise on effects and music discs.

Artistically, bass and top cuts can enhance the illusion of exterior settings. Slightly accentuated bass can assist the impression of size and grandeur in a vast interior.

Drastic filtering and distortion are effective in certain applications. In simulated telephone conversations, the "distant" voice is treated, this distort source being transferred by video switching as the picture changes.

Clearly, heavy-handed, inartistic mutilation has nothing to recommend it, while sensitive, intelligently-contrived modifications can put the finishing touches to any visual illusion.

Acoustical control

Natural sound quality varies with the acoustical character of the source's surroundings. If we have scenes depicting a cathedral, a cave, a small room, a packed theatre, a mountain-top, we cannot

114

expect the same sound quality to suit them all. Echoing, booming, hollow, smothered, flat, edgy, the acoustic quality takes many forms.

In television and film studios we have: the acoustics the setting suggests (e.g. depicting the bottom of a well); the real acoustics the setting actually has (e.g. a single painted curved flat of plywood);

Fig. 6.17. Pt.1.

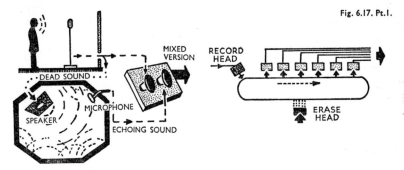

ARTIFICIALLY GENERATED REVERBERATION. *Left:* Echo Room. Sound from the dead studio is taken to the audio control console, some of it being tapped off and played over a loudspeaker in a special echo room, where hard, smooth walls cause the sound to echo and re-echo. The highly reverberant sound is picked up by a microphone and fed to a fader on the console. The direct sound and its reverberant version can be intermixed in any proportions required.

Right: Multi-headed Tape Recorder. The dead sound is recorded on a continuous loop of magnetic tape, then reproduced in rapid succession by a series of heads. The multiple version being combined and mixed with the original. The erase head wipes the tape clean for re-use each revolution.

Fig. 6.17. Pt.2.

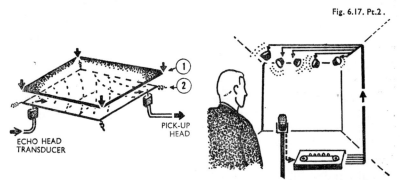

Left: Echo Plate. Audio fed to the echo head vibrates a large suspended steel sheet (2), these rebounding sounds being picked up by a second head and amplified. Adjustment of the distance between the steel plate and a parallel absorbent surface (1) varies reverberation time (over 1 sec.).

Simpler devices using steel springs as a reverberant medium are very compact, but can produce noticeable coloration of the sound quality.

Right: Multi-speaker Reverberation. The dead sound is fed into a multi-headed tape recorder, the delayed outputs of which are played over corresponding carefully positioned loudspeakers in the same studio. The reproduced sound reverberates within the studio and joins with the original source's microphone pick-up.

the studio's acoustics, giving us reflections from distant walls and ceiling.

By using dead studios and directional microphones, reverberation can be lessened. By drapes, carpets, upholstered furniture and limited acoustical treatment to scenic surfaces, reflection may be prevented from increasing excessively. But smooth, panelled, wooden, plastic, etc., surfaces will mean added sound reflection. They may resonate, too, at mid- and lower-frequencies.

Acoustical control is not easily achieved in television; although enthusiasts have provided astonishing realism by placing microphones in drain-pipes, dustbins, etc., to simulate confined surroundings. The ubiquitous acoustical screens of radio studios are rarely so usable in television to adjust local acoustics.

Reverberation can be simulated with arguable success by several means. The main methods are outlined in Fig. 6.17.

Practical sound problems

Many of the everyday problems for the sound mixer can be alleviated by cautionary warnings to the performers. To be able to control the audio level effectively, our mixer will, for instance, need to have sounds that are of more or less predictable volume.

Unexpected loud coughs, thumps, the crash of kitchenware on to a hard table-top, will send the volume-indicator reeling past its maximum limit. The listener will be treated to a tumultuous din. The sound transmitter may even overload and trip off the air. If the mixer quickly lowers the audio-gain to reduce the volume of the offending noise, all quieter sounds become inaudible by comparison.

Sources should preferably be of reasonably matched loudness. If two people talk, one whispering, the other shouting, the former is unlikely to be heard, for audio-gain is adjusted to accommodate the loudest sound. The microphone may compensate partially, by favouring the weaker source (e.g. to improve the separation between a soloist and a full orchestra a few feet away). Similarly, voices will be lost when talking against loud backgrounds of hammering, machinery, etc.

Extraneous noises should be kept to a minimum. Crackling paper, rustling taffeta, creaking wickerwork furniture, etc.

Sound pick-up should be kept to a minimum during wide movement, if possible, especially when the moves are fast or extensive.

Finally, the performer can help both the sound boom and the

116

cameraman by not talking down on to tables, or into a subject being demonstrated; by not moving about in rapid, erratic sprints or rising quickly from a seated position.

The last may lead not only to his being visually decapitated as his top half passes out of shot, but the odds are that he will bang his head on the microphone. This problem seldom arises more than once per speaker!

Recorded sound

Basic principles
Modern sound recording systems fall into three groups:—

Disc.
Magnetic tape.
Photographic film.

Each has its advantages and shortcomings.

Direct disc recording. Here a chisel-edged sapphire cutter engraves a fine, undulating V-shaped spiral in the cellulose lacquer on an aluminium blank. The swarf thread displaced is continually collected by suction or brush mechanism. The *pick-up's* reproducing needle is vibrated by the groove's undulations, and converted (as in the microphone) into an electrical audio signal.

Unfortunately, agreement is not universal among recording organizations on the extent of equalization necessary. Consequently, several standards developed concurrently. If we reproduce with the wrong compensation circuits, the audio-quality will be affected. We shall hear peaks and troughs, causing excess or loss in parts of the sound spectrum. Standardization trends have reduced this confusion (e.g. The National Assn. of Broadcasters' Standard), but a backlog of varied equalization still exists, and multiple switches are fitted to many professional reproducers to suit the disc being played.

Fig. 6.18.

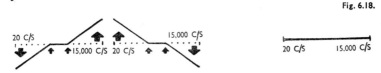

DISC EQUALIZATION. Recorded with bass roll-over around 250-500 cycles, with top pre-emphasis beginning around 500-2000 cycles (*left*), and reproduced with an exactly inverse characteristic (*centre*), the result is an even, overall response, avoiding bass overcutting during recording, and keeping reproduced disc surface to a minimum (*right*).

117

DISC REPRODUCTION TECHNIQUES. Most discs today, except archive material, are of microgroove form. Their potential high fidelity and quiet background noise is only maintained, however, by careful handling, lightweight high-grade disc reproducers, and scrupulous cleanliness. Suitable coverings, anti-static treatment, scratch avoidance, are essential for good disc maintenance.

All disc systems have certain inherent advantages: immediate access to anywhere in the recording—to extract a single sound is comparatively simple, or to make rapid jumps between sections; several discs can be used in quick succession without difficulty, even on the same turntable. Multiplay techniques from the same disc are practicable. Given the facility of a variable turntable, discs can be reproduced at various speeds for effects purposes (see page 409).

Against discs are their vulnerability to damage, susceptibility to wear, and background noise build-up.

The amount of information contained in a single groove is influenced by the disc's speed. Speech at 120 words per minute would result in $1\frac{1}{2}$ words per groove on a 78 r.p.m. disc, 3 words on a 45 r.p.m. disc groove, and nearly 4 words at $33\frac{1}{3}$ r.p.m.

This means that earlier techniques of dropping the pick-up into the prescribed groove of a moving disc are less practicable than for former high disc speeds. Wax pencil marking too is best avoided.

There are several techniques that enable us to select particular passages from any disc. A calibration scale can help here.

Playing it; fading up at the cue-point.
Playing up to the cue-point; stopping the turntable; restarting it on cue.
Playing to the cue-point; raising the pick-up there, poised over the required groove. Lowering the pick-up on cue, and fading up. On some equipment the disc itself is disengaged from the turntable instead.

A headset circuit usually allows the operator to pre-hear (pre-fade listen) the pick-up's reproduction, even when faded out, to facilitate setting-up.

Where provision is made to allow two pick-ups to reproduce one turntable's disc, we can obtain additional facilities:—

Continuous reproduction of a section or side of the disc.
Repeated effects.

Fig. 6.19.

BASIC TAPE RECORDER LAYOUT.
(1) supply spool; (2) take-up spool; (3) erase (wipe) head; (4) record head; (5) replay (reproduce) head; (6) tape drive roller; (7) idler pressure roller; (8) contact pressure pads or rollers. Tape may be spooled as shown or in cassette or in endless loop magazine forms.

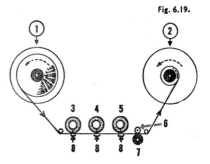

Reinforced effects (e.g. augmenting a crowd-noise disc by double-playing).

Echo (by playing two styli in one groove).

Magnetic tape. RECORDING AND REPRODUCTION. Unquestionably the simplest medium of all for recording and reproducing high-fidelity sounds, magnetic tape is suited to specialist and layman alike. Its applications are extensive. Its reliability and durability are high. Apart from avoiding over- or under-modulation through excessive or insufficient audio-amplification, the recording process consists of elementary switching. The rest is in the hands of those who are designing and maintaining the equipment, and the tape manufacturers.

The fundamental advantages of tape include: easy handling for recording and reproduction (particularly for cassettes); the recording is not worn by replay; accurate cueing is simple; the tape is readily edited.

The principal operational disadvantage of tape is the need to run the tape up to the playing point. Fast run enables us to reach various parts of the recording quickly by increasing the tape speed, but we cannot jump rapidly back and forth in a complete recording, as with disc. Instead, dubbing or editing is necessary.

Furthermore, some tape media do not store well, but buckle, or stretch, especially under high temperatures. Periodic rewinding and careful storage is desirable.

To record on the tape, we first neutralize it magnetically by an *erase head*.

Here a high ultra-sonic frequency is used to energize an electric magnet (occasionally a permanent or a D.C. electro-magnet, although this encourages higher background-noise.) We can bulk-erase the tape in a special unit beforehand, but more usually the tape is cleaned just before it approaches the record-head, as part of

119

Fig. 6.20.

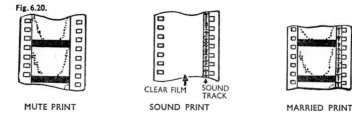

MUTE PRINT   CLEAR FILM   SOUND TRACK   SOUND PRINT   MARRIED PRINT

SOUND RECORDING ON PHOTOGRAPHIC FILM. The picture and the sound are usually recorded in separate cameras at the same speed, the picture at 24 frames per second (*left*) which is the same as 90 ft. per minute for the sound. This speed facilitates good recording and editing. The sound "picture" is a narrow track, fluctuating in density or area. They are then printed together on to one film, forming the combined (married) print, which is the version normally projected (*right*). The sound is printed 20 frames in advance of its corresponding picture to allow a loop to be formed in the projector between the intermittent movement at the projection gate and continuous movement at the sound head. This displacement creates problems if the combined print subsequently needs editing. We can also reproduce the two separate prints in synchronism, running them double-headed (duplexing) through the same projector or on individual picture and sound projectors.

the recording process. Any previous recording has now been wiped off; we can re-use the same tape almost indefinitely. Replaying the clean tape we should only hear the faint rustling of tape noise--if anything at all.

The neutral tape then passes against the record head, which is fed with the audio signal we wish to record. As each particle of the tape contacts the head it becomes magnetized. This magnetism corresponds in strength to the audio signal at that instant. This magnetic pattern is invisible, though, and we cannot tell by eye whether we are handling clean or recorded tape.

To assist in the recording action, the record-head is actually being fed with a little of the ultrasonic oscillation being used for erasure. This bias-current seems to "shake-up" the tape's molecules at the moment of recording (magnetically speaking), and helps to achieve better sound quality and lower tape-noise.

To reproduce the tape, we pass the magnetic pattern recorded on the tape over a re-play (playback; reproducer) head. Although very weak, this pattern causes an electric current to be generated in the head; the size and frequency of this current being proportional to the original audio signal.

Recording only takes place when the appropriate circuits are switched on by the record button. Some interlock device usually prevents our wiping a tape by its accidental use.

The duration of given tape footage depends on its running speed. Standard tape speeds range from 30 inches per second—15—$7\frac{1}{2}$—$3\frac{3}{4}$ —$1\frac{7}{8}$. A tape must be reproduced at the same speed as when

REPRODUCING THE OPTICAL PRINT.
Steady light from an exciter lamp is focused into a fine slit, and the area or intensity of this light interrupted by the passage of the film sound-track over it. The resultant light fluctuations are focused on to a photo-cell, which generates proportional currents—the audio signal.

MAGNETIC RECORDING ON FILM.
The sound is recorded upon a thin strip of oxide coating, replacing the usual optical track. This provides a print suitable for a projector having magnetic pick-up facilities.

Alternatively, the oxide coating can completely cover the film base. Multiple sound-tracks (e.g. dialogue, music, effects) can be recorded or reproduced simultaneously or individually, side by side. The sprocket-holes enable the mute and sound-track to be reproduced in exact synchronism.

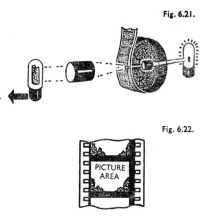

Fig. 6.21.

Fig. 6.22.

PICTURE AREA

recorded, or we shall get a marked pitch-change. The higher its speed, the higher the maximum frequency we can record and reproduce, and the simpler the editing process owing to the greater spread of the recorded sound pattern. The lower its speed, the greater the tape economy; but the greater the fall-off in higher frequencies. Background noise, too, is lower at greater speeds.

Fidelity depends too upon the head's design. One with a slit a quarter-thousandth of an inch wide will provide a response of 40 c/s to 14,000 c/s, with a tape-speed of $7\frac{1}{2}$ inches per second. A head of coarser slit lasts longer, but provides poorer high-note response.

Tape editing is simplicity itself. We cut the tape at the selected spot (preferably making a 45° splice) with non-magnetic scissors or razor-blade. The ends are rejoined with solvent, or an adhesive patch. The joint is inaudible. Problems arise where a tape has two or more side-by-side tracks, though, for any editing will involve them all.

One can edit by re-recording, or by cutting and rejoining. In dubbing, distortion, hum, background noise, and similar defects will become cumulative. How effective this approach is, therefore, must be influenced by the equipment being used. Editing by cutting is convenient, quick, and flexible, but necessarily disrupts the original recording. Where portions are to be selected from a complete sequence, it may be preferable to edit a dubbed copy. Non-magnetic tape is often interspliced as lace-up *leader*, and as *blanking*, during silent run-on periods between sequences, to overcome the need for intermediate stops and restarts.

To obtain additional copies of a tape-recording, one must resort to dubbing. High-speed duplication processes are used commercially for mass-production, but copying normally takes as long as the recording's duration, although several dubbings can be made simultaneously.

Audio effects

Most audio effects we meet will fall into two broad categories:

The individual *spot-effect*, usually made at a precise moment (e.g. a pistol-shot).

Continual background-sounds: of seawash, bird-song, traffic, etc.

Many spot-effects are quite successful when made by the performer himself, while others need the effects-man's skill to put over the right kind of sound illusion. The battering-ram forcing the castle door in our studio setting rarely produces a convincing impact. Better, perhaps, to dummy the noise by dropping a sandbag on to the ground, near a floor microphone. For brief sounds like door-bells, telephones, gun-shots, opinions are divided on the relative effectiveness of performer-operated, off-camera, and recorded effects. Each method has its advantages and drawbacks. Discs handle more flexibly than magnetic-tape, but are too easily marred by surface-noise. One solution lies in using push-button controlled loops of multi-track magnetic tape.

Some sounds can be simulated electronically. Gunfire effects, for example—ricochets and all. But many people feel that the ingenuity of these devices is more praiseworthy than their authenticity.

In everyday life, many a major disaster has taken place to surprisingly undramatic sound accompaniment. Cars crash, ships sink, fires rage with unimpressive cacophony. A concocted montage of sounds can frequently suggest the spirit of the occasion better than the true ones, although the sounds we use may be totally unrelated to the actual event (Table 17.II). And it is here that the choice and presentation of audio effects will so often come into their own.

Recordings are unquestionably the backbone of audio effects, especially for background-sounds. For, whether on disc or tape, they have the advantage of predictability. Their volume, duration and character are all known factors. Exact matching with the picture-action can be difficult, though. Just think of the comprehensive library needed to suit footsteps-effects alone.

Foldback

Playing sound-effects or music over studio-loudspeakers has the advantage that performers can synchronize their cues or action with them. But, against that, these sounds overheard by the studio-microphone can become coloured by studio acoustics and acquire a falsely hollow quality.

Miming and post-synchronizing

The business of having someone mouth silently the words of a recorded song as if themselves singing, needs little introduction. Many vaudeville acts have used this gimmick for comic-effect. This miming idea can help, too, when studio sound pick-up conditions are unfavourable, owing to balance problems, or the performer's exertions. The sound can be recorded first (pre-scoring), using that performer's or another's voice, and played-back over the studio-loudspeaker. Alternatively, the picture can be filmed first and sound provided live later (post-syncing).

Complete operas have been performed this way. But lip-sync is not easily maintained over long periods. Sound and lip movements all too easily fall out of step. The accurate synchronism that is possible when using these processes during motion picture making comes from repeated rehearsal. In live television it is safer instead to avoid the issue whenever possible.

Music and effects track

As television has spread throughout the world, it has become increasingly profitable to record programmes with an eye to potential markets. Clearly a show is more saleable if its duration fits neatly into typical commercial time-slots (e.g. 25 minutes) and permits intermediary "natural breaks" at appropriate intervals.

To extend sales, arrangements are now often made to supply with the picture not the normal total sound accompaniment but a "*music and effects*" track alone. Dialogue can then be added later in the language of the purchasing country (by post-synchronizing), and the original, appropriate music and effects included.

7

FILM REPRODUCTION

FILM televising equipment (Telecine) ranges in complexity from slightly adapted cinema projection apparatus, to electronic scanners transcribing the picture directly into a video signal. Both 35 mm. and 16 mm. systems are widely used (continual developments have markedly improved the latter), catering for optical and magnetic sound-tracks.

Generally speaking, the use of 8 mm. film (even when transferred to a 16 mm. print) has not proved satisfactory for television purposes, where good picture quality is the aim.

The principles of projection

If the film were drawn at a constant speed past a steady light-source, we should see only an indecipherable streaking image. We have to arrange, instead, to "freeze" each frame for an instant as it passes the picture-gate. This gives the eye time to discern successive pictures clearly.

This can be done by:

A mechanical shutter regularly interrupting a continuous light-source.

A regularly flashing light-source (pulsed by television synchronizing pulses).

Scanning the moving picture with an image of a television raster.

Optical dissolves between images held still by a prismatic or mirror system.

Productional requirements of film systems

Many shows will only require straightforward projection of a length or loop of film, at a constant speed. But there are several less conventional ways in which we can use film in television.

Holding a single film-frame for an indefinite period.

Running the film up from still-frame to full speed (on mute picture) in a short time (e.g. 5 secs.) without a flicker, frame-slip, moving phase-bars, etc.

Running at adjustable constant speeds, e.g. from 16 f.p.s. to 60 f.p.s.

Dissolving slowly from one frame to the next, in a film strip.

Transmitting colour and monochrome prints with equal fidelity.

Only the mirror-drum or prismatic-dissolve system satisfies all of these.

Picture projection rate

The standard projection-rate of motion pictures is 24 frames-per-second.[1] But standards of television systems are necessarily related to the frequency of the public A.C. power supplies. So, in Britain, the mains frequency being 50 c/s, the television systems produce 25 complete pictures per second.

Fortunately for such countries, it is possible to reproduce films using 24 f.p.s. standard film at this higher rate of 25 f.p.s. (i.e. 4% fast), without detectable speed-up.

Television systems based upon 60 c/s power supplies employ a 30 pictures/sec. repetition rate, and to attempt to project a 24 f.p.s. standard film at 30 f.p.s. would result in a 25% increase, which is unacceptable. Somehow, the 24 f.p.s. motion picture standard has to be adapted to the 30 f.p.s. television standard.

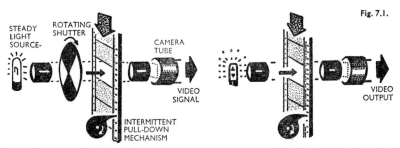

Fig. 7.1.

INTERMITTENT MOTION FILM PROJECTION SYSTEMS. *Left:* The steady light is regularly interrupted by a rotating mechanical shutter. Pulled down intermittently frame by frame, each picture is held and lit momentarily, and focused on to a camera-tube. The electron-image formed is subsequently scanned, producing a video signal. *Right:* With this method, each film frame is illuminated 2 or 3 times (while intermittently held), by a regularly flashing light from a gas-filled flash-tube. As previously, the image is focused on to a camera-tube (plumbicon or vidicon type), exposing during the vertical retrace (blanking) period.

[1] By covering and uncovering each *held* picture this rate can be effectively doubled.

125

CONTINUOUS MOTION FILM PROJECTION SYSTEMS. A plain white TV picture (raster) is displayed on the face of a special picture-tube. This is focused on to the transparency or film, which, according to its density, affects the brightness of the scanning spot when seen from the reverse side. A continuous video signal is generated by the photo-tube, corresponding to these light variations. This apparatus enables captions, slides, film-strip, etc., to be televised similarly.

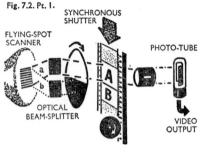

Fig. 7.2. Pt. 1.

The flying-spot image is passed through alternate optical paths to scan the continuously moving film. A synchronous shutter ensures that the *odd* scanning field explores the picture when the frame is at A, the *even* field scanning when the same frame has moved on to B, thus keeping the two in pace. The scanning raster has a 4 : $1\frac{1}{2}$ aspect ratio, so when A and B are interlaced the scanned area stretches to the standard 4 : 3 ratio, compensating for the film motion.

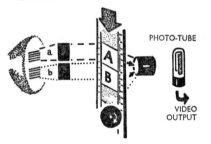

Fig. 7.2. Pt.2.

Another method is to have odd and even fields *displayed separately* on the flying-spot scanner. After odd scanning at A, the same frame is even-scanned at B.

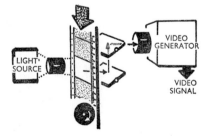

Fig. 7.2. Pt.3.

A succession of tilting mirrors or polygonal prisms cause the continuously moving film to appear stationary while being scanned. Each frame is held and then dissolved optically into the next. The light source may be constant, flashing, or flying-spot, and the video generator a camera-tube or (used with the flying-spot) a photo-tube.

CONVERTING FILM PICTURES (24 F.P.S.) TO TV (30 F.P.S.). This diagram shows the time relationships between the film pull-down and dwell time; the duration of its projection on to a storage-type camera-tube; and the scanning period. In the case of 35-mm. film an uneven pull-down rate and a 2–3–2 illuminating cycle is used; whereas for 16-mm. the pull-down is fast and even.

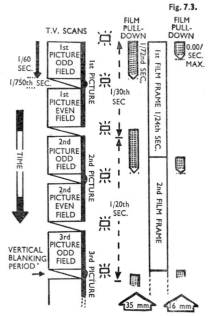

Fig. 7.3.

Ingenious intermediary mechanisms have been devised to deal with this situation. The methods usually consist of rotating shutters and electronic blanking-circuits, which sample the 24 f.p.s. projected film in a rhythmical 30 f.p.s. sequence.

Film print quality for television

Although any print acceptable for cinema screening will televise, we can often improve matters by careful attention to print quality by avoiding dense, flat, excessively high or low-contrast prints; shots that vary rapidly or widely in contrast or brightness; large dense areas of very light or very dark tone; pictures relying largely upon colour for their visual appeal (in monochrome television).

Fairly thin, well-graded prints are usually the most suitable.

Televising colour film on monochrome television

Televising colour film brings its problems.

Firstly, the colour of the *film-scanner's luminant* can modify tonal values. We shall notice this particularly with flesh-tones, which may be unduly light or dark. Where the light-source is white, however, and the pick-up device panchromatic, tonal quality will be naturalistic—providing the original colour film was satisfactory.

127

TABLE 7.1

FILM STANDARDS

35 mm. projection	16 mm. projection
Emulsion towards light source.	16 mm. B. & W. Usually emulsion to light. 16 mm. colour. Emulsion towards *lens*.
Sound-track on R.H.S. of picture.	Sound-track on L.H.S. of picture.
Projected at 24 frames per sec. (25 f.p.s. for some television systems).	24 f.p.s. sound copies (1 ft. = 1$\frac{2}{3}$ secs.). 16 f.p.s. silent films only (1 ft. = 2$\frac{1}{2}$ secs.).
16 frames to each foot of film.	40 frames per foot.
90 ft. projected per min. running time (18 ins./sec.).	36 ft. per min. at 24 f.p.s. (7.2 ins./sec.).
The duration of 1000 ft. of 35 mm. film equals 400 ft. of 16 mm. film.	
Sound track 20 frames (15 inches) in advance of corresponding picture.	Sound in advance by 26 frames is standard, but 20 frames advanced when print is a photographic reduction from 35 mm.
Standard reel sizes: 1000 ft., 2000 ft.	100 ft., 200 ft., 400 ft., 800 ft., 1200 ft., 1600 ft., 2000 ft.

An incandescent source, with a high red content, is more likely to pale-out reddish hues, while a flash-tube light-source having a blue hue, has the reverse effect; lightening blues, darkening reds. Colour filters may improve the conversion quality.

Most flying-spot scanners use blue or green phosphors, because of their higher actinic value and short after-glow period. So, remembering that a filter lightens its own colour, while darkening its complementary hue, we shall expect to find drastic colour-distortion. With poor colour film, conversion can be imperfect indeed.

A further variation can arise through distinctions between techniques when making colour and monochrome motion pictures. The lighting techniques particularly differ. Whereas black-and-white photography must express itself through tonal variation alone, colour media can employ both tonal and colour separation, to distinguish between planes. Colour separation may mean nothing when transformed into monochrome. And we shall lose too the psychological impact of juxtaposed colours.

In short, distortions and differing emphasis may cause a televised colour film to become entirely remote from the original.

Televising negative film

The film negative will generally be too precious to risk damaging through continual handling and projection. At least one positive rough-cut or rehearsal-print will usually be made; possibly with a show copy for on-the-air performance.

But where we must avoid the time or expense of making a positive-print, the negative itself can be reproduced. The video signal's polarity can be reversed by simple circuitry to phase-reverse the picture. This inverts all picture tones, turning a positive into negative, or a negative to positive.

Video equipment necessarily needs a minimum of light so that film shots of dark scenes may televise better in negative form (phase-reversed later in the video chain). For, in passing more light, they allow improved camera-tube performance.

The electronic control of picture quality has its complications when televising negative stock. The overall gamma of the video system, when reversed for negative pick-up, may differ from its normal positive characteristics, so the electronic positive may not be identical to printed positive tonal quality. Any print blemishes will be reversed in tone. Picture background (e.g. the phosphor-grain in flying-spot scanners) can become prominent. Electronic controls on the film-scanner will "feel" substantially different for the operator when phase-reversing negative, and can make rapid, unobtrusive compensations more difficult (e.g. when intercut shots have widely varying exposure).

The mechanics of film projection

We can project film:—

As a *reel* or reels of up to 20 mins., as a *length* of limited duration (e.g. 2 secs. 2 mins.). A series of lengths will often be joined by opaque *blanking*, and wound on to one reel. Individual isolating-leaders may be used for their cueing, as a continuously-repeated close *loop*, or as short *film-strips* made from a number of still-frames.

The leader

Any film is usually headed by a *leader*. In its full form, we have:

The *protective-leader* (6-8 ft. of raw, unused stock).

The 24-frame *identification-leader*, giving details of print-type, reel number, picture title.

The 20-frame plus 12 ft. *synchronizing-leader*, giving projection

129

information, and numbered footage marks from 11 ft. up to 3 ft. before the first picture.

This ensures sound/vision synchronism by aligning their respective start-marks, and facilitates accurately-timed changeovers between reels. Abbreviated and modified forms of the original A.S.A. leader have been developed for television use; providing, in addition, linearity and tonal-limit checks, etc. The Society leader meets both motion picture and television projection needs.

Cueing

Knowing the run-up time taken to attain full speed, we can set the leader with a suitable footage-mark in the projector's picture-gate. Then, by anticipating the cut-in moment into studio action by that period (e.g. 3-8 secs.), we cue the film. At the instant we require it, the first frame appears, and the picture is switched on to transmission. Where such precision is unnecessary, we can cue the film to "roll", and cut over to it when it appears. A few surplus frames are best left to "top-and-tail" any inserted film length, to avoid inadvertent cropping or run-out.

Change-over synchronizing

Change-over cues take the form of a small, white-circled black dot in the top right corner of the film-frame. The motor-cue at the conclusion of a reel lasts 1/6th sec. (i.e. 4 frames). This signals the operator to start the second projector's motor. After 10 ft. and 12 frames (7.1/6th secs. at 24 f.p.s.) another identical cue-mark appears, to signal him to switch vision to the new projector. He has 12 secs. in which to do this (18 frames) before the opaque run-out trailer. Reels will usually end with a fade-out, or cut to a new scene. Where changes are made during action, however—particularly when accompanied by sustaining sound—the transition can be unavoidably obvious.

Film loops

By joining the beginning and end of a length of film, we obtain a loop. Continuously projected, this gives endless repetition.

With such a loop, we can demonstrate a particular cyclic motion. The follow-through action in a golf-stroke; the complex operation of a lathe-tool. In television, the loop is widely used for continuous effects which have no marked rhythm, such as rain, mist, snow (see Visual Effects, Chapter 20).

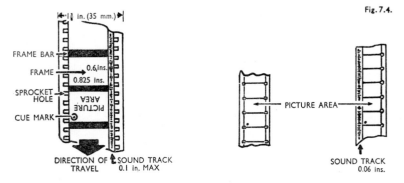

Fig. 7.4.

FILM STANDARDS. *Left:* 35 mm. sound. Projected frame area is .825 ins. x .6 ins., and camera frame area is .868 ins. x .631 ins. *Right:* 16 mm. mute and sound. Projected frame area is .38 ins. x .284 ins., camera frame area is .41 ins. x .29 ins.

By careful editing, it is normally possible to match the joined ends, so as to avoid any jump or jerk as the loop is repeated.

Maximum loop duration depends upon the projector's design. Whereas 10 secs. is a typical limit, 3 mins. or more can occasionally be accommodated. The film is then carried in take-up pulleys, or in a centreless spool.

Fig. 7.5. Pt. 1.

WIDE-SCREEN PRESENTATION. The standard television screen has a 4 x 3 shape, so we can only project a wide-screen film by fitting the film's height to the TV screen's height, thus wasting the side sections of the film, or by fitting the film's width to the screen's width, wasting the top and bottom of the TV screen. This is done by adjusting the film-scanner's frame height and width electronically; by suitable projector-lens adaptation; or by having a special 4 : 3 print made (identical compensations being made by the film labs. instead).

The first method (*top left*) disrupts the original composition, and widely-spaced subjects may be lost. The second (*top right*) produces a sawn-off picture with little pretence at a panoramic effect, but none of the original is lost.

Fig. 7.5. Pt. 2.

A convenient alternative is to use an electronic or optical zoom, adjusting the picture format to suit the situation, panning (*left*) by electronic shift in the scanning circuits. This will have to be carefully selected and controlled for optimum effect.

131

TABLE 7.II

FILM RUNNING TIME

Durations	35-mm. projection at 24 f.p.s.	at 25 f.p.s.	16-mm. projection at 24 f.p.s.
1 sec.	1 ft. 6 ins.	1 ft. 9 frames	3/5 ft.
5 secs.	7 ft. 6 ins.	7 ft. 13 frames	3 ft.
10 secs.	15 ft.	15 ft. 10 frames	6 ft.
30 secs.	45 ft.	46 ft. 12 frames	18 ft.
1 min.	90 ft.	93 ft. 8 frames	36 ft.
5 mins.	450 ft.		180 ft.
10 mins.	900 ft.		360 ft.
15 mins.	1350 ft.		540 ft.
30 mins.	2700 ft.		1080 ft.
45 mins.	4050 ft.		1620 ft.
60 mins.	5400 ft.		2160 ft.

Telerecording

Filming the image on the face of a picture-tube (kinescope) presents several technical problems.

The film-recording device must lock into the television system's scanning rate, or else we shall see horizontal banding (phase-bars; shutter bars) passing up and down the picture; or similar stroboscopic effects.

The picture's line-structure must be obliterated (by vibrating the scanning spot vertically at a high rate), otherwise there will be strobing effects (moiré patterning) on re-televising the recording, due to the interaction of the new and the original scanning-lines.

The relatively low light-level available from the picture-tube means that highly sensitive film stock is essential. Colour stock is slow, so colour kinescope recording is achieved by combining separate R-G-B records.

As laboratory processing is needed before the film recording can be checked or edited, delay and uncertainty are unavoidable.

Optical telerecording equipment ranges from continuous-motion cameras using flying-spot or mirror systems, to synchronized intermittent methods, using shutters and ultra-rapid film pull-down during picture-blanking (frame suppression) periods. The results currently vary from picture quality barely distinguishable from the original telecast, to picture quality that is barely recognizable!

Video-tape magnetic systems can record either monochrome or colour television, with its accompanying sound. Equipment ranges from large studio consoles, with full monitoring and many associated facilities, to one-man-camera back-packs of high mobility.

Recording methods, too, vary with design. For example, a single head may have a 1-in. tape circuiting it in a helical scan; or a rotating multi-headed recorder may be used transversely scanning a 2-in.-wide tape. Various ingenious methods have been devised to achieve fast, *effective* recording speed on a tape actually moving at some 10–15 in. a second. This has resulted, however, in a diversity of non-compatible equipment design approaches.

Video tape has the outstanding merit of providing immediate playback. By brushing on a suspension of carbonyl iron the tiny line pattern of fluctuating density can actually be made visible for editing or test purposes.

Reproduction simply involves using the recording head in a "replay mode"; the sound having been recorded simultaneously along the edge of the tape.

Editing processes can take several forms:

CUTTING AND SPLICING, where we run the tape up to the editing point, mark it, recheck it, and then paint the tape to reveal the synchronizing signals at which to make the cut. Unfortunately splices mechanically weaken the tape. As sound and picture are spacially displaced along the tape, "tight" editing can be difficult.

DUBBING involves switching between two or more reproducers to re-record their outputs on to a composite tape. But dubbing usually involves quality deterioration, and versions beyond third generation are seldom acceptable for transmission.

MODE CHANGE employs the deceptively simple concept of reproducing one's tape up to the editing point, then switching the machine instantly to the "recording mode", to take the new material. So the resultant tape presents electronically conjoined pictures. Even simple animation is possible this way.

Compared with film production, where edits are quickly and easily made, re-examined and modified, video-tape editing processes originally proved quite laborious and inflexible. Facilities developed, until with the CMX System extremely sophisticated video-tape editing became feasible. This permits random access, vari-speed reproduction, reverse-play, stop-motion, transition-selection and colour correction among its other features.

8

TELEVISION SETTINGS

THE design of television settings has developed from a rich inheritance of theatrical and motion picture practice. But the television designer or art director has many problems that are peculiar to his own medium. These can require a fresh and often extremely original approach. Limitations of space, time, man-hours and money have to be accepted.

And it is here particularly that we shall begin to appreciate not only the art, but the craft—in both senses of the word—that underlies the designer's work.

To be completely successful, settings should satisfy several requirements:—

Artistically, they must be appropriate to the subject; and to the programme's purposes and aims.

Studio mechanics must suit the studio dimensions and facilities, accommodate the production treatment, giving operational freedom to cameras, sound, lighting, etc.

Camera parameters. The tones, colours, contrasts and finish of settings should take into account the characteristics of the camera-tubes and, hence, the studio lighting techniques, if we are to achieve the optimum pictorial effect.

This done, the success of the designer's work relies not only on his own vision and interpretation, but on that of the lighting director, and the programme director himself.

Types of setting

"Setting" is a very broad term. It can include scenic arrangements ranging from a hung drape to a full-scale replica of a village. Whatever might be its actual form, we are able to classify its artistic aim as follows:

Neutrality
Non-representational. Without immediate association of ideas or locale, e.g. a plain, grey, background.

Realism
Replica. A faithful copy of the original scene. Accurate reproduction of detail, i.e. the setting looks like a particular place.

Atmospheric realism. A setting portraying a type of scene, e.g. resembling a typical Victorian drawing-room.

Symbolic realism. From apparent atmospheric realism, where we combine a few associative details to suggest or simulate (for instance, using a desk and chair before a plain background to suggest a typical office), to a single symbolic object that recalls a certain environment (e.g. a branch, or even its shadow, suggests a complete tree).

Fantasy (unrealistic settings)
Abstraction. Shape, form, and texture, arranged to express mood, thought, character, etc., without direct relationship to the real world.

Silhouette. A concentration on subject outline.

Bizarre. Deliberately distorted reality. Covering a field from vaudeville backcloths and cut-outs, futuristic settings, to the lopsided houses of the madman's world in *Dr Caligari*.

Studio layout

The studio plan (ground plan; floor plan; staging plan)
A bird's-eye scale drawing of the studio floor, generally drawn on $\frac{1}{4}$ in. = 1 ft. squared paper, shows the layout, proportion, and distribution of settings. Later, the productional mechanics are indicated upon it (standardized symbols represent setting-units, furniture, etc.). After mutual acceptance, copies of the plan are distributed to the programme director, art department, technical and lighting directors, etc.

Elevations
A $\frac{1}{4}$ in. scale side view, showing the surface features and dimensions of all vertical scenic planes. Larger scale versions provide

135

information to scenic artists, carpenters, etc., on detailed design or construction.

Scale model

Sometimes a utility scale-model of the studio arrangements is prepared by erecting cut-out elevations upon the studio plan. Scale details can be added if required. This helps those concerned with direction and staging, and the performers themselves, to visualize the completed set-up.

With imagination and experience, one seldom requires models, but where the setting is complicated and precise visual treatment is intended, the scale-model can save considerable time and misunderstanding. Using a miniature viewfinder, we can see directly the shot obtained from any viewpoint.

Sketches

The designer will frequently supply sketches. These may be key-shots, depicting the programme director's visual treatment; atmospheric sketches showing the feel of the setting; scaled architectural sketches; or simply rough scribbles. Whether used for operational planning or just to bring to life the atmosphere that stark, clinical, scale-drawings cannot convey, these sketches can inspire and coordinate the rest of the visual team.

Methods of setting construction

The studio setting is normally built from a number of separate, prefabricated scenic-units, positioned and fastened together. Subsequently, they are *dressed* with appropriate furnishings, properties, drapes, etc., to create the total scenic effect.

By using articulated units, sets can be erected (set-up) or struck (pulled apart) with the minimum time and trouble. Most of the component units will have been designed so that they can be taken out and re-used over again, in different combinations, for many productions.

Although their appearance can be altered considerably by resurfacing, these stock pieces (modular units) have several basic generic forms

The flat.
The solid-piece (rigid-units).
The cut-out.
The background.

136

Fig. 8.1. Pt.1.

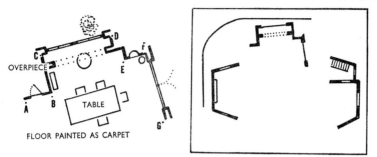

STUDIO LAY-OUT. *Left:* The setting plan. *Right:* The complete studio plan.

Fig. 8.1. Pt.2.

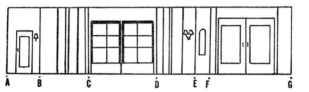

The extended elevations. (Details of finish, etc., have been omitted for simplicity.)

The flat

The flat is built up of a plane surface of fireproofed hessian, $\frac{1}{4}$ in. plywood, or prepared boarding, on wooden frames. Sizes range from 8 ft. to 14 ft. high, and 6 in. to 12 ft. wide.

Fig. 8.2. Pt.1.

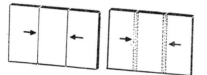

FLATS. *Left:* Where flats are set edge-to-edge, their joins show. *Right:* When surface pattern does not hide joins, they may be covered with pasted-on paper or fabric stripping, but this precludes later re-angling of flats and may still not hide joins in plain, light surfaces.

Fig. 8.2. Pt.2.

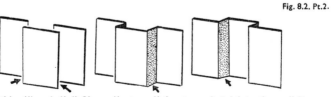

Where possible, "breaks" (*left*), or "returns" (*centre* and *right*) in the wall-lines are a more effective disguise, visually more interesting, and provide better acoustics.

137

Fig. 8.3.

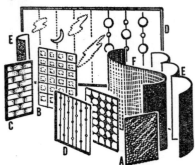

SCREENS AND DECORATIVE FLATS.
(A) Decorative screens of expanded metal, wire mesh, perforated metal, etc.; (B) tracery of plastic, plaster, metal, wood; (C) woven screens of cardboard, wood, metal, wicker, slats, cord; (D) open screens of stretched wire or cord—supporting small motifs, profiles, decoration—spaced poles, slats, etc.; (E) curved flats (small cycloramas); (F) flexible screens.

Many surface treatments are possible. A surface colouring medium is commonest; using a distempera, or casein prepared "cold water" paints. Wallpaper provides an enormous selection of patterns, tones and textures. Finally, there are the less usual coverings of fabric, patterned or expanded metal, plastics and cording.

Any surface-modelling can be simulated by painting, or with suitable photographic or printed papers, or modelling may be built up in wood, papier mâché, canvas-covered wire, plaster, or thin plastic relief modelling attached to flattage.

The solid piece

Solid pieces (rigid units, built pieces) fall into two categories:—

(i) Those that are direct imitations of architectural features. This includes such features as doorways, fireplaces, windows. These are built as hollow wooden or plastic shells, and fitted as plugs into contoured flats, or set up as free-standing units.

Fig. 8.4.

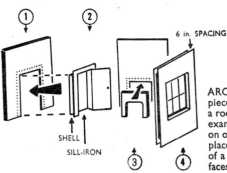

ARCHITECTURAL UNITS. With solid pieces we build up the main features of a room. (1) contoured flat or frame (an example of a single-sided unit—viewable on one face only); (2) door plug; (3) fireplace plug; (4) window unit (an example of a double-clad unit—viewable on both faces).

138

The field here is endless: staircases, columns, balconies, lamp-posts, garden-pools, statuary, rocks, etc.

(ii) The second group includes units that serve useful mechanical purposes, but have no directly imitative associations, such as rostra and ramps.

Fig. 8.5.

BUILT PIECES. (1) Pillars: cylinders or half-shells of plywood, 6 ins. — 2 ft. diameter, up to 15 ft. high; (2) Staircases (steps, stair-units): groups of two or more treads, usually with 6-in. risers, matching the heights of stock rostra; (3) Rostra (platforms, parallels): variously shaped level platforms, on permanent or folding frames with sides boarded-in as necessary and painted; (4) Ramp: sloping plane surfaces; (5) Arch-way: a typical solid piece, resembling masonry; (6) Drape frame: a light framework in single or hinged units, used to carry draperies, instead of hanging them from a bar or barrel; (7) Cove: a shallow, concave surface, used to merge horizontal and vertical planes without a join; (8) Block (step-box): wooden shells, from about 6 ins. x 1 ft. to 2 ft. x 2 ft., to provide half-steps, to raise the height of actors, furniture, etc.

The cut-out or profile-piece (set-piece). See Figure 8.6

The cut-out consists of a flat, vertical profile in plywood, compressed card, prepared board, etc., often attached to stock flats to modify their outline. We shall meet the cut-out applied:

As isolated decorative pieces, e.g. a flat "Christmas tree" cut-out representation;

As wings masking off the edges of an acting area;

As ground-rows, concealing a background floor join; or

To suggest scenic planes, e.g. representing a skyline of hills, roof-tops, and the intermediary terrain.

Because cut-outs are invariably flat-faced, they seldom bear close scrutiny where they are supposed to represent solid subjects. Only at a distance does a cut-out tree, for example, still look like a real tree. Viewed at an angle, the illusion fails completely. Surface painting is seen to be flat. Where they are only decorative non-realistic planes, this will not matter.

Suitably painted and built, though, cut-outs can be cheap, very effective, substitutes for elaborate scenic arrangements. As the camera moves, we see parallactic movement between planes, achieving a surprisingly realistic, three-dimensional, impression.

Fig. 8.6. Pt. I.

THE CUT-OUT OR PROFILE-PIECE. *Left:* Profile pieces painted in perspective help to create the illusion of distance for a street scene. *Right:* Profiled ground rows suggest progressive planes.

Fig. 8.6. Pt.2.

Ground row (*left*); wing (rear and front views) (*centre*); decorative piece (*right*).

Backgrounds

Strictly speaking, any surface seen behind a subject is its background. Some people use the term "backing" synonymously. More precisely, *background* is a scenic surface in front of which action takes place; *backing* applies to parts of a background seen beyond windows, doors, voids, etc. of our setting, usually implying more distant planes.

The neutral background

A non-associative background. At its simplest, a plain, even-toned surface, formed by a run of edge-to-edge flats, a backcloth, cyclorama, translucent screen.... More decorative effects can be obtained by uneven lighting (shading, stippling), or surface texturing (puddling, spattering, etc.), decorative screens, and so on.

Drapes have wide applications for neutral backgrounds, from cottons and linens, to lush velours and velvets. (Plasticized or glazed materials, remember, may reflect troublesome flares.)

Stretched, pleated, swagged, their fluid form makes them, perhaps, the most adaptable of scenic elements. A warning, though, against the prominently-patterned material. Not only does it distract, and tend to make foreground faces appear defocused, perhaps causing moiré effects; it soon becomes recognized by the regular viewer.

The flat pictorial background

There is obviously a limit to how elaborate a set we can build in the studio. Fortunately, the camera cannot distinguish easily at a distance between flat and solid subjects, so, even when seeking complete realism, we can simulate it convincingly on a flat pictorial background.

Ideally, it will need to be: free from blemishes, spurious hot spots or shadows; matched to the perspective, proportions, tones, etc., of the foreground; viewed straight-on, to avoid the distortion of angular viewing.

In practice, surprising deviations from the ideal will still deceive.

These flat scenic backgrounds are provided in television by the following means:

Painted cloths.
Photographic enlargements.
Back projection.
Electronically inserted backgrounds.

Painted cloths (backdrops, backcloths). Extensively used, these large painted sheets are often creative masterpieces of the scenic artist; at times almost indistinguishable from photographic reproduction. The cloth may be rolled around the pole that weights its lower edge, or flown by being raised out of view. Of canvas or twill, cloths are normally hung on pipes, battens, or frames.

To reduce the painted-look that even the best cloths are liable to have, a black or white gauze net is sometimes hung over them to lend greater distance.

Photographic enlargements (photo-murals, photo blow-ups). Enlarged photographs represent the ultimate realism obtainable from backcloths. Enlargements are made on sections of sensitized paper, stuck on to a flat, or a canvas support. More elaborate translucents are enlarged on to sensitized fabric, when they are often rear-lit.

Initially expensive, the photographic enlargement can be stored and, like a painted cloth, be re-used many times; particularly where only portions of the whole are seen, as when backing a window.

Back projection (B.P., rear projection, process projection). See Chapter 20. Here, a still or moving picture of the background subject is projected on to the rear of a translucent screen, before which action takes place.

Electronically inserted backgrounds. See Chapter 20. By special electronic equipment, we can place one camera's performers and/or

Fig. 8.7

SKELETAL SETTINGS. May be decorative effects based directly upon natural subjects (*top left*); may use fragments of realistic settings (*top right*); or may be abstract pattern (*bottom*).

scenery in the background picture provided by another picture source. Unseen by the performers themselves, this background can come from any camera, film, slide projector, etc. The action appears as if actually happening within this artificial background, derived from a photograph, drawing, or model, etc.

Skeletal settings

A product of modern theatre, skeletal settings have found some application in motion pictures and television, for musicals, dance spectacles, and the like. They are usually formed from isolated scenic units, or from a framework of metal or wood ribs, with profiled areas attached to this assembly.

The effect is a spatial pattern; purely decorative, or with abstract associations.

Primarily of value for non-realistic spectacle, the skeletal setting is seen to best advantage in long shots. In closer viewpoints it loses its pictorial appeal.

Surface formation

Apart from any contouring built into them, most scenic units will have smooth, even surfaces. Where we need texture, surface relief, or figuring of any kind, this can be added by several standard methods.

Surfacing

By careful surface painting, we can imitate the finish of many materials with surprising realism.

Wood graining, by combing and brushing a painted surface, is a familiar everyday example. Many allied techniques enable us to transform a flat lay-in (even-toned, textureless, surface painting).

Dry-brush work. Overpainting a dry, flat surface with a nearly dry brush, to leave a sparse pattern of brush-marks across the ground colour. Suggests metal, wood, stone, or fabric.

Stippling. A series of small, close dots or patterns of colour on a different ground-tone. Applied by a coarse brush, sponge, wrinkled paper, cloth. The mottling suggests stone, cement, or earth.

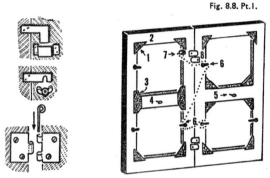

Fig. 8.8. Pt.1.

METHODS OF FIXING AND HOLDING SCENIC UNITS. Most scenic units are not self-supporting. Quick-fix arrangements enable us to clamp (*left*), clip by L-plate, wing-nut, or loose pin hinge (*centre*), or lash units together in a stable assembly (*right*). (1) corner block; (2) rail; (3) keystone; (4) toggle; (5) brace eye; (6) brace cleat; (7) lash eye; (8) stop cleats to align flats.

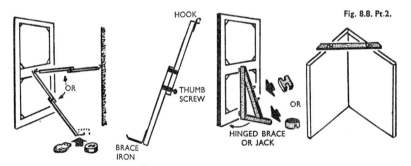

Fig. 8.8. Pt.2.

The unit can be braced to the studio wall, or supported by a brace weighted with a stage weight or sand-bag (*left*). Occasionally, an extensible stage brace is used with the brace-iron (*centre-left*) fixed to the floor where a very solid support is needed.

Other methods include a weighted "jack" (*centre right*), or especially where double-cladding prevents normal bracing, bracing struts across the tops of sets (*right*). In certain circumstances (columns, trees, poles, etc.) suspension lines, or bottom weighting, may be the only practicable methods of support.

143

Puddling. Wet colours, allowed to flow together, intermixing to give random variations suggesting ageing plaster walls, earth, etc.

Daubing. Dabs of colour, applied with a rolled rag, patted irregularly over a surface, giving varying density of tone.

Scumbling. A translucent coating, usually of a darker tone over a lighter one.

Glaze. A transparent dry-brushed application of a lighter tone on a background, e.g. to suggest highlight sheen.

Scuffing (dragging). Brushing so that painting skims a surface, leaving textural depressions untouched.

Wash. A thin coating of lighter or darker tone, over a background (body) colour, simulating highlights or shadows.

Rough-cast. Sprinkling a powdered material (sawdust, sand) irregularly on to a freshly painted surface, to introduce random changes in tone or texture.

Spattering (dottling). Mottling with random brush-thrown splashes.

Surface detail and contouring

Because one can readily suggest solidity by suitable drawing and shading, we can provide an illusion of brickwork, panelling, moulding, etc., by brush alone. Realism will depend upon the artist's skill; upon how closely the camera scrutinizes it, and whether oblique viewpoints or cast shadows give away the true contours.

Various photographic wallpapers available, enable brickwork (of varied type and age), stonework, wood-graining, rough-cast, pebble beaches, etc., to be laid by the yard. Little skill or labour is needed, and surface details can be painted on as required.

Papier mâché, plaster, fibreglass, plastics. These can all produce real works of art in the hands of skilled craftsmen. From spectacular statuary and architectural features, to tree-trunks, brick walls, and cobbled paving, surfaces can be formed that completely bear the closest scrutiny—even up to fingernail scratching distance. Each material has its own particular advantages and disadvantages concerning cost, weight, durability, ease of working, etc.

Wooden contours. Can provide detail such as carvings, panelling, mouldings, etc., attached to stock units, and removed and stored as needed. Glass-fibre or plastic reliefs are now increasingly used instead.

Floor treatment

The studio floor surface is generally in the middle tones, serving as a neutral ground for most programmes. However, to leave it like this would be to throw away many scenic opportunities.

Painting

Is the most economical and comprehensive method of floor treatment, for water paints enable temporary washable surfaces to be applied and removed at will. As well as overall tonal changes, we can suggest paving, floor-boards, carpets, decorative motifs, abstract design, and so on. For simple decoration, plastic tape and stencilling have been found hardwearing, and easy to lay down.

Painted floors have the advantage of allowing free camera-dolly movement, but they are liable to become dirtied all too easily. Black floors are particularly vulnerable, showing all tyre and foot marks.

Scattering

Scattering of innocuous materials like peat, sawdust, cork chips, leaves, can transform the floor surface quickly for naturalistic effects. They do tend, however, to stray around the studio, especially where wind machines are in use. Materials such as sand or salt should invariably be avoided, for they foul-up studio equipment.

Covering materials

Further convenient methods of floor treatment include the use of carpets, raffia matting; heavy duck, canvas, or tarpaulin floor cloths (plain or pre-painted); photographic or patterned wallpaper; panels of prepared board, plaster; or turf. All merely require laying. On the other hand, they may impede camera movement.

Ceilings

There have been designers who contended that a ceiling is worth more to an interior's authenticity than the most extensive care in its dressing and surface treatment. Be that as it may, a ceiling certainly can add a psychological "something" in appropriate circumstances.

Unfortunately, ceilings bring their productional problems, too. They often restrict lighting treatment badly, and can provide hollow,

Fig. 8.9. Pt.1.

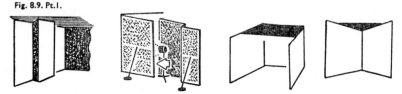

CEILINGS. A complete ceiling is feasible, if action and treatment are planned to allow lighting through windows, doors, by lamps behind furniture, and by "practical" lamps, etc. Translucent ceiling material may help but backlight projected through it produces hot-spots. *Left:* Also by false returns in walls, so that a seemingly-normal side wall from the front, allows concealed lighting from behind. *Right:* Partial ceilings, positioned to suit shots that necessitate a ceiling.

Fig. 8.9. Pt.2.

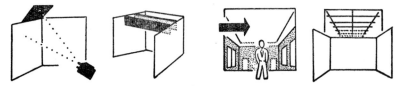

Left: Overhanging flats or (*centre left*): cutting pieces (e.g. beams, vaulting) and suspended fittings (e.g. chandeliers) entering the top of the picture, suggest a ceiling. *Right:* ceiling detail painted on the background (suitable for limited action and static camera), or on a cloth hung behind the set.

Fig. 8.9. Pt.3.

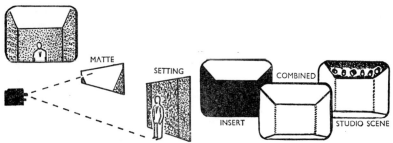

MATTE

SETTING

COMBINED

INSERT

STUDIO SCENE

Left: A camera-matte between camera and settings. *Right:* electronic insertion; fitting a picture of the ceiling, from another source, into the studio camera's shot.

"boxy", sound quality. More often than not the viewer will have to imagine the ceiling, just as he does the non-existent fourth wall.

But where the camera takes low viewpoints in an interior scene, we can no longer ignore the ceiling dilemma. If we see interminably high walls, or meaningless void, instead of a ceiling, the illusion will be unconvincing. Numerous solutions have been devised, and these are summarized in Figure 8.9.

Studio mechanics and set design

As we have seen, the most obvious approaches to set design are not necessarily the most practicable, either economically or operationally. In this section we shall meet many routine design problems, and their typical solutions.

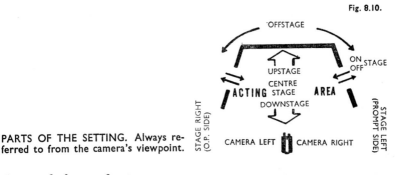

Fig. 8.10.

PARTS OF THE SETTING. Always referred to from the camera's viewpoint.

Size and shape of sets

The floor area taken up by a setting will be influenced by such factors as:—studio size and construction methods; the number and size of other sets; the kind and extent of action required; the type of setting we are depicting.

Naturalistic settings will rarely be much larger than their original, if they are not to look too vast and empty in long shots. Frequently they will be much smaller. Often their true size and perspective will be enhanced by using wide-angle lenses, so that quite a sawn-down version appears spacious on the screen. Lastly, sets may be provided by partial construction, special effects, and other scenic devices, when the size of set will be determined chiefly by the amount of performers' movement we require.

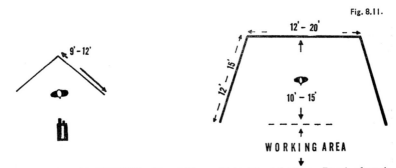

Fig. 8.11.

TYPICAL SIZES OF SETTINGS. Two-fold set. *Right:* Three-fold set. Depth of working area in front of set is about 15 ft.

147

Fig. 8.12. Pt.1.

PROPORTIONS OF SETTINGS. *Left:* Cross-shots will shoot-off easily. The set lacks depth and solidity, but occupies little room (can be piled or nested). *Centre and left:* Slightly improved, but setting looks disproportionately wide in cross-shots.

Fig. 8.12. Pt.2.

Left: Very satisfactory proportions for general use. *Centre left:* Absence of splay on side-walls tends to crowd cameras towards centre, but prevents exaggerated room-size effects in cross-shots. *Centre-right:* Any cross-shooting is restricted. Cameras and sound-boom congest towards the centre. *Right:* Little deviation from straight-on shooting is possible. Sound pick-up may be "boxy", and lighting tends to be steep.

Fig. 8.12. Pt.3.

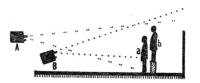

The height of a setting depends upon the longest shot to be taken, A (for long shots background height must be increased); the lowest camera height, B (to avoid tilting over the set); and the height of the subject (where action takes place on a rostrum, staircase, etc.). A 10 ft. high flat is adequate for general use, a 12 ft. one being needed for larger settings. High backgrounds should be avoided to prevent steep lighting.

Shooting-off. Wherever we inadvertently shoot-off past the confines of the set, and see another setting, or the studio beyond, the solution is usually self-evident—to readjust the camera's viewpoint, or to add masking.

But, by anticipating shooting-off, we can save needless last-minute rearrangement. Problems mostly arise when: cross-shooting on a shallow set (Figure 8.12); shooting downstage from upstage parts of the setting; cross-shooting on inadequately backed windows and doors; reverse-shooting through a window or door, into a set; using low-angle shots; seeing unwanted reflections in mirrors, pictures, etc.; or using adjacent splayed settings with communicating openings (Figure 8.13).

Fig. 8.13. Pt.1.

SHOOTING-OFF. *Left:* On composite sets, cameras and booms on one part of the set may be seen through adjoining windows by cameras working in another part. *Right:* The camera may shoot-off over the top of a setting.

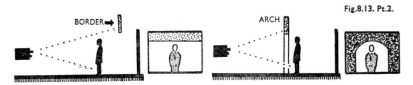

Fig.8.13. Pt.2.

Left: A cutting-piece (border) may be introduced as an artificial limit. *Right:* An architectural feature restricts the line of vision.

Height and depth in floors. Where we need elevated floor areas (e.g. landings, hillocks), vari-height rostra are used. Any irregularities are added with blocks, framework, tightly-packed sandbags, etc.

Holes in the ground are obtained by building up the overall floor level and leaving the depression where required. This way we can simulate graves, bomb-craters, trapdoors to cellars, wells, etc., without resorting to holes in the studio floor.

Where cameras are to track along elevated floor areas, the flooring will, of course, need to be level, flat, non-skid, and suitably reinforced. For heavy sets or gear, proper tubular scaffold is a must.

Rostra can provide sound problems, particularly in exterior scenes. Footsteps are likely to sound hollow and unnatural as they resonate. Surfacing with felt or foam-plastic sheeting, plus internal packing, can improve matters considerably in most cases.

Space problems

Space problems will always be with us, however large the studios, where built sets are used. Economics apart, we still have operational continuity to consider; time and space to move performers, cameras, booms, from one place to the next.

In any complicated, quick-moving show we cannot have sets too widely spaced, therefore, if we want to make good use of a small camera complement.

Many cunning arrangements have become commonplace, to enable us to condense space and provide operational freedom.

149

Fig. 8.14. Pt.1.

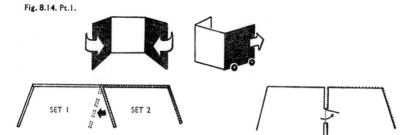

SPACE PROBLEMS. Numerous standard solutions have been devised. *Top left:* Side walls hinged to fold within the set when not required (folds, flippers). *Top right:* Walls may be wheeled or manhandled into position when needed (wild walls, floats) or hung (flown) and raised or lowered (mechanics and lighting permitting).

Bottom: Nearby sets may be joined (although unrelated) by a common dividing wall swung to splay as required, or with communal features (e.g. a connecting door).

Fig. 8.14. Pt.2.

The set may be re-vamped—its appearance changed by leaving the basic structure in position, and modifying the dressing, etc. (*top*), by re-arranging the scenic elements (*left*), or by lighting changes (*right*).

Fig. 8.14. Pt.3.

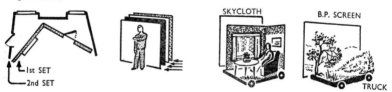

Left: Nesting sets may be placed one within the other, the inner one being struck to reveal the outer. Or packed sets, consisting of individually decorated flats laid in a pile, the top one being removed to reveal the next in turn.

Right: Wheeled platforms (trucks, floats, stage wagon) allow a small setting, or parts of a large one, to be wheeled into places. For elaborate settings this enables rapid set building, more economical use of space, and greater re-arrangement of settings for camera manoeuvres.

The three-dimensional world

When we look at a photograph or drawing of the three-dimensional world, more often than not we accept this flat image as a reasonable representation of the real scene. We see in it the solidity, space, depth and proportion of the real subject—or imagine we do. We are reconstructing, in our minds, an impression built up from various visual clues before us.

These visual clues are important. Many of them are obvious and commonplace. Others are less familiar. But they underly many of our visual techniques. Subconsciously, perhaps, for we mostly appreciate them by habit. We shall take advantage of them in lighting and camera treatment but, chiefly, as we might expect, in scenic design.

If we do so deliberately, we can: create a more convincing illusion of depth and solidity in our picture; avoid ambiguity, and, what is more important, conjure up false visual clues so as to deceive the viewer about the true size, proportions, etc. We can also provide clues that temporarily mislead his judgment, for dramatic or comic effect.

We can summarize these clues as:

Subject detail. Linear perspective—converging lines.
Relative sizes. Overlapping (masking). Foreshortening.
Surface tones. Surface contours and shadow.
Vertical displacement. Aerial perspective.
Parallactic movement. Colour.

Further practical applications

As the reader will have appreciated, there are endless uses to which we can put these visual clues. Two aspects particularly have been widely explored in television and motion picture production—*scenic distortion* and *multiplane techniques*.

We might have discussed them later as visual effects, but they are such a regular aspect of scenic design that we can more conveniently deal with them here.

Fig. 8.15. Pt.I.

THE THREE-DIMENSIONAL WORLD. Because the amount of detail the eye can see decreases with distance, fine detail suggests closeness; coarse detail, distance. A factor in the arrangement, distribution and texturing of scenery.

(Fig. 8.15 continued on page 152.)

Fig. 8.15. Pt.2.

Left: The convergence of parallel lines on a distant point with progressive compression of depth is a familiar indication of depth and distance (linear perspective). Objects at regular intervals from the eye particularly appear to contract in this way. **Right:** By depicting this in two dimensions, depth or distance can be deliberately falsified (see False Perspective).

Fig. 8.15. Pt.3

Comparing relative sizes of familiar objects gives us a strong sense of depth. **Left:** Where distant background is plain, there are no visual clues to depth. **Centre:** Where distant objects have no recognizable size, the clues are indefinite. **Right:** By comparing the relative sizes of recognizable subjects, we assume the picture's scale.

Fig. 8.15. Pt. 4.

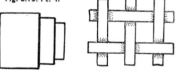

Left: Where one plane partially overlaps (masks) another, the half-seen plane appears further away. **Right:** Suggesting planar relationships between surfaces that would otherwise appear conjoined.

Fig. 8.15. Pt.5.

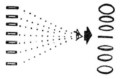

As we look at a surface from an increasingly acute angle, its shape and apparent depth alter, the amount of distortion (foreshortening) influencing our impression of its form and position. This illustrates the value of camera movement; how horizontal planes emphasize vertical movement and vice versa, enabling us to reduce or emphasize these effects by set design.

Fig. 8.15. Pt.6.

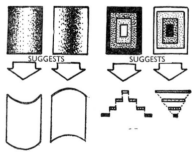

We saw how surface shading affected apparent shape and how surface brightness affected apparent distance and size. (Fig. 5.12. Pt. 2). Useful for suggesting, emphasizing or hiding shape and distance.

(Fig. 8.15 continued on page 153.)

Fig. 8.15. Pt.7.

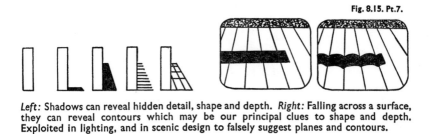

Left: Shadows can reveal hidden detail, shape and depth. *Right:* Falling across a surface, they can reveal contours which may be our principal clues to shape and depth. Exploited in lighting, and in scenic design to falsely suggest planes and contours.

Fig. 8.15. Pt.8.

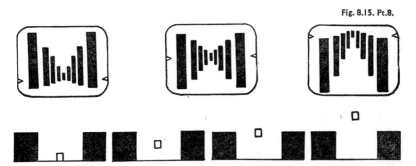

(See "Position of Subject in Frame"). *Top:* In general, objects at the bottom of the picture look nearer than those at the top. *Bottom:* So we get a greater impression of depth by placing the furthermost planes high in the picture, providing the camera is not high enough to shoot over, rather than through, progressively distant planes. Similarly, we can emphasize height in an elevated shot.

Fig. 8.15. Pt.9.

While near objects are usually sharp and well contrasted, the haze of distance causes far objects to become softened in outline, with less pronounced tonal and chromatic contrasts (aerial perspective). Achieved in scenic design by using gauzes, and less contrasty lighting for "distant" planes, and by restricting camera depth of field.

Fig. 8.15. Pt.10.

CAMERA
MOVEMENT

The rate at which planes move aside as we pass depends upon their distance from us. Whenever we alter our viewpoint, the relative positions of planes will change (parallactic movement). This is a reminder of the value of lateral camera movement, the shortcomings of flat scenic surfaces, and the advantages of multiplane design.

Warm colours (red, orange, yellow) appear nearer than cool ones (green, blue, violet), and therefore suggest greater depth when used in foreground areas of a colour picture. Pastel shades appear more distant, softer and with less detail than hard primaries, which look closer and more clear-cut.

153

Fig. 8.16.

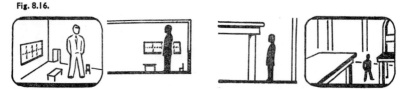

SCALE CHANGES. *Left:* If a set is built with everything proportionately scaled down, people assume gigantic proportions. *Right:* If everything is scaled up, people are dwarfed.

Scenic distortion seeks, by deliberately changing the normal size or proportions of the setting, to suggest that it is larger or smaller than it really is; this is scale distortion.

By means of false perspective, the set can be made to appear more or less extensive than it is in reality.

Multiplane techniques. Our impressions of depth are, as we have seen, chiefly comparative. We compare the appearance of progressive planes. The more planes between our viewpoint and the horizon, the more information for us and the stronger the illusion. A

Fig. 8.17. Pt.1.

FALSE PERSPECTIVE. Although only strictly accurate for one pre-arranged viewpoint, reasonable variation is possible. *Left:* The simplest of all false perspectives is a painted or photographic flat background which merges with the painted floor in a continuous effect; but give-away shadows must be avoided. *Right:* Alternatively, the whole setting may be built in false perspective, but this necessitates keeping action downstage to avoid accidental scale changes.

Fig. 8.17. Pt.2.

▶PERSPECTIVE MODEL◀

A *perspective* model background allows a certain amount of parallactic movement to be obtained.

Fig. 8.17. Pt.3.

In many settings, by deliberately arranging to have large objects in the foreground, and progressively smaller items further from the camera, perspective can be cheated.

mountainous skyline, for example, will give us less sense of distance if only flat fields separate the horizon from us, than if there were a series of intervening ridges. Since it is principally from the foreground and the middle-distance that we derive this feeling of depth and solidity, we can play upon this in our staging techniques.

In an exterior setting, we might move through low walls, gates, bushes, etc., that have been arranged in the foreground. In interiors we can interpose furniture, architectural features (e.g. columns, statuary), drapes, between the camera and its subject. Even in decorative or abstract settings this principle can be followed. Where the foreground has been completely cleared to leave plenty of room for camera moves, we lose much of the illusion of completeness; the impression of space, solidity, and movement that foreground planes offer.

Against this, one needs to be warned against overdone extremes. Constantly forcing the camera through foreground debris, and peering through bric-a-brac, can be visually disturbing. In the studio, it becomes a hazardous game of skittles.

In a carefully contrived production, prominent foreground planes appear to come naturally into picture as the cameras move about the setting. They just happen—or so it seems. They form a definite part of the compositional treatment; of which, more later.

Fig. 8.18. Pt.1.

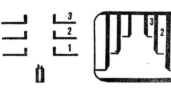

MULTIPLANE TECHNIQUES. *Left:* Intermediate planes between foreground and background enhance the illusion of depth, even when quite simply contrived.

Right: Where foreground subjects will obstruct the camera, and cannot be moved while out of shot, we can use break-away pieces (for doorways, arches, windows), which part as the camera passes through them.

Fig. 8.18. Pt.2.

Foreground subjects for multiplane effects can be arranged "naturally" in a carefully contrived production.

155

Although the viewer accepts various abnormal viewpoints unquestioningly, foreground subjects that one normally associates as being against a wall, invariably prove distracting. Shots through shelves, from behind mantelpieces and fireplaces, have been explored, but usually found wanting.

The illusion of four walls

It is rarely necessary to have four visible walls to an interior setting to imply completeness. Filmic convention has got the viewer used to the idea of three, or even one, background plane, as sufficient; the other invisible walls being imagined. But occasionally we find ourselves obliged to establish the complete room. In motion pictures, individual camera set-up allows the whole set to be built and struck to suit each viewpoint. In television, we have to resort to other subterfuges.

Fig 8.19.

ILLUSION OF FOUR WALLS. A low-wall arrangement can be used to provide a four-walled set. Although cameras and booms are unimpeded, we may run into editing reversals.

We could actually build a four-walled set, arranging to raise, hinge, or slide one or more wild walls when needed, to give access to cameras and booms. But there exist many operational hazards, especially for lighting.

It is simpler to provide peep-holes or hiding places for cameras. Holes behind pictures, concealing curtains, hinged wall-flaps, gauzes, see-through mirrors, all have their merits. Sometimes an existing doorway or window may suffice. Even a three-walled set may look entire by these methods.

Where only elevated shots of a small room are needed, the arrangement of Figure 8.19 has been effective before now, while in the last resort, we can even split the room into two adjacent half-rooms, and then, by judicious cheating and careful cutting, imply that they are continuous.

Partial settings

Strictly speaking, most realistic settings are only partially built. We are unlikely to reconstruct a complete cathedral where a church door will suffice. But in partial settings we meet the deliberately

Fig. 8.20. Pt.1.

PARTIAL SETTINGS. An environment can be implied by building a complete but abbreviated portion. *Left:* a shop interior can be suggested by shelves attached to a flat, and a trestle table. *Right:* A brick flat, stairs, ash-can, drainpipe and poster, become an alley.

Fig. 8.20. Pt.2.

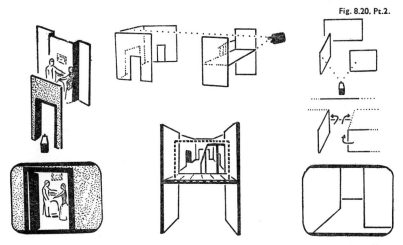

A setting can appear more extensive by the use of scenic foreground pieces. *Left:* A room interior shot through a foreground door. *Centre:* The impression of a staircase head conveyed by three flats and a handrail. *Right:* Careful abbreviation (*top*) implies solidity (*centre*) in the camera (*bottom*).

Fig. 8.20. Pt.3.

Another method is electronic insertion, with part of the studio scene (*left*) inserted into a separate background picture.

contrived approach to set design, in which only what the camera must see is built. Admittedly, this does not allow the director much opportunity to change his planned shots once he has agreed upon the abbreviated set-arrangements, but such a design approach does make for considerable economies in expense, space, etc.

Carefully controlled, this technique permits greater elaboration and scenic variety, especially for short scenes. Unskilfully used, it may lead to a cramped, portmanteau quality in the staging.

Fig. 8.21. Pt.1.

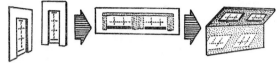

MULTIPLE USE OF UNITS. Individual pieces of scenery can be completely transformed by re-arrangement. Two glazed doors become a long window, or a decorative ceiling light.

Fig. 8.21. Pt.2.

Ingenuity transforms railings into window bars or an interior decorative grille.

Fig. 8.21. Pt.3.

Similarly a low window can be turned into a throne, and an arch into a bookcase.

Partial settings fall into two general groups:—those that imply the whole, by showing a complete but localized part, and those that suggest that the setting is more extensive than it really is, by judiciously positioned parts.

Realism and falsity

In the theatre, the artificiality of all settings is accepted by the playgoer as a convention of the medium, even where realism is inferred. In television and film, this is less true. Here the setting is more usually judged by realistic standards. We cannot accept here obvious crudity as a reasonable representation of reality. The artificiality of cut-out tree-wings, for example, does not mix convincingly with natural foliage. But where artificiality creates a genuine-looking effect, deception can go to any lengths.

The situation is paradoxical. On the one hand, we may successfully use a cotton-actuated curtain to suggest billowing winds, a few leafy shadows to suggest a nearby tree; perversely, a carefully constructed stone wall may have some indefinably fake air about it, or

158

a genuine antique tapestry may look less realistic than a dye-painted canvas copy.

Where realistic settings are concerned, scenic design is based upon skilful deception; upon presenting the characteristic flavour of the scene. It is not enough for an actor to wear the right clothes for the part; the clothes may need to look worn-in to the part, too. Many settings lose their impact through looking too fresh.

Deception must be discreet. All too often we are asked to believe in: newspapers with pasted-on headlines or photographs that would deceive no one, or painted seascapes, showing foaming breakers that never move.

Very occasionally, one can accept still backgrounds showing subjects that are normally moving. Motionless trees, even static people, may survive convincingly for a while, but sooner or later they appear suspiciously inert; particularly where sounds of associative movement (e.g. soughing winds, milling crowds) accompany the shot.

Realism is sometimes best achieved by using the real thing, sometimes by substitute materials, or a mixture of both.

A country scene may be built up by the use of real turf, heather, tall grasses, and defy the closest scrutiny. In longer shots, grass matting (suitably draped and discoloured) may get by. Artificial flowers and foliage (especially plastic reproductions) may look livelier and more convincing than real cut branches and potted plants that have wilted under hot studio lights. Real water pools may be no more natural-looking on camera than faked-up glass sheeting.

Gauzes

Thin cotton netting, known as scrim or gauze, has extensive applications. In black or white, with a mesh around 1/16 in. to 1/8 in. diameter, it provides scenic effects that are not readily achieved by other means. (See Figure 8.22.)

It may be used as glass in windows, to lend aerial perspective to planes beyond. Laid over scenic backgrounds, it softens their contrast and texture. By painting or attaching solid shapes to gauzes, one can suggest forms suspended in space; either frontally lit or silhouetted.

Gauze will need to be of sufficiently open weave to prevent excessive light absorption, or the obscuring of subjects beyond. And, unless deliberately draped, it will need to be hung without visible wrinkles or seams.

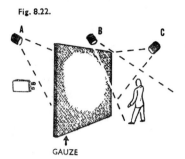

Fig. 8.22.

GAUZES. (Scrims). Lit from the front by A alone, the gauze appears a solid plane, surface painting or decoration shows up brightly against a plain white or black background. Unlit subjects behind the gauze are invisible.

By reducing A's brightness and illuminating the subject with B (behind gauze), the subject and setting are revealed with outlines and contrast softened, surface painting having almost disappeared.

If C is added to rear-light the gauze, it will increase the mistiness over the scene, while silhouetting details on the surface of the gauze.

GAUZE

The cyclorama

An old theatrical device, the cyclorama (Cyc.) helps to create the illusion of unrestricted space. (See Figure 8.23.) Its shallow, C-curved surface may be a semi-permanent structure of plywood or hardboard, or temporarily formed from a cyc-cloth (of duck, canvas, gauze, or velour) hung on a curved batten, or fixed cyc-rail.

Small cycloramas, using a standard curved flat, occasionally provide backings outside windows.

With a matt mid-grey finish, the plain cyclorama can serve the small studio as a maid-of-all-work, for its neutral background can be lit to varying values, decorated with projected shadows, or become a "sky-cloth" at choice.

Permanently built cycs can be given excellent surfacing, but tend to localize staging to wherever they are, and cause hard, reverberant, sound quality.

Cyc-cloths can be hung where needed, but may not be free from blemishes, even when carefully stretched and weighted.

Mobile scenic units

See Moving B.P.; Moving Shadows—Chapter 20.

Moving vehicles

Where live action is to take place in, or on, a "moving" vehicle, we have to somehow convey this idea of movement to the viewer—even when the automobile interior is, in fact, only a sawn-off rear-end. Wavering the camera head hopefully is seldom sufficient.

We get our impression that a vehicle is travelling from several clues: from passing background scenery; lights and shadows passing over the interior; movement of parts of the vehicle (e.g. jerking

Fig. 8.23.

THE CYCLORAMA. The cyc. can be merged with the floor by a *merging curve* or *a ground cove*. The former is a concave facing, merging the cyc. imperceptibly with the floor so that painted designs (especially perspective) appear continuous. The latter is a concave ramp, behind which ground-lamps can be hidden to light the lower part of the cyc.

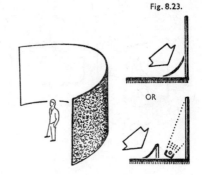

reins, revolving wheels), signs of its progress (e.g. dust-clouds, billowing clothes, etc.), or seeing people being swayed, jogged, etc. We may hear the vehicle's progress (e.g. engine, hoof-beats, wheel-noises, etc.) or the sounds of other subjects passing by.

Mounting it on a suitably manipulated rostrum (Figure 8.24) provides realistic movement for most situations. Even for modern, smooth-running methods of travel, a slight exaggeration of movement is more convincing.

Where the vehicle is too large to rostrum-mount, excellent effects have been achieved, even without Moving B.P., by having the quay-side, railway platform, or whatever, move past instead (Figure 8.24).

Movable scenic sections. Mobile units are invaluable when we want to move around a subject in restricted space. A rotating turntable enables us to see it from various viewpoints, without tortuous camera moves. Wheeled-units enable parts of the setting to be moved out of the way, or repositioned to form new compositional patterns (Figure 8.24). Groups of singers, orchestral sections, can be slid aside for scenic effect. Facilities here will range from manual towing or pushing by hidden stagehands, to remotely-controlled, electrically driven trucks.

Set dressing

Dressing—the business of furnishing and decorating the built set, is obviously an enormous field for discussion, but we can mention a few pointers here.

The televised scene easily acquires a filled-up look, so that, even

Fig. 8.24.

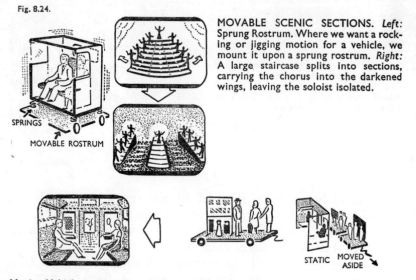

MOVABLE SCENIC SECTIONS. *Left:* Sprung Rostrum. Where we want a rocking or jigging motion for a vehicle, we mount it upon a sprung rostrum. *Right:* A large staircase splits into sections, carrying the chorus into the darkened wings, leaving the soloist isolated.

SPRINGS

MOVABLE ROSTRUM

STATIC   MOVED ASIDE

Moving Vehicles: where the vehicle is too large for a movable rostrum, we can sometimes put the exterior moving subjects on trucks instead.

when we want to suggest excessively crowded surroundings, surprisingly little dressing will often do the trick. It is all too easy to clutter the setting with knick-knacks that are hardly seen, and contribute little. Discreet selection, rather than a profusion of properties is what should be aimed at.

We seldom find an over-furnished setting, because space has to be left for performers and camera movement. The reverse more frequently happens. In long shots we then see wide expanses of floor, with a few scattered pieces of furniture.

The camera-crane can shoot over low furniture, but for pedestal cameras we shall have to strike and reset any foreground units whenever we want to move into the set for upstage angle-shots.

Furniture design can present its problems. Deep, well-sprung settees and armchairs invariably place the seated persons so low that they have to bend forward awkwardly as they get up, and appear leggy when sitting down. Sometimes the distance between people seated and standing is too great for good sound and composition. Skilful propping-up with cushions and/or furniture blocks, may improve matters. (Furniture blocks are 4 ins. square or cylindrical blocks, recessed on one face, universally used to raise furniture to a better working height. Peculiar looking in long shots, they may be disguised by cylindrical cloth gaiters.)

162

If there can be said to be any set dressing tool, it must, surely, be the ubiquitous *staple-gun* or the *staple-hammer*, which can shoot or knock open-ended wire staples through cardboard and soft wood up to 1/8 in. thick. To turn their ends over, for securing materials to a thin surface, a flat metal plane is held at the rear-side. These tools are most versatile and can be used for attaching pictures, drapes, posters, foliage, wallpaper, to flats; felting rostrum tops, constructing animated captions, etc.

Realistic substitutes

Natural materials cannot always be used in the studio scene. They may be too heavy, cumbersome, impracticable, or dangerous. Substitute materials have to be used, and in skilled hands these can look more realistic than the real thing. Most of these dodges have been inherited from motion picture design.

Rocks, Caves. Built up from wood and wire (expanded metal) frameworks, covered with stiffened and painted canvas, papier-mâché, plaster. Expanded polystyrene and glass fibre are outstanding.

Burning logs. Logs formed from chicken-wire, covered with an asbestos/fireclay compound, are "burned" over several concealed gas jets. Flames can be produced, too, using paraffin wax firelighters.

Glass. To avoid smashing during storage or transportation, tracing linen, gauze, or thin plastic sheeting are often used as substitutes.

Frosted-surface plastic can be rendered semi-transparent by a clear-varnish coating. Useful to simulate ornamental frosted glasswork.

Frost patterns can be imitated by stippling glass with a solution of stale beer and Epsom salts. Spray-on plastics are rather more reliable.

Sugar-glass or brittle plastic are used where panes of glass have to be smashed during action.

Gravel. Real gravel is usually excessively noisy. Fine cork-chips are preferable. Plastic chips and expanded mica can also be used.

Dust and dirt. Coloured sawdust, fuller's earth; peat.

Rain. Superimposed film-loop of rain, or a black-background water spray to reinforce or replace "practical" water.

Suspended perforated metal pipes, hose-spray, even watering-cans. Floor protection (e.g. low tarpaulin troughs) and drainage are imperative. Absorbing troughs of sawdust may suffice.

Water. Shallow, black-bottomed water tank. Plastic sheeting pools. Remember that water forms a heavy mass, and leaks are catastrophic. Glass or plastic sheet on black may be more convenient.

Wet surface. Surfaces sprayed with glycerin, varnish, or water-soluble glue. Plastic gloss sprays are quick for local use.

Ice. Clear or opal plastic sheet laid on light-toned cloth.

Snow. Most materials have shortcomings (inflammable, or fire hazards). For falling snow, finely shredded paper is effective, fire-proofed confetti may substitute. On the ground, formaldehyde-based plastic, sawdust, expanded polystyrene is convincing. Avoid salt or sand.

Film-loops of snow, superimposed on shots of exterior scenes are convincing enough, while for night-scenes, moving light patterns have proved effective under low light conditions.

The traditional snow-bag is a suspended hessian sheet, anchored along one edge, the other being hung from a rope-support. Tugging the rope causes the snow-material to seep through slits cut in the bag. More sophisticated dispenser machines drop "snow" at a controlled rate.

Ageing sets

In all its pristine newness, the realistic setting lacks, somehow, the "lived-in" or worn appearance that lends conviction. To remedy this, a process known variously as antiqueing, blowing-down, dirtying-up, is carried out on it. With an air-brush (spray-gun), parts of the setting are toned down, shaded with a fine film of dark water-paint, in places where time and usage would cause grubbiness or discoloration—around door-frames, in corners and crevices, near fireplaces, at picture-edges, by wall fittings, by door handles. If over-enthusiastically applied, it will look as if dilapidation has set in. Aerosol sprays are very useful for rapid, localised treatment.

Pre-damaged sets

Occasionally, the story demands that settings become damaged. Damage can be simulated in many ways by: filming all sequences, or those in the undamaged setting; providing duplicate sets, undamaged and damaged; or by having the damage occur during the pro-gramme.

We shall usually supplement the physical change with overlaid sound effects (dust, fire, smoke, etc.) to heighten the impression.

Wreckage. (Collapse—earthquakes, shelling, bombing, fire, etc.)

Pre-broken walls. Jagged sub-sections, pre-cut and camouflaged, are withdrawn or pushed into the setting on cue. Parts of the set likely to fall on to performers are made of balsa, cork, canvas-covered wire, cardboard, etc. Suspended or balanced debris (timbering, dust, etc.) is held out of shot, being dropped in by pulling a release-pin. Expanded polystyrene blocks have wide applications.

After-effects

Fire. Discoloured walls and charring, by two-colour painting. This appears homogeneous under first lighting, but contrasts when the colour of lighting, is changed or the camera filtered (see Samoiloff effect). Defocused shadows cast on walls may suggest discoloration.

Flooding. Same method to simulate discoloration. Peeling wallpaper is obtained by tearing, and then holding the paper flat against the wall by cotton threads or tape. Releasing these causes the paper to peel.

Furniture

Furniture that is to collapse can be pre-damaged by destroying or sawing, then lightly glued, or by joining with matchsticks or tooth-picks.

Balsa or papier-mâché furniture will avoid injury to actors.

Electronic and lighting considerations in scenic design

Good staging practice places a number of restrictions upon set design. Occasionally we can ignore them, and by careful calculation (or good luck) still produce superb pictures—but only occasionally.

One cannot give precise formulae for design; but some tips based on experience may prove useful. Many of these practical findings have been grouped together here, including relevant points raised in other specialized chapters.

The final effect

Although colourful design may delight those in the studio, it has no significance to the viewer in monochrome television. This is easy to forget. Visual appeal can inadvertently hinge on colour relationships that are lost on transmission. Compositional line, tonal balance, tonal transition, are the monochrome alphabet. Colour harmonies, colour balance, and colour association, essentially require colour transmission. Design features that make for good colour staging can become meaningless on a black-and-white screen. Dissimilar hues may reproduce as the same grey-scale value.

A similar situation arises when a setting has been conceived only as a complete effect, and not designed with the camera's shots specifically in mind: pictorial effects that are sensational in very long shots can become disappointingly meaningless when dissected into closer shots.

Tonal values

Marked tonal contrasts may give bite to a long shot, but panning past strongly contrasting tones in close-shots, or intercutting between different viewpoints, can cause visual havoc. Pictures will not be of consistent or matched quality (Chapter 4). Shots tilting quickly from the open sky to the ground demonstrate this problem well, for the correctly-exposed sky-shot leaves the succeeding ground reproduced as a dark, underexposed, mass. At the other extreme, we find that with insufficient contrast, subjects' tones can merge all too readily with their backgrounds. That is normally a matter for co-ordinated setting, lighting, make-up and wardrobe treatment.

Surface detail

Plain walls, unbroken by detail or modelling, have limited pictorial appeal, although shading or decoration with light may enhance their appearance. They are liable, also, to show up camera-tube blemishes.

Conversely, striking detail can distract, or create conflicting compositional lines.

Small detail, on the other hand, will not only be lost when defocused, but may dilute into an overall tone of a different value entirely. A crisp, black-and-white striped pattern, for instance, can defocus into quite an unexciting grey ground. Fidgety action near to such a background is likely to emphasize this effect as sharpness of focus varies.

Close horizontal stripes are an anathema to television staging. Whether from distant scenic-lines, or close fabric pattern, the violent line-beating that arises distracts one from all else.

166

Surface finish

Bright surfaces of any size leave the camera-channel at a disadvantage wherever a full tonal range is to be reproduced. Varying picture-brightness, clogged shadows, burned-out highlights, are typical results. Consequently, highly-reflective surfaces of all kinds are fraught with problems, particularly when they are curved.

Most designers shun highly-reflective materials where possible. Some sheen can give an undeniable sparkle of vitality, but the borderline between attractive sheen and obtrusive flare is sadly narrow. Little wonder that it is many a designer's lot to have glassless pictures, windows and furniture, imitating surface reflections with dry-brush highlights. Tantalizingly enough, he may find mirror-finish floors, metal foil wallpaper, and the like, quite successful in certain controlled situations, where lighting and camera treatment permit. Haphazardly used, they just frustrate.

First-aid treatment. Where a surface does cause flares, we can often alleviate the trouble by tilting or re-angling it. Coating or daubing it with various materials is sometimes effective, although we may modify the surface's appearance in the process; for example: flour paste for varnished paper and wood, glossy paint, plastics; putty, plasticine for small glass or metal surfaces; metal polish for light-toned surfaces that are not to be handled; furniture wax (spray or solid) for most glass, metal, wood, etc.

Where parts of a setting are over-bright, one's natural reaction is to reduce the light falling upon it. But this may be impracticable, or it may rob nearby surfaces. Instead, short of completely repainting, the only expeditious solution may be to spray or spatter the surface lightly with black water-paint. Overdone, this procedure produces dirty, lifeless results, interfering with surface drawing. In skilled hands, the transformation can be perfect. Over-dark surfaces can often be improved by similar light-toned treatment.

Within the setting, over-light surfaces are always avoided. Any white dressing or properties (like the costume), should be very much off-white. Tablecloths, sheets, papers, china, drapes, etc., will be light blue, green, grey, or yellowish. Items such as books and newspapers may have to be blown-down to the required tone. Fabric may need dyeing. Carefully arranged, these non-white subjects will still reproduce as white, but without the distressing lack of modelling, or degraded picture quality, that lighter tones would have given. But if we overdo it, we may finish up with people in dirty shirts, reading greyish newspapers!

167

9

TELEVISION MAKE-UP

Types of make-up treatment

MAKE-UP treatment follows three general forms; these are straight make-up, corrective make-up and character make-up.

Straight make-up

A basic treatment affecting the performer's appearance to a minimum extent. Straight make-up falls into two groups:

Skin-tone adjustment (flesh, lips) to provide a good tonal balance in the picture: darkening pale faces; lightening swarthy complexions.

Routine improvements: subduing blotchy skin tones, shiny foreheads; strengthening under-prominent eyebrows; removing beard-line (blue chin); darkening over-light ears; lightening deep-set eyes, etc.

Corrective make-up

This seeks to reduce less pleasing facial characteristics, while enhancing more attractive points. Corrective treatment can range from slight modifications of lips, eyes and nose, to strapping sagging skin or outstanding ears, or concealing baldness.

Character make-up

Here, emphasis is upon the specific character or type the actor is playing. By facial re-shaping, re-modelling, changes in hair, etc., the subject may be so transformed as to be unrecognizable as himself. Film characters like Fagin and Frankenstein's Monster are supreme examples of this.

But most character make-up is less spectacular. Subtle changes in the actor's disposition can be introduced. The down curled lip, studious frown, neglected cheek, are hackneyed instances.

168

The conditions of television make-up

The principles and practice of television make-up are almost identical with those of motion pictures, except that the pace and continuity of television usually prevent the elaboration and refurbishing possible in film. Theatrical make-up methods appear rather too broad and crude under the camera's scrutiny. They are confined, therefore, to more stylized treatments; e.g. clowns, ballet, pierrot, etc.

Make-up routines vary slightly between organizations, but for a production of any size, the following is typical enough.

At a preliminary meeting with the programme's director, the supervisory make-up artist will discuss such details as character interpretation, hair-styling, special treatments, any transformations during the programme (ageing, etc.), and so on. Performers who need fitted wigs, or trial make-up, are then contacted. Any film-insert sequences involving make-up are duly serviced.

The next stage is the studio rehearsal, Here, either of two approaches is common.

In the first, the performers are made-up beforehand as experience suggests. This forms a basis from which the make-up artist can judge more exactly the treatment and tones needed. It allows, too, the lighting director and video control operators to finalize their aspects, i.e. tonal balance, contrast, and exposure.

The supervisory make-up artist watches camera rehearsals on a picture-monitor, noting any changes that seem desirable. These notes are then passed on to individual make-up artists, who handle the performers.

The other approach to treatment (especially for straight or corrective types) is to see them on the screen first, and then to service the performers as time and facilities allow.

A straight make-up for men may take around 3 to 10 minutes; women require 6-20 minutes on average. Elaborate needs can double, or even treble, these times.

After a few hours, the cosmetics will usually have become partly absorbed, or dispersed through body-heat and perspiration. The surface finish, texture, and tones, will have changed somewhat from when they were first applied. Consequently, fresh make-up or refurbishing will then be necessary.

Apart from on-the-spot retouching and freshening (mopping-off and applying astringents), performers are normally treated in make-up rooms near the studio. Miracles of makeshift quick-changes are regularly achieved on the studio floor amidst the turmoil of production. But, unless time-limits necessitate this (as for rapid ageing, wounds, character transformations, etc.) a more leisurely procedure is preferable.

There will, too, always be occasions when performers cannot have make-up, owing to allergy or temperament; or must have "blind", off-camera treatment, which is seen for the first time when the show is on the air. The wise director keeps such occasions to a minimum.

Whereas film make-up can be suited to individual shots or scenes, the continuity of television production generally prevents such refinements. A long shot ideally requires more defined, prominent, treatment than a close-up. (A similar situation, in fact, to that found in lighting techniques.) But one may not even be able to do anything about such distractions as perspiration, or dishevelled hair, when the player is on-camera for long periods.

Apart from artistic considerations, many technical factors affect make-up treatment:—

lighting intensity, balance, direction, etc.;
scenery, relative face/background contrasts;
video adjustment (picture control): exposure, gamma, black-level, colour response, etc.;
costume: relative face/costume tones contrast.

This helps to explain why treatment of the same performer may need to vary from one show to the next. Sometimes an astringent and light powdering may suffice, while at others, more particular make-up is essential.

The principles of make-up

The broad aims of facial make-up and lighting are complementary. Occasionally one or the other will predominate, but they must always be in concord, or they will simply cancel each other out. One can sometimes compensate for the other's shortcomings—e.g. lightening eye-sockets to anticipate shadowing cast by steep lamps. But, whereas the effect of lighting changes as the subject moves, that of make-up remains constant. This distinction is important when we consider corrective treatment.

Make-up achieves its effects through:—

Large area tonal changes—e.g. making the entire face lighter.
Small area tonal changes—e.g. darkening part of the forehead.

170

Broad shading—e.g. blending one tonal area into another.

Localized shading—e.g. pronounced shading to simulate a jaw-line.

Drawing—e.g. accurately lineated lines and outlines.

Contour changes—e.g. built-up surfaces.

Hair work—e.g. moustaches, wigs, etc.

Many of the illusions produced by make-up are similar to those of other graphic and plastic arts. Localized highlighting will increase the apparent size and prominence of an area, while darkening reduces its effective size, and causes it to recede. By selective highlighting and shading, therefore, we can ostensibly vary surface relief and proportions considerably.

We can reduce or emphasize existing modelling, or suggest modelling where none exists; remembering, though, that the deceit may not stand up to close scrutiny.

A base or foundation tone serves as a covering agent that will obscure any blotchiness in the natural skin-colouring, blemishes, beard-shadow, etc. This can be extended, where necessary, to block-out the normal lips, eyebrows, or hair-line, before drawing-in another, different, formation.

Selected regions can be treated with media a few tones lighter or darker than the main foundation, and worked into adjacent areas with fingertips, brush, or sponge. After this highlighting and shading, any detailed drawing is done, using special wax lining-pencils and brushes.

We cannot summarize the craft of television make-up in a few pages. Its principles are uncomplicated, but its practical applications are numerous.

We can show diagrammatically, however, a few examples of elementary treatment (Fig. 9.1).

Make-up materials

Current make-up media used in television and motion pictures include:—

(i) A dry, matt cake of compressed powder.

(ii) A non-greasy base of creamy consistency, in small stick-containers, or jars.

(iii) A grease-paint foundation, contained in tubes.

(iv) Powder and liquid bases.

Before applying a foundation (base), the skin is first prepared by

Fig.9.1. Pt. 1.

EXAMPLES OF BASIC MAKE-UP TREATMENT. *Left:* Hard hairline; forehead lowered by shading; face width reduced; cheeks depressed. *Centre:* Softened-off hairline; forehead width reduced by shading; deep-set eyes lightened; cheek and chin modelling increased. *Right:* Hairline reshaped, extended; forehead height reduced; chin made to recede; "apple-cheeks" highlighted.

Fig. 9.1. Pt.2.

UNTOUCHED

Variations in nose shape achieved by shading (side shading, ridge highlighting).

UNTOUCHED

Eye make-up. *Left:* Untouched. *Left centre:* Area above eye protrudes as a result of lightening, eyebrows raised and thinned, making eye seem smaller. *Right centre:* Eyebrows thickened and lowered, and area above eye shaded, to open up eye and make it appear larger. *Right:* Method of lengthening eyes.

Fig. 9.1. Pt.3.

Left: Method of broadening eyes. *Left centre:* Distance between eyes reduced by shading. *Right centre:* Distance between eyes increased by shortening brows. *Right:* Detailed example of eye treatment: base of top lashes underlined thinly from centre to just beyond outer corner, end being upturned. A short upturned line is occasionally drawn under lower lashes, but this tends to reduce eye size. Eye shadow light towards the inner corners, slightly heavier at outer corners. A white pencilled line along inside edge of lower lid.

thoroughly working it over with cold cream, to remove any traces of existing cosmetics. Wiping this off with paper tissues, an astringent (e.g. Eau-de-Cologne) is patted on, to close skin pores, reduce absorption and perspiration, and generally freshen.

(Following an after-transmission clean-off, the skin is washed with soap and water.)

Which of the above media is used, is dependent upon the effect required, the degree and nature of the treatment, and personal preference. Each material has its particular features.

172

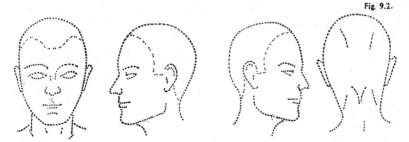

TYPICAL BLANK OUTLINE FACE CHART FOR MAKE-UP DETAIL. These charts record detailed notes on facial treatment, colour, materials, hairwork, etc.

(i) The **cake-foundation** is worked up with a moist sponge, and applied thinly over the whole face. Its covering power (ability to overlay other tones, blemishes, etc.) is excellent. Its finish is predominantly matt. When too thickly applied, it can produce a mask-like appearance; flattening out facial modelling. Highlights and shading can be introduced with a sponge-edge, using shades two to four tones different from the foundation colour.

(ii) The **cream-base** is dabbed on to small areas, and worked by fingertips evenly over the regions being treated. Localized highlighting and shading with different tones is pat-blended or merged with flat-topped sable brushes. Surface finish depends upon any subsequent powdering to set the foundation. Unpowdered, the skin has a soft sheen. Powdering reduces shine, leaving a smooth, satin finish. The cream has fair covering power, and may be retouched easily. Leaving the skin's natural texture visible, it permits lighting to reveal more subtle half-tone modelling.

(iii) **Greasepaint** is similarly spread from small dabs. Although it has little covering power, and may need refurbishing sooner than a heavier foundation, it is easily worked and retouched (by brush or fingertips). Powdering modifies its distinct shine to a silky gloss.

Such after-powdering, when applied to any completed make-up, softens its contrasts, and binds its materials. Any excess can be removed with a powder-brush.

(iv) **Powder-bases** are supplied in compacts, and resemble reinforced face powder. Applied with a puff, their covering power is fair. They have general use for broad shading and tonal improvement, emergency repairs, shiny foreheads, and the like. Most

173

liquid-type media have poor covering power, and tend to become streaky and patchy.

Several accessory preparations we shall find common to the beauty salon and the make-up room alike, e.g. mascara, eye-shadow, false eyelashes. A few, like rouge, have less value on-camera, being better simulated by other means; while materials such as tooth enamel (to blacken or whiten teeth), artificial blood capsules, are more suitable to the television and film studios.

The make-up artist's tools range from brushes (for modelling, lining, applying lip-colour, etc.), sponges, wax pencils, to palette knives. But for many workers, their fingers are their most-used aid, with which they blend foundation media into a homogeneously moulded complexion.

Surface modelling (prosthetics)

Where we wish to change the physical contours of the flesh, we can do so:

(i) By manipulating the subject's own skin.
(ii) By sticking on new formations.

(i) To produce scars, ridges, etc., *non-flexible collodion* may be brushed on to the skin. This mixture of pyroxylin, alcohol and ether, contracts the flesh, the painless contraction increasing with successive layers. Although readily removed, collodion can irritate sensitive skins. In such cases, quick-drying, liquid plastic sealers may be preferable.

Fish skin can be used to contort the flesh within limits. The selected area of flesh is drawn into its new position, and the fish skin spirit-gummed to hold it in place. Apart from imparting an oriental slant to eyes, this material is mostly used to emphasize or flatten out folds of superfluous flesh.

(ii) We can build up surface contours by several methods. Wax nose-putty, mortician's wax, plasticine, can all be moulded into shape and stuck on. A latex rubber-base solution, *flexible* collodion (resin, castor oil, ether, alcohol), or various plastic equivalents, can be coated over a selected area until it is sufficiently extended. Where large protruberances are wanted, pads of cotton-wool or sponge can be attached first by spirit-gum. Surfacing is then carried out over this sub-structure. Because normal foundation media do not take to these surfaces, plastic sealers are finally brushed over to key them on, and prevent any interaction.

Alternatively, special foundation media can be used.

The various methods we have just outlined form the basis of nose-modelling, eye-bags, warts, wounds, etc.

For more drastic changes, partial or complete masks are moulded and attached. They provide us with anything from double chins to grotesques. The advantage here is not only that more extensive alterations can be introduced, but that they can be prepared beforehand, and fitted more quickly and easily.

Hair

Hair may be treated and arranged by the make-up artist, or by a separate specialist. Such hair-work includes:—

(i) Alterations to the performer's own hair.
(ii) The addition of supplementary hair-pieces.
(iii) Complete wigs, covering existing hair.

(i) In television, a certain amount of restyling, resetting, or waving, may be carried out on the performer's own hair, but where extensive alterations, such as cutting or shaving are needed, complete wigs are more popular. Hair colour is readily changed by sprays, rinses, or bleaches. Hair whitener suffices for both localized and overall greying. Over-light hair can be darkened to provide better modelling on-camera, while dark hair may need gold or silver dust, or brilliantine, to give it life.

(ii) Where the performer's natural hair is unsuitable for treatment, hair-pieces or full toupees can be attached to the scalp, and unified with existing hair. Where scalp masque cannot, by staining, hide baldness, these made-up hair-pieces may be necessary. For women, postiche work (pinned-on hair) may provide flowing tresses, buns, ringlets, etc., augmenting a short coiffure.

Beards, moustaches, side-boards (side-burns), stubble and the like, may be "home-grown", preformed by using prepared hair-goods, or built up from cut hair-lengths. Prepared hair-goods are, undoubtedly, most popular. These are obtained prefabricated by wig specialists. Here, hair has been tied strand-by-strand to a fine silk or nylon netting (lace), and can be dressed to suit beforehand. This method demands less time and skill of the make-up artist.

Treatment built up from cut hair, on the other hand, is a lengthy, skilled business, but more versatile. Human and yak hair are the most-used materials. (Crepe hair has limited application.) Hair is laid in spirit-gummed sections, until the required area has been covered. Finally, the hair is trimmed, waved, and fixed (lacquer-sprayed), as necessary.

(iii) Wigs need little introduction. They cover the performer's own hair entirely, and are again formed from hair tied to a shaped foundation net. The front and sides of this net may be stuck to the forehead and temples and, where necessary, hidden by overlaying it with the base medium.

10

PRODUCTIONAL ORGANIZATION

Technical planning

The need for planning

IN all art media, the time eventually arrives when the smoke from one's pipe-dreams clears. Delicate atmospheric sketches become timber, paint, and economics. Plot treatment has to be transcribed into hard facts of studio facilities, floor space, and camera-angles. Only when faced with the actual problems of the physical effort, time, space, and expense involved, do we begin to get down to the bare bones that underly television production techniques.

One can only use facilities to their utmost. Ingenuity may make them seem extensive on the screen, but they have their limits, and it is here that the whole production team's co-ordination pays dividends. The firmer the director's understanding of the studio mechanics, their limitations, and their potentialities, the less are these inherent restrictions going to frustrate his creative ambitions. The more last-minute modifications prove necessary during rehearsals, the more will the show lose its freshness and cohesion, and become just another workable routine.

Effective pictorial and sound treatment can just happen by accident, but somehow these accidents seem to occur more often to the creative artist who appreciates the characteristic properties of his craft. Spontaneous inspiration is well enough for a solo worker, but television production necessarily involves teamwork, and one's ideas have to be interpreted by other people in the team. Last-minute changes can be more far-reaching than is immediately apparent. In short, the more definite the director's work-out of his programme ideas in terms of the mechanics at his disposal, and the more complete his shot-by-shot planning, the greater his chance of success, and the better will his team's efforts be co-ordinated to the same end. Any precise treatment or special effects always take more time to rehearse and co-ordinate than straightforward routines, but this time and effort can be reduced considerably by careful and accurate planning.

Even the novice director can point his cameras, and intercut between varying viewpoints, without much difficulty. The principal difference between his work and that of a skilled director lies in the significance of their respective treatments. The latter is more likely to evoke a meaningful audience reaction, through the way he has chosen to present his subject.

This is the ideal. Some allowance has to be made for the un-expected. An actor out of position, a late entrance, can necessitate off-the-cuff changes. A workable, planned, show can stand such deviations, for there is a reference framework to which operations can be steered back. But situations where the studio operations crew hardly knows what it is supposed to do next, or which of sundry rehearsed versions is finally to be used, are not unknown. Usually, they get by thanks to the sheer experience and competence of the crew.

The technical planning process

Some time before zero date, the programme director will have gathered together his general ideas about his forthcoming production, and discussed them with his set designer (three to six weeks beforehand seems typical). The designer then transforms their agreed ideas into sketches and rough floor-plans. Whether the director initiates scenic treatment, or gains inspiration from an environment created by the designer, varies with individuals. Often it is a little of each.[1]

Procedures vary, of course, but the one we are going to outline is widely followed. Having blocked-out the general production treatment he is hoping for, the director, joined by his designer, then arranges a technical planning meeting. Here, they meet together with those responsible for technical operations and organization of the show—normally the technical and lighting directors. They will discuss the intended treatment, any hazards of the design and lay-out, and agree upon the practicability of the whole scheme. The agreed positions of cameras and sound booms are marked, and, after any corrections, copies will be sent to all departments involved in studio staging and operations.

[1] At about the same time, many other arrangements are being decided, too:— casting, rehearsal schedules, wardrobe, make-up, publicity, costing, etc.

But this is a very abbreviated version of proceedings. Let us look at them more closely, for here is the gestation period of the production the viewer will eventually see.

The designer will have distributed the settings about the studio floor, in a pattern that he hopes will suit the running-order of the production, allowing ample working space for cameras, booms, and lights. Clearly, we want to avoid unnecessary congestion, fast or complex moves for equipment. Pushing scale silhouettes of camera-dollies and booms around this floor-plan, the director begins to transcribe his treatment ideas into operational mechanics (Figure 10.1). Do camera cables cross? Will a camera kill that floor lamp?

Reminiscent of a game of chess, problems continually present themselves:—

"I'll need a high-shot from the crane here. That'll mean releasing it from the previous set before that scene's over, and another camera taking those shots. The quick reaction shots we need at the end will require three cameras, so the next scene will have to be shot on a single camera. There's only a pedestal available; that will prevent the long tracking shot I'd hoped for. Perhaps Camera 2 can turn its head quickly and get an establishing shot of the next scene from its present position ..." For continuous production, such moves are inevitable.

At last, it seems to be a workable proposition, and time has come for technical and operational vetting.

(i) **Camera positions.** The *script breakdown* (Table 10.II) shows us the sequence of scenes throughout the show. Upon it will be marked exactly which facilities are in use, and how they are distributed, as the script treatment is worked out. Knowing the amount of action in each scene; how he wants to treat it; the scene's duration; the director can allocate his cameras and booms accordingly. He can see at a glance his plan of campaign.

Few studios will have more than four cameras; although networks may use up to six. Some studios use pedestals or rolling tripods exclusively. Others include a small crane or two. (Large cranes are found only on biggest shows.) The more limited the facilities, the tighter the planning, if we are to achieve fluid, imaginative, camera treatment. Stop/start recording may ease matters here.

As we saw in Chapter 3, each type of mounting has its peculiarities. A pedestal can be rolled quickly to a new position, but provides fairly restricted adjustment when there. The crane is highly flexible, but is more bulky, repositions less readily, and so on. So, for example, we cannot end one scene with a crane-shot, and begin

Fig. 10.1.

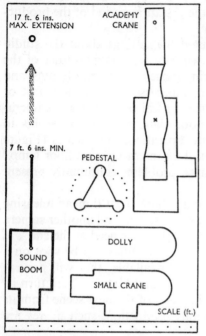

STUDIO EQUIPMENT STENCILS. Scale silhouettes showing floor space occupied.

the next similarly, if we have only one crane. Such operational factors eventually influence artistic treatment.

A simple production may involve only a series of straightforward camera moves from one set to the next. But in a complex show, with numerous scenes, intercut shots, and elaborate visual treatment, difficulties naturally arise. To have the required number and type of mountings in position when we want them, may necessitate calculated moves, countermoves, and subterfuges. If we need three cameras throughout most of an important scene, and we have only three available, we have to do some quick thinking about the treatment of the previous and succeeding scenes. The obvious solution is to begin and end that scene on two cameras. But where time or space prevent our moving the third sufficiently quickly, either less ambitious treatment, or some time-occupying device, may be inevitable. A caption, film-insert, additional business, may be needed to get the cameras and booms repositioned.

Apparently innocuous scripts can provide quite thorny problems. We may, for example, have a succession of short-duration scenes, or need to cut frequently between two or three settings. This is likely to leave the cameras spread around the studio. With few facilities, that means limiting the camera treatment for each set; no opportunity now for reaction-shots, or intercut viewpoints. We may overcome this dilemma, perhaps, by using rapidly replaceable backgrounds, multiple-settings, and the rest. But here is just one hidden aspect of live studio production. Edited recording overcomes such dilemmas.

Camera positions having been decided, the floor-plan is marked with camera-outlines, numbered, and lettered in sequence. In the studio, these positions are duplicated by drawing or taping corresponding marks on the studio floor. Graduated tracking lines or wheel marks are chalked by the cameraman during rehearsal, to record any variations about these positions. In this way, providing planning has been accurate, very predictable results can be estimated beforehand on the scale-plan, long before entering the studio. For less difficult, or less organized productional approaches, the cameras will only be allocated loose positions, their shots being worked out and positions marked-up during rehearsals. A common procedure!

Camera positioning is quite a simple process. Knowing the shot we want, the transparent lens-angle is laid on the plan, and then slid away from the subject, until the appropriate width-of-scene is taken in. Its lens position marks where our camera's lens will have to be. Conversely, if our camera distance is restricted for any reason, we lay the protractor where the camera is positioned, and see which lens-angle gives the required shot (remembering perspective). A camera set-up graph can help us to solve this sort of problem that careful planning involves, in the shortest possible time. A few of the many ways in which it can be used include:—

(i) To work out camera distance required for a particular shot with any given lens, find type of shot on vertical scale. Follow across to lens angle (sloping lines), then vertically down to the horizontal scale for the answer.

(ii) To find lens angle for any shot from a given distance, follow type-of-shot line across from the vertical scale; take a line

Fig. 10.2.

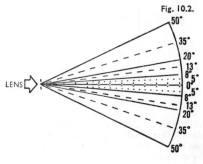

CAMERA LENS PROTRACTOR. The horizontal lens angles available by lens changes are marked on a transparent protractor. Angles of 50°, 35°, 20°, 13°, 8°, and 5° are widely used. Vertical angles (for use on elevations) are $\frac{3}{4}$ of the respective horizontal angles. As a general guide, remember the width taken in by each lens at 10 feet. The shot width for other distances will then be proportional.

181

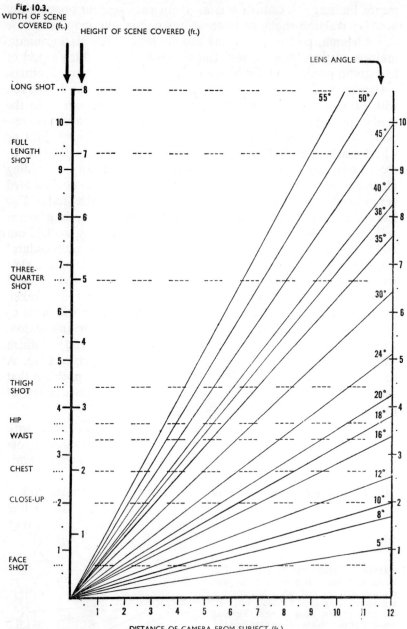

Fig. 10.3.
WIDTH OF SCENE COVERED (ft.)

HEIGHT OF SCENE COVERED (ft.)

LENS ANGLE

LONG SHOT

FULL LENGTH SHOT

THREE-QUARTER SHOT

THIGH SHOT

HIP

WAIST

CHEST

CLOSE-UP

FACE SHOT

DISTANCE OF CAMERA FROM SUBJECT (ft.)

UNIVERSAL CAMERA SET-UP GRAPH. The graph shows direction and staging details that would otherwise need continual measurement. It is calibrated in feet, but can be interpreted in inches. To use it for greater distances multiply the scale readings by a similar factor (e.g. times 5).

vertically up from the camera's distance on the horizontal scale. Where they intersect shows the required lens angle, or the nearest available that is suitable.

(iii) To see the kind of shot obtainable with any lens at a given distance, take a vertical line upwards from the distance scale. Where it cuts the lines of lenses available, look left to the vertical scale to see the nearest shot type.

(iv) How much can we see? A vertical up from the distance scale to cut the lens angle being used. From here, look left to the height and width scale, for the area taken in.

(v) What is the maximum scenic height taken in by the camera? For a level camera: half height found on graph, plus the height of the camera. For an upward tilting camera: add half vertical angle of the lens to angle of tilt. Find width taken in by twice this combined angle, at the appropriate distance. Add camera height to half this answer, and the result is the maximum height seen.

(vi) What distances give the same kind of shot (for the main subject) using various lenses? As for (i) page 181. Repeat with each lens angle.

(vii) Proportions. A TV screen has a 4 : 3 format. Knowing the height or width of the area being shot (e.g. a caption card 22 ins. wide), its other dimension is on the vertical scale beside it (in this case 16½ ins. high).

(viii) A subject is to fill 1/3 of the screen height (or width). Multiply subject height (or width) by three (i.e., we must take in three times the subject height). Look up this figure on the vertical scale; across to lens being used; down from intersection point to camera distance needed on horizontal scale.

The Graph is also used to estimate shooting-off possibilities; setting heights needed; available shots; likelihood of cameras in shot; perspective; special visual effects, etc.

The corresponding vertical and horizontal angles covered are:

Horizontal:
5   8   10   12   16   18   20   24   30   35   38   40   45   50   55
Vertical:
3¾ 6   7½   9   12   13½ 15   18   22½ 26   28½ 30   34   37½ 40

(ii) **Sound pick-up arrangements.** Visual treatment having been decided, sound pick-up must be arranged. The extensible sound boom is the most flexible microphone mounting, and is the work-

Fig. 10.4.

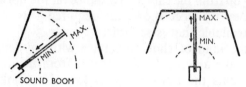

THE SOUND BOOM'S COVERAGE OF THE SETTING. This will vary with its position, being greatest in the central area.

horse for most studio sound pick-up. There will usually be two, occasionally three, per studio. Where a setting is large or awkwardly placed, or concurrent sound sources are over 6 ft. apart, two microphones will probably be needed. Two booms on one set, though, will mean disproportionately greater lighting difficulties and compromises. It may prove better to use a static microphone for the second source; a stand, slung, or hidden microphone. These may be necessary, anyway, where booms cannot reach, or would come into shot. Quick cuts from close to long shots may reveal the boom.

As with cameras, timing has to be allowed for when planning. Time to swing, time to move to a new position, possibly with the need to release the boom to get to the next set, or "fill-in" before it arrives. Sound, we see, has its continuity problems too.

(iii) **Lighting.** The lighting director becomes involved in most aspects of production treatment and staging. During the technical planning meeting he discovers both the artistic framework of the programme, and the circumstances in which he is expected to achieve it. There are many other factors, though, that are too varied for us to mention here—such considerations as fire and safety regulations; union rules; etc. And these are likely to differ widely between organizations.

The pre-rehearsal

Preliminary planning having been completed, the production will then, in all probability, go into rehearsal. Small shows may have to be content with a talk-over in the director's office; finalizing during the studio pre-transmission run-through. But productions of any complexity will undergo two to three weeks' outside rehearsal beforehand in rehearsal rooms, so freeing studio-time for camera rehearsals and transmissions.

A full-size plan of the studio settings is taped-out on to the floor. And in this imaginary environment, made a little more tangible with makeshift furniture and properties, performers rehearse their parts. Chairs and odd scenic-pieces provide landmarks such as doors, windows, and fireplaces. The director, when he is not pre-meditating on the strategies of delivery, action, business and other niceties of cast-direction, will watch proceedings from his cameras' anticipated positions. There he can align shots and adjust perform-ers' positions, using a viewfinder as a substitute for the camera lens.

Once rehearsals have begun to "gel", the specialists who earlier attended the technical planning meeting (perhaps with members of the camera and sound crews), will arrive to watch a run-through. Now they will be looking for any changes since the earlier planning, and any operational snags that the rehearsal may reveal.

Now we can see that a camera will not have time to change a lens here; some mute business will be needed there to provide time for a boom swing. The speakers here are too far apart for a single micro-phone; shall we tighten the shot by pushing them closer together? Perhaps two booms will be preferable?

By anticipating snags at this stage, valuable camera rehearsal time and last-minute alterations can be saved when the show reaches the studio.

Studio rehearsals

In the studio, the actors encounter the real setting for the first time. The production crew meet the director, and find the part they have to play. Studio rehearsal-time runs from around one to ten times the show's transmitted duration, according to its complexity and to studio availability.

Directors use this precious time in a fairly regular pattern. Occa-sionally, rehearsal begins with a preliminary *dry-run* or *walk-through*. Here, the cast go through the action of each scene in turn, while the technicians look on, the director explaining treatment and continuity. By now the *professional* director has firm ideas.

The first camera rehearsal will normally be a *blocking-out* or stopping run-through, with the crew at their operational positions. From the production control room, the director instructs and guides the crew through his treatment, pausing, where necessary, to correct, reposition performers, rearrange mechanics, etc.

Next, the complete run-through, having as few stops as possible, for final touches.

Lastly, the dress rehearsal, or final run-through, which should be an exact replica of the show which will reach the public.

Some directors stop during earlier rehearsals, to correct every defect. Others feel that this disrupts the flow and timing of the show. They prefer to stop only for major mishaps, giving corrective notes at the end of each rehearsal. On balance, this is probably the better scheme, and proves sufficient if the show has been properly planned.

The script

The television script will eventually contain basic detail of all the camera and sound direction. But in its initial form, it is often arranged in single column format like this:—

TABLE 10.1

THE REHEARSAL SCRIPT

Interior: lounge. (Night.)
(GEORGE enters, walks to table, switches on table-lamp.)
GEORGE *(calls)*: The lights are all right in here. It must be the lamp, dear.
(GEORGE takes a revolver from table-drawer: slips it into his pocket His hand withdraws, holding a telegram. He looks towards the door: holds up telegram and reads:
    "SORRY CANNOT COME WEEKEND. BRIAN SICK. WRITING ... JUDY."
He screws it up and throws into fire.
    Door opens: EILEEN enters briskly.)
EILEEN: Really these people are too bad. Judy promised faithfully that they would be here to-night.
GEORGE: Maybe the storm has delayed them a bit. It's pelting down outside. They'll be here all right.
(EILEEN sits on settee: lights a cigarette. GEORGE joins her.)
EILEEN: If she knows you're here, it'll take wild horses to keep Judy away.
GEORGE: How many more times must I tell you ...
EILEEN: Why do you have to keep pretending?
GEORGE: I've warned you before, Eileen, you've been going too far lately.

The exact layout varies with organizations, but this form of *rehearsal script* is common. It contains location, stage instructions and action, characters, dialogue. (The margins will later have camera and sound treatment typed in on the duplicator-stencils.)

A *running order*, prepared from this script after the technical planning meeting, will contain a breakdown of the entire studio operations throughout the show. We can now see at a glance the moves and timing involved.

TABLE 10.11

SCRIPT BREAKDOWN

Page	Shot Nos.	SETTING	Timing	Cam. 1	Cam. 2	Cam. 3	Cam. 4	Film	Boom A	Boom B	Various	Grams	NOTES
1		Film	40 secs.								S.O.F. (O)		
1	1	Titles (Roller)	20 secs.				4	Super				Disc 1. (Rain)	Superimposed after 40 secs.
1	2–9	Dining rm.		1			4						
3	10–27	Lounge			2 Centre	3 RHS	4 LHS		A			Disc 12. ("radio") Disc 8. band 2 (thunder)	Night (lightning) Release 3 after Shot 26
10	28	Garden				3 LHS		Rain Loop Super		B		Disc 3. (wind)	Day
10	29	Street		1					A			Disc 7. (traffic)	
11	30–43	Cafe		1	2	3 RHS	4				Slung	Disc 5. (crowd)	
15	44	Lounge								B			Evening
		Etc.											

S.O.F. = Sound On Film.   (O) = Optical.

187

Fig.10.5.

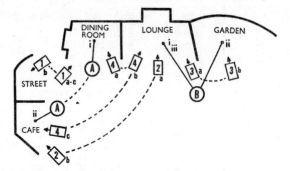

ROUGH SKETCH OF OPERATIONAL LAYOUT. (A version of this may accompany the camera cue-cards.) (1) (2) (3) and (4) are cameras; A and B are sound booms. Successive camera positions are marked a, b, c. Boom movements are marked i, ii.

TABLE 10.111

THE CAMERA SCRIPT

Shot No.	Vision	Sound
	FADE UP	*Interior: lounge (Night).*
10	CAM. 2. (20°) DISC. 1. RAIN. DISC 12. RADIO.	*(GEORGE enters, walks to table, switches on table-lamp.)*
	L.S. panning on entrance to mid-shot.	GEORGE *(calls):* The lights are all right in here. It must be the lamp, dear.
11	CUT to CAM. 3 (13°) C. Up.	*(GEORGE takes a revolver from table-drawer: slips it into his pocket. His hand withdraws, holding a telegram. He looks*
12	CUT to CAM. 2 (20°) Mid-Shot.	*towards the door: holds up telegram and reads:*
13	CUT to CAM. 4 (13°) B.C.U. Pull focus on fire.	"SORRY CANNOT COME WEEKEND. BRIAN SICK. WRITING ... JUDY." *He screws it up and throws into fire.*
14	CUT to CAM. 3 (35°) L.S.	*Door opens: EILEEN enters briskly.)* EILEEN: Really these people are too bad. Judy promised faithfully that they would be here to-night.
		GEORGE: Maybe the storm has delayed them a bit. It's pelting down outside. They'll be here all right.
	track in to mid 2-shot.	*(EILEEN sits on settee: lights a cigarette. GEORGE joins her.)* EILEEN: If she knows you're here, it'll take wild horses to keep Judy away.
		GEORGE: How many more times must I tell you ...
	DISC 8. BAND 2. LIGHTNING FLASH.	EILEEN: Why do you have to keep pretending?
15	CUT to CAM. 2 (13°). C. Up. George. THUNDER CLAP.	GEORGE: I've warned you before, Eileen, you've been going too far lately.

Once outside rehearsals have enabled the director to finalize his treatment, and decide on productional detail, the *camera script* can be prepared. This will have, added to the rehearsal script, the camera shots, lenses, camera movement, editing, visual effects, and sound arrangements. They may get modified during camera rehearsal, but this version serves to co-ordinate the whole production team.

Finally, the cameramen, who are too preoccupied with their viewfinders to follow a detailed script, are given *camera cards* (*run-down sheets*). These contain their particular camera's shots throughout the show and, perhaps, a sketch of its moves about the studio.

TABLE 10.IV

CAMERA CARDS (*Run-down sheets*)

CAMERA 3

Shot No.	Setting	Lens		Shot
11	LOUNGE R.H.S.	13°	C.U.	G's hand takes gun from drawer, to pocket, pulls out telegram.
14	LOUNGE R.H.S.	35°	L.S.	EILEEN enters. *Pan her, tracking* in to MID-SHOT as she sits.
19	LOUNGE R.H.S.	20°	C.U.	GEORGE: Reaction shot of eyes.
23	LOUNGE R.H.S.	13° B.C.U.		EILEEN'S mouth on scream.
25	LOUNGE R.H.S.	20°	C.U.	G's hand puts gun in drawer.
28	*QUICK MOVE TO GARDEN.* GARDEN	35° MID. SHOT		Follow GEORGE carrying body: ¾ back-view. Depress to *low elevation shot*. Past G's legs to C.U. of hand.
44	*MOVE TO R.H.S. LOUNGE.* LOUNGE	13°	L.S.	INSPECTOR enters: pan into MID. SHOT as 'phone rings.

Production control

Who controls the production? Throughout this book we shall call him the director, for simplicity, but his title and functions can vary.

(i) There is the one-man producer-cum-director, who is responsible for all the business and artistic arrangements of the production. Its origination, interpretation, casting, staging, and treatment, may all be his concern, as well as actually directing operations from the studio control room.

189

(ii) The producer may be the business-head of the programme, responsible for programme organization, finance, policy, etc., while his director is concerned with interpretation, staging and directing its production.

(iii) In this set-up, the producer's role is similar to the first arrangement we discussed above, but he is assisted by a director, who directs productional mechanics by proxy for him, in order that the producer can more readily detach himself from these chores, and evaluate his own treatment. By this scheme, one producer may control the preparation of several productions simultaneously, leaving the routine organization to his director.

As we have seen in earlier sections, the programme director's talkback microphone enables the whole production crew to hear him. Those in the studio itself wear head-phones; other areas (e.g. Telecine) have loudspeaker talkback. In some cases, they may have reverse talkback, and be able to reply. The technical director generally has his own parallel circuit, sometimes with private wires to individuals in the crew.

During rehearsal, the programme director may occasionally give instructions, over the loudspeaker talkback, to the whole studio, though normally he will use his floor manager as his contact-man on the studio floor. The F.M. will liaise between the director and the performers; passing instructions (couched, perhaps, in more diplomatic terms), cueing action, checking their positions; generally ensuring the smooth running of non-technical aspects of studio routine. Having an on-the-spot view of activities, he may anticipate hazards not visible to the director himself.

Cueing systems

To ensure that action begins and ends at the instant it is required, reliable cueing systems are essential, especially if accurate timing and continuity are to be maintained. Performers may be cued to enter, speak, start, stop, and so on, according to circumstances.

Most cueing is done by the floor manager, on the director's instructions. He will either make hand-signals to the performer himself, or throw a cue to someone nearer, who will relay it. Visual cues may be taken from shots on a nearby picture monitor.

Word-cues are widely used, the performer taking his cue from dialogue or action around him. Small, portable, cue-lights can cue performers from switches on the director's desk.

Stand by; go ahead.

Cut it; stop; finish; omit rest of item.

You are cleared. You are now off camera and can move, or stop action.

Volume up; louder.

Volume down; quieter (sometimes precede by "Quiet" signal).

Quiet; stop applause.

Tighten-up. Get closer together.

Open-up. Move further apart.

Come nearer; come downstage.

Go further away; go upstage.

You're on that camera, play to that camera. (Sometimes preceded by "Turning actor's head" gesture.)

Play to the light indicated. (When actors are shadowing, point to light source and to area of face shadowed.)

Turn around (in direction indicated).

Speed up; faster pace; quicker tempo. (Movement's speed shows amount of increase.)

Slow down; slower pace; stretch it out. (Indicated by slow "stretching" gesture.)

O.K.; you're all right now; it's O.K. (Confirmation signal.)

OR

We're/you're on time.

Are we on time? How is time going?

2 MINS.   OR   ½ MIN.

You have ... time left (Illustrated—2 mins. and ½ min.)

Wind-up now.

To audience: *you can applaud now.* (May be followed by "Louder" signal.)

Stop. (For applause, widespread action, etc.)

The incongruous effects of early or late cueing are all too familiar, as is the performer who finds himself still on the air long after a friendly farewell. So is sound hangover from one scene, when we are on the next. But these examples do point the necessity for clearing action after it is completed. A solution to many such situations is to have, not only prompt cueing, but bridging business as well. Then the character can be arranging papers, lighting a cigarette, etc., to cover the raw edge of the cueing.

TABLE 10.V

INSTRUCTIONS TO CAMERAS

Instructions	Meaning
Stand by two: ready two: coming to two, ... on two, shot 53, three next.	General cueing procedure for cameras, switcher, video operators, sound, etc.
Pan left (right)	Fig. 3.5.
Tilt up (down)	
Centre up (frame up)	Arrange subject in picture centre.
Focus up	To a camera standing-by defocused. (Chapter 3).
Lose focus on ...	Defocus subject mentioned.
Split focus	Focus evenly on main subjects; (where differing distances prevent sharp focus on both).
Follow focus on ...	Keep sharp on ... (a moving subject).
Focus forward (or back)	Refocus nearer to (further away).
More (less) headroom	Fig. 15.18.
Cut at ...	Compose picture to place frame at subject mentioned.
Lose the ...	Compose picture to avoid subject mentioned.
Stand by for a "rise"	Warning that subject will get up, stand.
Single shot	Shot containing 1 person.
2-shot, 3-shot, etc.	Shot containing 2, 3 persons.
Close-up, mid-shot, $\frac{3}{4}$ shot, Long shot, etc.	Fig. 14.2.
Get a wider shot	Use wider-angle lens from same camera position.
Widen the shot	Track back to get a slightly longer shot.
Looser shot: fuller-shot: Not so tight	Leave more space between subject and frame (i.e. longer shot).
Tighten the shot	Leave less space between subject and frame (i.e. closer shot).
Track in: dolly in:	
Track back: dolly back	Fig. 14.15.
Creep in (back)	Track very slowly in or out.
Crane left (right): also tongue, slew or jib	Fig. 14.18.
Crane up (down): also boom or tongue	Fig. 14.17.
Depress (elevate)	Equivalent of crane down (up) for pedestal mounting.
Tongue in (back)	With crane base at right angles to subject, swinging boom to and from it.
Zoom in (out)	Adjust zoom-lens for larger (smaller) image.
Clear on two	Camera 2's shot now finished. Camera may move, to next position, etc.
Clear two's shot	Remove obstructions from camera 2's shot.

TABLE 10.VI

INSTRUCTIONS TO SWITCHER (*vision mixer*)

Instructions	Meaning
Take One: *Cut to* One: *Cut* One	Switch to Camera 1's picture.
Fade-up Two	Turn Camera 2's video-fader up from zero to full.
Stand by to fade Two: *Ready to fade* Two	Prepare to fade Camera 2's picture out.
Fade: fade out: fade to black	Turn transmitted camera's video-fader down from full, to zero.
Stand by to mix: dissolve Two	Warning before mixing-cue.
Mix to Two: *dissolve to* Two	Mix from present camera being transmitted, to Camera 2's picture.
Superimposing Three: *ready to super* Three	Warning before superimposition-cue.
Superimpose: add Three	Fade Camera 3's picture up, adding it to existing sources.
Take Two *out: lose* Two	Remove (usually fade) Camera 2's picture from transmission; leaving the rest.

TABLE 10.VII

INSTRUCTIONS TO SOUND-MIXER (*Audio Control*)

Instructions	Meaning
Fade up sound: fade up boom	Fade up to full from zero, sound pickup in general, or source named.
Stand by grams (or music, effects, etc.)	Warning before cue.
Cue grams: go grams	Go-ahead cue for disc (tape) reproduction.
Creep in music: sneak in music	Begin music very quietly; gradually increasing volume.
Down music	Reduce volume of music.
Music under: behind, or to background.	Keep music volume low relative to other sources.
Hit the music	Begin music at full volume.
Up music	Increase volume of music,—usually to full.
Sound up (down)	General instruction to increase (decrease) overall sound volume.
Kill music (cut the music)	Stop the music.
Fade sound	Fade out all programme sound.
Fade grams:	Fade out grams: usually leaving other sources still audible.
Cross fade: mix	Fade out present source(s) while fading in the next.
Segue (pronounced "Seg-way")	One piece of music immediately follows another without a break.

Note: For Grams/music above, read any identified sound source.

TABLE 10.VIII

GENERAL INSTRUCTIONS

Instructions	Meaning
To *CAPTIONS:*	
Flip, wipe, animate, etc.	Cueing caption changes.
change caption, go roller	
To *FILM:*	
Cue telecine: roll film	Start film-projector.
To *FLOOR MANAGER:*	
Stand by to cue. Cue action.	Warning and go-ahead cues to performer.
Hold it!	During rehearsals, when director wants to stop all action, to make a correction.
Give him 2 minutes	Performer has 2 minutes left.
More voice (or less) from ...	Louder (or softer) delivery from ...
Tighten them up	Move them closer together.
Wind him up	Signal speaker to conclude.
Kill him	Stop the speaker or his action.

(and instructions respecting signals in Figure 10.6).

Fig. 10.7. Pt.1.

PROMPTING DEVICES. *Left:* Held-up cue card, idiot board, goof sheet. *Centre:* Cue flipper on camera. *Right:* roller containing the entire script, attached to camera, or concealed in the setting. May be remotely controlled by special operators or by the speaker. These have the disadvantage that the speaker looks off-camera when reading.

Fig. 10.7. Pt.2.

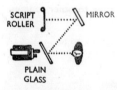
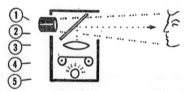

To avoid looking past the viewer, optical systems are used. *Left:* Double reflector. *Right:* Single reflector; (1) camera lens; (2) half-silvered mirror; (3) magnifier; (4) script roller; (5) light.

All such systems have the advantage that the script need not be memorized, but the delivery may lose naturalness and spontaneity, the gaze become fixed, with obvious lateral eye-movements, and the speaker and caption may get out of step.

Prompting

Subjected to the continuity of live production, even the most experienced performer can deviate from his script, dry up, or cut lines, so some method of prompting is advisable if the show is to continue smoothly.

There are several standard systems:

To have a verbal prompt from a script-following prompter nearby. To avoid the viewer hearing this off-stage voice, the person prompting may have a cut-key, which he presses when speaking. This temporarily breaks the sound circuit; a cure that can be more obtrusive than the prompt it conceals.

There are the personal reminders. These are legion, and range from palmed notes, inscribed shirt cuffs, and headlines stuck beneath the camera, to reminders scribbled on the setting itself. More than one singer has written out a lyric on the handrail of a staircase.

More extensive visual methods of prompting are shown in Figure 10.7.

A further system that has had uses for prompting and instructing is the *radio-talkback*. Here, a concealed earpiece worn by the actor, reproduces messages picked up by the tiny radio receiver he carries.

The experienced performer takes a prompt easily, without outward sign, and will need little assistance. The inexperienced may be so put off by a prompt as to dry up altogether.

In the modern TV studio, visual prompting systems are regularly used. So extensively, in fact, that many programmes and performers would be considerably hampered by their omission or failure. Certain types of production are firmly based on prompters for their success, enabling announcers, "anchor men" and introducers, to present even unread last-minute scripts with panache. No longer need speeches, lectures or dialogue be learned, for they can be read with ease from the prompter as the performer faced his audience, the camera. With experience, the presentation can be so convincing that even "ad-lib" jokes read entirely from the prompter-script appear quite spontaneous.

In advanced camera-prompter systems, an inbuilt TV picture-tube displays a remote vari-speed typed roller caption. This is seen by the speaker as a bright, large size script at the camera-lens position (the picture-tube replaces the script-roller and light in Fig. 10.7, Pt. 2, right).

11

THE TELEVISION MEDIUM

The background of television production

WHEN television began, its workers could do no more than guess at its needs and potentialities. So for the most part they adapted the diverse skills they had learned in film, theatre and radio, to this new, composite medium.

Innumerable staging and operational problems emerged, too, requiring fresh and original techniques, until, gradually television, as we know it to-day, evolved.

As the complexity of production has grown, it has become increasingly difficult for individual specialists to appreciate more than a few of its facets fully. And yet co-operative teamwork is essential. In addition, it is becoming harder for us to discern which aspects of production have evolved through convention, usage, or error, and which are inherent in its very nature.

Individual members of the production team are so preoccupied with their own ephemeral contributions to the show, that it is not always easy for them to appreciate the significance of their labours for the viewer, though this is all that really counts ultimately. Here, filming and telerecording are invaluable aids; for lessons can be learned and applied next time. Film can be analysed at leisure. Repeated viewing of any programme will inevitably reveal its shortcomings. Live production provides only a lingering memory —a most unreliable means of judging merit.

Despite the variations of production treatment, it is possible for us to trace a number of general axioms and styles that make for good television.

Techniques may still, perhaps, be regarded as at an embryonic stage of development. Some are progressive and exploratory; some mere technical operations; others a shadowy imitation, converted from other media. A parallel, in fact, to the way motion pictures have developed.

The development of motion pictures

Film presented its users with a tool of such potentialities that it was only by experiment and patient enquiry that it began to yield up its secrets. At first, the very existence of movement was sufficient to satisfy its beholder, but soon, seeking to tell a story, to convey emotion, film-makers began to demand more effective means of expression. They sought this through the means they already knew: they aped the techniques, contrivances, and conventions, of the theatre.

Gradually, film began to part company with mock-theatre and to formulate lines of its own. An untried range of devices and treatments were taken up and used; sometimes to good effect, sometimes with indiscretion. The film developed its own conventions, clichés, and routines.

Some methods became almost universal, others seemingly died—to be resuscitated later, perhaps, in a different form. Certain workers were not content to accept these findings at their face value, and were spurred on to explore this new terrain; to discover its aesthetics and the nature of its persuasion. But for the most part, the ready gimmick and the proved box-office success satisfied most makers of films. A flickering toy was now a powerful medium, capable of serving the heights and depths of Man's aspirations.

Television, too, has striven to find a mode of expression of its own, and has followed a similar course.

Comparing theatre—film—radio

In the theatre, it is the performer, and what he has to say and do, that is of primary importance. The setting—except for spectacular displays—takes a secondary place to dialogue and delivery. The whole stage presentation revolves around the actors: small, remote figures within a lighted frame, opening on to a highly conventionalized make-believe world.

Although the theatregoer's viewpoint is fixed, the filmgoer's seems limitless. At one minute he can peruse detail invisible to the human eye; at the next, he rises to stratospheric heights, to be whirled away in time and space. His freedom is apparently endless.

But the limits are there—and very strictly controlled, too, for they have been selected for him by the film-makers. He is only seeing the aspects and interpretation of the subject and its scene that they have chosen for him. The filmgoer is normally oblivious of the fact, however, and accepts the director's blinkers unquestioningly. His attention may wander a little within this framework, but he will seldom be free to concentrate upon aspects of his own choosing (as in the theatre). If skilfully persuaded, he will be inclined, instead, to follow the general line of thought suggested by those who made the film.

In film, the performer is no longer all-important. He can be subjugated, or removed altogether. His style of performance must almost always be naturalistic. Any ill-concealed falsity or overemphasis will be revealed under the camera's scrutiny, as gross exaggeration. Reaction and interaction are more important than the performer's action.

Everything the camera sees, it necessarily interprets. It is innately selective. By the very way in which it portrays a subject, its comment upon that subject will vary. The director has therefore to discriminate in his visual treatment, if his work is to convey his particular ideas to his audience.

However extreme the director's aspirations, time, space, dimension, can all seemingly be conquered by ingenious manipulation of this medium. He can conjure up visual and aural relationships that have no natural counterpart; effects that stimulate the imagination and stir the emotions.

Radio has its lesson to offer, too, for it is a medium that has turned its inherent shortcoming into an incomparable virtue. Its absence of vision makes a continual demand upon its audience's imagination. Some of its programmes, moreover, rely so strongly upon this fact that they are virtually untranslatable into visual terms without losing something of their original appeal. Furthermore, the listener's image will be personally satisfying, for it is his own. Where vision is superfluous to the significance of a programme, radio scores. Where vision is paramount, it is at its weakest.

Radio's lesson to television is clear. Unless carefully handled, visual media can provide a surfeit of information. Consequently, the viewer is not stimulated into thinking very positively about what he sees. Instead, he witnesses a series of familiar situations, presented according to an increasingly familiar pattern. His mind becomes relaxed into a state of inactive acceptance.

The characteristics of television

If we analyse briefly the nature and appeal of the television medium, we shall meet some interesting, if paradoxical, points.

Through the medium's mere accessibility, the viewer will willingly watch programmes he would not otherwise consider. The occasion having involved little personal inconvenience or expense, he is rather less expectant than when visiting the cinema or the theatre. But he is also less determined to enjoy what he sees.

Comfort and informality mean that the viewer is less inhibited, and can respond more freely than in public. But the infectious influence of a nearby audience is missing and he can take a more detached attitude towards the programme.

It was once argued that television is a medium that cannot become subservient to other activities. It demands the full attention. But in point of fact people do manage to combine viewing with an astounding variety of pursuits. With what success, is another matter. If he is not interested the viewer can indulge in conversation or other diversions, so the TV programme must capture his attention early on. Uninterrupted, however, television viewing can allow considerable concentration, and a correspondingly more critical attitude.

Within the studio, live presentation has its limits. Imperfections cannot be remedied, so standards normally average below the optimum. But, because artists and crew are tensed for their once-only performance, standards may, paradoxically, prove higher than when filmed "cold". Limitations of space, settings, mechanics, and rehearsal time, are all restrictive. The speed and variety of treatment are, therefore, constrained. Finally, the quality of the picture the viewer is watching cannot be relied upon to represent what is being transmitted.

The actuality telecast (remote; outside broadcast) brings to the viewer life as it is being lived. He gets a vantage-point at public events, games, etc., often seeing more than if he were there himself. But actuality is double-edged. Sometimes restrictions of viewpoint can mean that he sees less than the flexibility of a film coverage would have offered. The engineering problems for remote telecasts can be enormous, but once the viewer has got used to the idea of being able to look direct from a helicopter, down a coal-mine, or from Mars, his standards re-assert themselves. He begins to judge the programme on its intrinsic merits, irrespective of operational and engineering hazards. And, owing to the greater degree of compromise such shows usually involve, this can be frustrating from the producer's point of view.

Live television can rarely provide directly the flexibility and perfection of technique possible in film. However, as the director can draw upon either live or recorded presentation as the occasion requires (economics permitting), this is not an ultimate restriction. The dictum that television is essentially a live medium dies hard. After all, it is the value and quality of the product that is the final gauge, and recording offers many advantages.

An interesting factor here is the inconstancy one can find in audience attitude and criticism. Any individual's judgment is surprisingly flexible, and liable to be adjusted to the occasion. Setting aside situations where emotional bias or prejudice will condition the audience's attitude, other subtle issues seem to affect one's evaluation. One tends to expect a higher standard of a filmed production than one known to be live, and, accordingly, becomes more critical.

Conversely, where a live show has the characteristic polish of a filmed programme, an audience will start to apply more stringent standards of criticism. A programme is likely to be judged according to the standards at which it seems to be aiming. Amateur-night and Grand Opera are viewed with entirely different appraisal. Where a show is pretentious, and fails to reach its aspirations, criticism can be harsh. Where aims appear slight, viewer-attitude can be generous.

Furthermore, an actuality transmission (remote) containing such obvious blunders as visual lens-changes, or wrong cutting, can be received with remarkable forbearance, while in a studio show involving greater difficulties, they would be intensely resented.

This varying audience-sympathy throws a little light, perhaps, on the tolerance with which lower-grade programmes and presentation are often accepted.

Immediacy

A foremost advantage quoted of live television is its immediacy. Watching an event as it is happening is believed to give some indescribable fillip to the show, quickening audience concentration and interest, giving them a sense of participation.

When the viewer knowingly watches a live performance, he is aware that the progress and conclusion of events are not completely decided (unlike a film record), and that anything unpredictable may happen, an accident, a diversion, a candid-camera shot, etc.

As a result of this inherent suspense, he tends to experience various mixed feelings: curiosity, anxiety, anticipation, hope, and so on. These are not so much a direct result of the action, as his imaginative interpretation of its possible consequences.

A great deal of entertainment relies upon this sort of audience-reaction for its success. Many shows, in fact, would have distinctly less appeal if the drama of the unpredictable was completely absent. Sport, deeds of daring, from knife-throwing to parachuting, gain tremendously from such immediacy. The scene-stealing ability of children and animals lies partially in the uncertainty of their behaviour. Similarly, for interviews, talks, quizzes, experimental demonstrations, etc., we may derive a stronger impression of spontaneity through knowing that the presentation is live. There is always the chance of the unexpected reply, or the unprepared comment.

But, for a number of programmes, it is arguable whether immediacy contributes much to success. In drama, dance routines, and musical recitals, for instance, any signs of hesitancy or fault seldom please. We realize that such programmes are supposed to be rehearsed displays, not extemporized, and our appraisals are based on this premise.

Recorded programmes can have an extremely strong pseudo immediacy. Although we realize that all action is completely decided, we soon become absorbed in the performance. The chair about to crash on to the hero's head, the aircraft spinning earthwards, the whitewash bucket poised above the door, still stir our feelings. Production treatment heightens the occasion. We do not disinterestedly dismiss events because they are recorded, but react as eagerly and emotionally as if they were live. Even when we know the outcome, we can still watch the situations leading up to it with interest. If this were not so, flashback techniques would be abortive, and most dramatic films would lose their point altogether—particularly when we already know the plot.

Immediacy, then, is a real enough factor. But well contrived writing and production techniques convey a strong impression of immediacy to the film audience, anyway, and the distinction between the appeals of this pseudo immediacy and true actuality is not all that marked.

Intimacy (presence)

Television's greatest appeal lies, perhaps, in the intimacy or presence that comes from its being within the home. So strong is this impression that, before now, viewers meeting a familiar television personality for the first time have been disconcerted to realize that their acquaintance has been purely one-sided! This situation is far removed from the fan-worship of film and stage addicts.

Watching a television programme, we feel not so much that we are being taken out into the world, as that the world is being brought

201

to us. Here is an important distinction between the cinema and television.

The cinemagoer's attitude towards the screen is the broader one of surveying; a feeling that easily culminates in participation. The television viewer's attitude is a more detached one of perusal, inspection, scrutiny, or observation.

Most of us watch television in our home, where we are used to meeting people and things we know and like. We are liable to be more receptive on that account, and the presentation appears more personal. If, though, the programme should clash with our taste, we are more likely to take exception to the intrusion, which suggests that we might consequently form more violent likes and dislikes regarding individual television shows than towards entertainment in public places.

The size and distance of the screen undoubtedly influences our response to what we see there. Although a close, small, picture may have the same effective size as a large screen further away, our impressions of size, space, and time will not be the same. Speeds of camera and subject movement, and editing, that are suitable for large screen presentation, can produce a different "feel" when transferred to the smaller screen.

When our eyes converge and focus upon any subject, we subconsciously gauge its size and distance. The television screen's images, our eyes tell us, are much smaller than ourselves. We tend, therefore, to adopt a superior disposition. On the large cinema screen, our eyes inform us that the people imaged there are many times life-size, so we are inclined to a feeling of inferiority relative to them (although, through self-identification with these supranormal images, one may occasionally experience the reverse). Recall how one reacts in daily life to people considerably larger or smaller than ourselves. As we shall discuss later, our basic responses to such factors are very deep-rooted.

12

PRODUCTION TREATMENT

"It is the programme material that matters—production treatment is incidental."

Here we have a misleading half-truth. A poor script or performer remains intrinsically poor, however brilliant the presentation, but with skilful treatment, audience-impact will be improved. The average can appear good and first-class programme material can become memorable, given suitable production treatment. Conversely, potentially good material can be ruined when badly treated.

When we televise or film an event, we do much more than just show what is going on. The camera and the microphone are essentially selective tools. We cannot use them without some degree of selectivity. How the subject is arranged, and how we look at it, may have been fortuitous, but our tools have selected their material from what is there.

And our audience will be affected accordingly. This arranging and selection process constitutes production treatment.

The purpose of production treatment is clear. It is devised in order to arouse calculated thought-processes in our audience. More exactly, we are arranging and relating a series of visual and aural stimuli. We are not just providing variety and continuity of action, although that is many directors' interpretation of the term. Whether we choose these stimuli carefully, or haphazardly, they will evoke responses in our audience. From skilfully devised stimuli, we can expect effective, reasonably predictable influences. From random or erroneously chosen stimuli, we shall get unpredictable results.

Must we rely here on chance inspiration, or is there any rational basis for our choice? There is plenty of evidence to suggest that logical principles underly production treatment,—not rules, but maxims that time has shown to be valid. Not a strait jacket to originality, but a foundation of understanding from which our

Fig.12.1.

OPTICAL ILLUSIONS. In each pair of drawings, the parts marked A are identical sizes, but we shall all see the same distortions.

techniques may develop—consciously or unconsciously. And these we shall gather together and discuss in this second half of the book.

Audience reactions

Human reactions are far too complex to be easily analysed, however convenient a stimulus-catalogue might be. But within limits, we can form a shrewd estimate of audience-reaction to a great many stimuli. Psychologists are divided about causes, but the effects are definite enough for our purpose. Optical illusions illustrate this situation perfectly. Looking at the drawings in Figure 12.1, almost all of us will have the same primary reactions to what we see. How we then go on to think and feel about what we are seeing, is another matter. One cannot control idea development, but we can usefully influence the general audience response.

Many stimuli will have an almost universal effect. Horizontal scenic-lines, for example, have a more restful, settled "feel" than leaning diagonals. A movement towards us attracts our attention more readily than a recessive movement. And, whatever our audience, they are likely to feel that way about such stimuli.

Various stimuli will have predictable influences for certain national, social, cultural groups. To a Frenchman, for example, "La Marseillaise" is his national anthem, and has national, emotional, significance; while to a Tibetan, it may simply be strange, orchestral music.

Black suggests mourning in the Western world while in China, for instance, white serves the same purpose.

Some stimuli will hold different meanings for different individuals. We each have prejudices and associations of ideas, arising from our personal characteristics and experience.

How strongly we manage to influence our audience will depend,

204

initially, upon how forceful the stimuli are which we present. Effects can be deliberately cumulative, and build up to a combined strength; or, if we are not careful, they may nullify one another. The underlying influences of our production treatment are there, and we cannot run counter to them without conflict (e.g. we cannot get a rising audience reaction from a depressive camera movement). But we may knowingly create conflict by contrasting opposing associations, achieving piquancy as a result (e.g. a picture of starving children singing, "We're in the Money").

The eye and the camera

Why not set up a fixed camera overlooking a scene of action, and leave it to televise what goes on? Surely that is what a spectator would see, if he were there himself. We have just extended his vision for him, so that he can watch from his armchair. Why the need for multi-camera treatment? Pan the camera when there is a chance of the subject going out of the frame, and cut to another viewpoint when it is getting too far away.

This is certainly one possible way of televising or filming. Some directors use it. The early film pioneers did so, too: they wanted a faithful record of the events that took place before them, and this appeared the best way of getting it.

Although this may seem the most straightforward, naturalistic, approach, it overlooks important factors.

Given a constant viewpoint, our audience will find themselves individually selecting what they want to watch. Their random, unguided choice will encounter varying stimuli, which will provoke a variety of reactions. Satisfactory enough, you may argue, where all we want to do is to see what is going on; when watching a televised tennis game, for example.

But with a small screen, concentrated viewing of localized areas becomes difficult, and fatiguing. We must provide variety to keep the eye's interest by providing closer views—so giving a larger image of the subject; by providing movement (of subject or camera) to vary the eye's centre of attention; by changing the camera's viewpoint, or by changing the subject seen.

And here we have the beginning of guided selection.

The impressions we gain from seeing with our eyes are not those we get when seeing with a camera.

The camera's picture has sharply-defined borders. The size and clarity of its image will vary with the characteristics of the lens, and according to how it is used. Changing the lens-angle will alter the field of view, and the effective size and perspective of the subjects in it.

We mostly imagine that our eyes are seeing sharply over an enormous angle and depth, but this is an illusion, for the eye can only detect detail over about $1\frac{1}{2}°$, the peripheral vision being poor, both in detail and colour pick-up. However, by unconscious scanning actions and rapid readjustments of the eye's positions and focus, we obtain this sense of freedom.

Our gaze can flick instantly from one spot to another without our being aware of the blurred intermediary image; only the starting and finishing points are noticed. We could imitate these actions on the screen to some extent, but the result would be unsatisfactory. The absence, too, of true three-dimensional vision and colour, and the relatively coarse detail and tonal gradation, all modify the nature of the screened image.

There are other, less definable, distinctions between true and screened versions of the visual world.

Walking around the real world, we are conscious of our bodily sensations; of our personal physical contact with it. The environment is still; we are moving within it.

On the screen, the world is moving towards and away from us; spreading-out as it approaches our viewpoint, diminishing as it recedes. We are still, and this section of the environment is moving with respect to us. When we move the camera, immobile subjects in the scene before us are imbued with life and movement quite as much as the performers.

We can experience something of this in everyday life, when looking from a moving vehicle. But the absence of real depth and solidity on the screen causes movements and forces to be suggested between planes that will not arise naturally (e.g. the tilting-floor effect that one sees on raising or lowering the camera). But more of this later. ...

The need for techniques

Our appreciation of the external world comes to us through both intellectual and sensory responses. Some situations we enjoy primarily through their appeal to the mind (e.g. reading, painting, music), while others mainly appeal to our feelings (e.g. swimming, eating, dancing).

The camera and microphone only provide us with an intellectual link with the viewer. We can encourage him to recall bodily sensations but, obviously, we cannot convey them to him directly. Think of the feelings that come from being in a fast-moving elevator, peering over a cliff-edge, or wading through mud. Clearly, just showing someone else experiencing these things will not capture the feeling of the occasion for our audience.

We cannot, strictly speaking, convey the majority of physical concepts directly. We cannot literally show the viewer an orange, fire or water. We provide images in sound and vision that reproduce some characteristic features of these things, and their effects. We show him the shape, texture, movement, or whatever it may be, to recall for him what these things are like.

The orange is a meaningless globe—until light reveals its shape and texture; a knife slices through it, its fleshy interior glistens, juice oozes and drips. We then feel that we are seeing an orange.

Factual statements alone are not enough. We have to convey an impression to our audience, so that they feel them to be true. Whether it is a commercial or a train-crash, an interview or a dramatic play, the axiom still holds.

Imagine we are filming a mountain climb. Straightforward shots would show what was going on, but they would impart none of the thrills and emotions of the situation.

But a steep camera-angle emphasizing the slope of the mountain-side; threatening overhangs; straining fingers; slipping feet; dislodged stones; labouring breath; slow ascending music—these can do it. The difference in audience-impact between the two approaches is immeasurable. Even though the viewer himself may never have climbed in his life, he receives something of the excitement and tedium of the occasion. By selective presentation, we can awaken associative feelings.

Sometimes the audience can be so strongly moved that sympathetic bodily reactions set in (e.g. mouth-watering, dizziness, seasickness ...). Even experiences that are outside the viewer's personal knowledge (e.g. the elation of free-flight; the horror of quicksands) can be conveyed to some degree. By seeing certain carefully-chosen familiar stimuli viewers can build up for themselves an idea of what this new, unfamiliar, situation must be like.

The director must extract from a scene the essentials that are suitable for his purpose; or, if needs be, provide them.

If this choice has been wise, the result will seem natural and convincing, while conveying the impression he intends. Paradoxically, more naturalistic results can come from deliberate techniques and devices than from true reportage of actuality (as we see on comparing everyday conversation with the sort of dialogue we find written in naturalistic plays and prose).

The artificiality of production treatment

The artificialities we encounter daily can become so much part of our lives that we are liable to forget that they are contrived. They appear natural developments. They seem the logical way of doing a particular thing. Perhaps, even, the only way of doing it. Are monochrome photographs an obvious, natural representation of the scenes we snapshot? Apparently not. Our ability to interpret and appreciate them is something we have to learn. To isolated peoples who have never seen such things before, photographs are meaningless until explained.

A great many of the configurations around us that appear natural are really the product of acquired techniques. It is as natural to the Oriental's ear that his music should use microtonal scales and free rhythm as it is to the African musician that his music should contain highly complex, interwoven, rhythmical patterns; or to the Westerner that he should use an octave scale, and relatively simple rhythm structure. Their music may sound peculiar and unnatural to us. Ours may sound mannered and inexpressive to them.

Westerners have been puzzled at Eastern art which ignores perspective and proportion. Its figures often appear to us to be walking in mid-air. But our use of localized shading to depict contour has suggested to their eyes that our subjects must surely be piebald!

We accept and use artificial conventions to express ourselves and to communicate our ideas to others. Eventually, we can become oblivious that we are using conventions at all.

Some conventions or techniques are automatically understood by our audience, through subconscious associated experiences. Most camera-work is in this category.

Others will require the audience's conditioning, so that they learn to recognize and appreciate those conventions as having a particular meaning (e.g. wipes, split-screen effects).

Remember, to us a close-up face is a commonplace shot, whereas to early film audiences, the sight of a talking, decapitated head, appeared bizarre and distasteful.

We can apply techniques so that they seem to arise naturally. Supposing we want to look upwards at an actor suddenly, to give him an overpowering aspect. We could have him seated, shooting from chest height. Then, as he got up, the camera would tilt up with him, thereby giving a low-angle shot.

Alternatively, the technique can be introduced as an obviously contrived device. The camera is looking at him from eye-level, then suddenly depresses to a low-angle shot. Or we could cut from a level camera to another camera of lower height.

Where techniques seem to occur naturally, they are unobtrusive and, in consequence, usually more effective. But where a technique does not appear as a logical development, or is unfamiliar, it tends to become over-prominent. When this happens, by drawing attention to itself instead of its purpose, it largely defeats its own ends.

As conventions become recognized and established, they can sometimes be extended, so that even flagrant changes become readily accepted by the audience. Examples of this are: extreme changes in camera viewpoint, a complete absence of sound-perspective, or excessive filmic time.

So, devices that a few years earlier would have been puzzling or obtrusive, can become regarded as a normal means of expression. But where a director deliberately thwarts strongly-established conventions, and tries to give them a new significance, he treads a thorny path. A clock made with hands that rotate counter-clockwise would not show welcome originality. For most of us, it would be a tiresome conflict with established procedure.

In any medium we shall encounter conventions. Those born from the limitations of the medium, helping us to express an idea concisely that might otherwise be less conveniently conveyed (e.g. dissolves, superimpositions). And those accepted amongst practising craftsmen as interesting or attractive gimmicks (e.g. ornamental wipes). Numerous conventions come and go with the rise and fall of fashion. Others remain as the most suitable ways of transcribing thoughts and sensations into visual and aural terms.

It has been argued that conventions reduce the need for the audience to concern itself with the nature of an event, leaving them free instead to concentrate upon its significance. But conventions can lead all too readily to restrictive stylization, and require, therefore, to be used with discretion.

Most of the conventions of television have come from the established cinema. We can list only a few:

Visual—Most editing principles and camera control. The extensive use of backlight. The presence of light in "totally dark" scenes.

Aural—High-pitched voices for small creatures (e.g. mice). The echo accompanying ghostly manifestations. Background music.

13

PRODUCTIONAL IMAGERY

WHEN we want to convey information to somebody, we can do so either through a relatively uncoloured statement of fact, or by deliberately choosing language or action that will imply something more than the information alone—setting out to influence emotions, usually by inference. Television programmes can have either intention (e.g. a newscast will attempt to be factual; drama seeks to be influential).

In this chapter, we shall investigate the nature of aural and visual images, and enquire how they can be arranged influentially. The field is vast—and controversial. But here is the basis of persuasive presentation.

The sound image

Sound is much more than mere accompaniment to the picture. Sound can be used to strengthen the picture's impact. It can form the focal point of our interest, making the picture subsidiary.

Generally speaking, the aural memory, although less retentive, is more imaginative than that of the eye. Most of us have a more perceptive, discriminatory attitude towards what we see than towards what we hear.

Not only does the ear accept the unfamiliar and unrealistic more readily than the eye: aural repetition is not quite so quickly recognized. A sound-effects disc can be re-used many times, whereas a dress or wallpaper pattern can become familiar after a couple of viewings. We can summarize the kinds of sounds we use as:

Factual sound

Random natural sound pick-up, e.g. overheard street conversation.

Selective pick-up of natural sounds, enabling a particular source to be heard.

210

Atmospheric sound

Realistic. Deliberately selecting certain naturalistic sound(s) from all those present, to suggest a particular, realistic environment: e.g. the wail of a foghorn to suggest a fogbound ship.

Fantasy. A deliberately introduced distortion of reality, stimulating one's imagination through allusive associations: e.g. a Swannee Whistle's note, to suggest flight through the air.

Abstraction. Sounds stimulating ideas and emotions (by their pitch. rhythm, etc.) without direct reference to naturalistic phenomena The listener's own associated thoughts lead to personal interpretation: e.g. musique concrète.

What functions can the sound have?

Factual	Conveying information directly (e.g. normal speech).
Environmental	Establishing location (e.g. traffic noises suggesting a street scene).
Interpretative	Of ideas, thoughts, feelings (e.g. a slurred trombone note, to proclaim derision).
Symbolic	Of places, moods, events (e.g. an air-raid siren, to denote an attack).
Imitative	Of the sound, aural character, movement, etc. of the subject (e.g. music imitating a cuckoo's call).
Identifying	Associated with particular people or events (e.g. signature-tunes, leit–motiv).
Recapitulative	Recalling our acquaintance with sounds met earlier.
Coupling	Linking scenes, events, etc. (e.g. bridging music carried over from one scene to the next).
Montage	A succession, or mixture, of sounds, arranged for dramatic or comic effect (e.g. a bassoon and piccolo duet).

The visual image

What kinds of picture are there? There are factual images, showing subjects in a familiar, undistorted view, without predominantly emotional appeal; and atmospheric images. These may be realistic, with subjects arranged in apparently naturalistic surroundings, but given a particular significance by selective presentation. Alternatively, they may be fantastic, when the visual arrangements deliberately distort reality in order to stimulate the imagination through associatory ideas.

In abstraction, images stimulate ideas and emotions by using line, form, texture, movement, tone, etc., without direct reference to naturalistic phenomena, and are subject to personal interpretations.

What functions can the picture have?

Factual	Conveying information directly.
Environmental	Establishing location (e.g. a shot of Big Ben to suggest London).
Interpretative	Of ideas, thoughts, feelings, etc., by visual associations (e.g. plodding feet, to suggest weariness).
Symbolic	Of places, moods, events, etc., by associative symbols (e.g. showing Old Glory to symbolize U.S.A.).
Imitative	Of action, appearance, etc. of the subject (e.g. the camera "staggers", as a drunk reels).
Identifying	Associated with particular people, events, etc. (e.g. trademarks; Napoleon's hat).
Recapitulative	Recalling our acquaintance with subjects met earlier.
Coupling	Linking events, themes, etc. (e.g. panning from a toy boat, to *water*, to a ship at sea).
Montage	An interplay between a succession of images, or a juxtaposition of subjects.

Productional rhetoric

Rhetoric is the art of persuasive or impressive speech and writing. How does it differ from daily speech? In the first place, it stimulates the imagination. It does so in many ways, through style and technique. By inference, by allusion, instead of direct pronouncement. By appealing to the inward ear and eye. The rhetoric of the screen has similar roots; and we shall find the work of the foremost film directors studded with powerful examples.

These techniques do not classify too neatly, for they spring from relationships that are themselves unsystematic. Often the demarcations are slight. And, depending upon how we lay the emphasis, so a situation's category can vary. But the real value of such analysis is in the way it can trigger off our imagination, to help us think up new ideas.

The dry bones of any technique can make dull reading, and it is only when we actually translate them into living illustration that one begins to realize their exciting potentialities.

212

As a foretaste, let us have a couple of examples. The actor can conjure pathos from a single gesture; the director can do so through the way he selects and arranges his treatment. A veteran performer ends his brave but pathetic vaudeville act amid gibes and cat-calls. He bows and smiles hopefully; we hear a single pair of hands clapping. The camera turns from the sad, lone figure, past derisive faces ... to where his aged wife sits applauding.

An old man dies. The ticking of his bedside clock ceases; the billowing curtains at his window now lie limp.

Our analysis must be brief, but from it can be derived endless original situations. Let us first summarize the devices we can use. Then we can go on to illustrate them in action. Better still, work out an example of each device in turn, for yourself.

Summarizing the devices used in productional rhetoric

1. Directly contrasting two ideas, situations, etc.
(A) Visually. By means of:
  (a) Picture Quality.
        Changes in (i) Brightness.
                   (ii) Clarity.
                   (iii) Tonal Contrast.
  (b) Editing.
        Contrasting (i) Shot duration.
                    (ii) Transitions (speed, accentuation, rate).
                    (iii) Cutting rhythm.
  (c) Camera.
        Contrasting (i) Subject size (larger, to smaller subjects. Closer, to distant shots).
                    (ii) Elevation.
                    (iii) Camera movement.
                    (iv) Perspective and composition (Contrasting differently composed scenes); (Changing composition to alter a shot's meaning); (Readjusting the centre of attention).

213

(d) Subject.

    (i) Subject change—revealing new information which changes the picture's significance by: subject movement, camera position, editing, lighting.

    (ii) Type of subject movement—contrasting the action, speed, direction, etc., of one subject with another.

    (iii) Kind of subject association—contrasting the mood, qualities, properties, etc., of one subject with another.

    (iv) Subject form.

(B) Aurally. (i) volume; (ii) pitch; (iii) quality; (iv) reverberation; (v) speed; (vi) rhythm; (vii) duration; (viii) methods of transition; (ix) composition; (x) sound movement; (xi) sound association.

(C) Picture and Sound contrasted. By relating any aspect of (A) with an aspect of (B).

2. Direct comparison between two ideas, situations, etc.
(A) Visual comparison. See 1(A), 1(B), 1(C) above.
(B) Aural comparison.
(C) Picture and sound.

3. Showing identical subjects with different associations.

    (i) Identical or similar subjects having different purposes, values, significances, etc.

    (ii) The original purpose (or associations, etc.) of a subject has become changed.

4. Linking a variety of subjects, through common association.

5. Juxtaposing apparent incongruities.

6. Implication. Hinting at a situation without actually demonstrating it. Examples range from filmic-time to censorable innuendo.

7. Unexpected outcome.	(i) Climactic build-up to an unexpected outcome. (ii) Anti-climax, following a build-up.
8. Bathos.	A fall in significance; from the sublime to the ridiculous.
9. Deliberate falsification or distortion.	"Accidentally" causing the audience to misinterpret.
10. Imitative interpretation.	(i) Between subjects. (ii) Between mechanics and subjects.
11. Associative selection.	(i) Direct, using part of a subject to represent the whole. (ii) Recalling a subject by referring to something closely associated with it. (iii) Symbolism—using a symbol to represent a subject.
12. Deliberate overstatement.	Excessive emphasis on size, effort, etc., for dramatic strength.
13. Deliberate understatement.	Preliminary under-emphasis, to strengthen the eventual impact of size, effort, etc.
14. An unreal effect seeming to evolve naturally.	
15. A natural effect introduced through obviously contrived means.	
16. Repetition.	(i) Of sound. (ii) Of picture.
17. Sequential repetition.	(i) A series of sequences, all beginning with the same shot, associations, etc. (ii) A succession of similar circumstances.
18. Successive comparison.	(i) Showing the same subject in different circumstances. (ii) Showing the same subject in different manifestations.

19. Pun.	Play on a subject's dual significance.
20. Irony.	A comment with an inner, sardonic, meaning; often by stating the opposite.
21. Modified irony.	An ironic modification of the real significance of a subject, situation, etc.
22. Dramatic irony.	The audience perceiving a fact that the character involved is unaware of.
23. Personification.	Representing an inanimate object as having human characteristics.
24. Metaphorical transfer.	Transferring the properties of one subject to another.
25. Flash-back.	Jumping back in time, to a point earlier than the narration has reached.

26. Referring to future events as if already past or present.

27. Referring to the absent as if present.	Usually relating concurrent events, by montage.

28. Referring to the past as if still present.

29. Cut-away (See Chapter 18).	Deliberately interrupting events to show concurrent action elsewhere.
30. Fade-out on climax.	Fading at the crucial moment in action: (i) To leave the audience in suspense. (ii) To prevent the climactic peak being modified by subsequent action.
31. Double take.	Passing by a stimulus casually and then returning to it quickly, as if not realizing its significance until it had passed.
32. Sudden revelation.	Suddenly revealing new information that we were not previously aware of, immediately making the situation meaningful.

216

33. Incongruity.   Where a character:
          (i) Accepts an incongruous situation as normal.
        (ii) Exerts disproportionate effort to achieve something.
      (iii) Displays disproportionate facility (i.e. exaggerated speed, etc.).
      (iv) Is unable to perform a simple act.
       (v) Imitates unsuccessfully.
      (vi) Does the right thing—wrongly.
     (vii) Caricature.

Examples of productional rhetoric

1.A.a.i. A sudden mood-change, by switching from bright gaiety to macabre gloom.

ii. Contrasting a soft-focused dream-like atmosphere with the hard, clear-cut state of harsh reality.

iii. Contrasting the airiness of a high-key scene with the restrictiveness of heavy chiaroscuro treatment.

1.A.b.i. Contrasting the leisurely pace of prolonged shots with fast-moving, short-duration shots.

ii. Contrasting the peaceful effect of a series of fades with the sudden shock of a cut.

iii. Contrasting the jerky staccato of rapid cutting with the deliberation of a slower cutting-rhythm.

1.A.c.i. Contrasting the size of a giant aircraft with its diminutive pilot.

i. Contrasting an individual's dominance in close-up with his relative insignificance within his surroundings.

ii. Contrasting the subject-strength from a low-angle shot with its inferiority from a high-angle viewpoint.

iii. Contrasting a forward aggressive move with a backward recessive move.

iv. Contrasting the restriction of limited depth of field with a spacious deep-focus shot.

v. Contrasting the normality of a straight-on shot with the instability of a canted shot.

1.A.d.i. The only refuge on a storm-swept moorland is revealed by a lightning flash as a prison.

1.B.i. Contrasting the busy crowds' noise by day with the hush of the empty street at night.

ii. Contrasting realistic with unreal sounds.

217

2.A. A pair of lovers embrace and the camera tilts up to show a pair of lovebirds.

B. A soprano's high C merges into the scream of a factory siren.

C. A helicopter hovers to the sound of a bee buzzing.

3.ii. A favourite disc, played at a gay party is later used to cover the sounds of a murder.

4. Shots of sandcastles, rock pools, deck-chairs; sounds of children's laughter, sea-wash, suggest the seaside.

5. Shots of a massive French locomotive end with the shrill, effeminate toot of its whistle.

6. Suspecting that he is followed, a fugitive leaves a café; the sound of feet joins his own in the empty street.

7.i. A thief grabs a valuable necklace—it breaks—the pearls scatter.

ii. It is Spring. Migrant ducks arrive and land on a lake but skid on its still-frozen surface.

8. After ceremonial orders, a massive gun is fired—producing a wisp of smoke and a pop.

9. South American music, a striped blanket on sun-drenched stone, a gay straw hat, cactus—but only a sunbather in a suburban garden, listening to the radio.

10.i. Someone is speaking to a deaf person; we see his lips in close-up, but without sound.

ii. An upward movement, accompanied by rising-pitch sounds.

14. A moving picture—on which a large hand appears and turns over the picture like the page of a book.

15. Watching a street fight as a distorted reflection in the chromium hub-plate of a nearby car.

16.i. A searcher shouts the lost person's name; it echoes and re-echoes.

ii. An angry crowd closes round a central figure. Close-shots cut alternately between the accused and individuals in the crowd.

19. A parrot at an open window whistles for food; a passing girl turns at the "wolf whistle".

21. A newspaper advertisement shows extraordinary bargains; then we see it is an old copy, used to line a drawer.

24. An elephant that flies ("Dumbo").

26. Looking at the projected plans of a ship, we hear the launching festivities.

27. A superimposed montage showing a missing man surrounded by headlines; radio announcers telling of his disappearance.

28. A derelict ballroom echoing to the sounds of bygone dances.

32. Entering a room, we see someone sitting reading;—a close viewpoint reveals the dagger-hilt protruding from his back.

33.i. A man takes off his hat—and eats it.

14

CAMERA CONTROL

Its principles and purpose

NOTHING could be easier than placing a subject in a setting, and shooting it with a couple of cameras. One takes an overall covering-shot; the other shows close-up detail. We move the subjects and cameras about a bit for the sake of variety, and to link areas. This is television (or film) at its lowest.

Some directors rely largely upon the appeal of elaborate settings, dramatic lighting, and period costume to sell their work; instead of developing persuasive production techniques. But elaboration and expense are no substitute for skilled presentation.

What is the principal difference between good and bad techniques?

Good techniques are meaningful and apt. They add persuasively and significantly to the raw programme-material.
Poor techniques are mechanical, meaningless, ambiguous, unstimulating.

The selection of the right techniques at the right time marks off the creative artist from the hack.

A fundamental that is frequently overlooked, is the intrinsic distinction between moving the camera, and moving the subject. Whether the audience's viewpoint is nominally subjective (participant), or objective (observer), the effect of movement is basic.

Camera movement becomes "our behaviour"—by proxy—towards the subject. *We* are moving. The camera is our viewpoint. *We* go up to look at the subject. If the director moves us as we do not want to be moved, we shall respond accordingly. (But remember that the skilled director persuades us to want or welcome a particular camera viewpoint or movement.) If this sounds improbable, try watching a distant, mute, television screen in a dark room. The subjective illusion is so strong that we feel our peep-hole vision

Fig. 14.1. Pt.1.

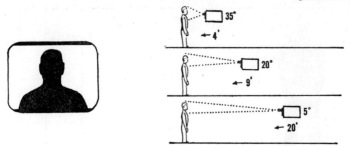

LENGTH OF SHOT. This depends on the lens-angle and the subject distance. A close shot is obtained with a close camera using a wide-angle lens; a medium distance camera with an average lens-angle; or a distant camera and a narrow-angle lens.

Fig. 14.1. Pt.2.

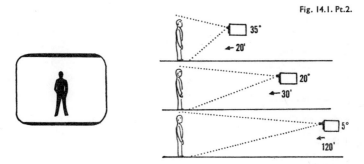

A long shot is obtained with a distant wide-angle lens; a very distant average lens-angle; or an extremely distant narrow-angle lens.

being moved around—almost physically—selecting from the complete scene.

When the subject moves and we are still, we become the recipients of its actions. The subject's movements become an "attitude" towards us; especially for advancing and receding action. Again, this effect is easily verified on a closely-viewed screen in a darkened room.

This distinction between camera movement and subject movement may result in subtle differences; for example, a performer having to walk the length of a corridor to come to us produces a viewer-superiority effect, whereas if we have to move the corridor-length to reach him where he sits waiting, the effect is one of viewer-inferiority. Where both subject and camera move, their relative effects combine.

221

Fig. 14.2. Pt.1.

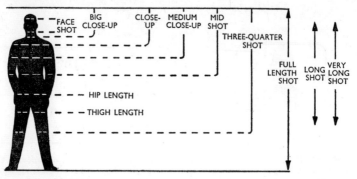

SHOT CLASSIFICATION. Shots of people have to be classified so that we can describe them in planning and production.

Fig. 14.2. Pt.2.

Left: Extreme Close-up (E.C.U., detail shot). *Centre:* Face Shot (V.C.U., very close-up). *Right:* Big Close-up (B.C.U., tight close-up, large head, full head); the head occupying $\frac{2}{5}$ of the screen.

Fig. 14.2. Pt.3.

Left: Close-up (C.U., head and shoulders); the head occupying $\frac{2}{3}$ of the screen. *Centre:* Medium Close-up (M.C.U., bust shot, chest shot); the head occupying $\frac{1}{2}$ of the screen. *Right:* Mid-shot (M.S., waist shot, close medium shot); the head occupying $\frac{1}{3}$ of the screen.

Fig. 14.2. Pt.4.

Left: $\frac{3}{4}$ Shot (medium shot, knee shot); the head occupying $\frac{1}{6}$ of the screen. *Centre:* Full-length Shot (F.L.S., M.L.S., medium long shot); the head occupying $\frac{1}{8}$ of the screen. *Right:* Long Shot (L.S.); body fills $\frac{1}{3}$ to $\frac{3}{4}$ screen height.

222

Lens and camera position

Camera set-up

Classifying the length of shot provided by camera set-ups is a somewhat arbitrary business, for although shots of people can be categorized readily enough, other subjects are less conveniently sub-divided. Terminology has been inherited from motion picture production, but has its variants.

The characteristics of various lengths of shot

The very long shot. Cramped studios usually preclude true very long shots; although they are readily simulated by process backgrounds or other special effects. Some directors shun this shot length, believing that the smallness of people results in undue visual fatigue. But this does not happen if the shot is appropriately used.

The very long shot has certain particular values. It establishes broad location, creates an overall atmospheric impression of an environment, and can co-ordinate several small action groups as well as accommodating widespread action.

The environment necessarily dominates in the very long shot.

Fig. 14.3.

THE DRAMATIC USE OF THE VERY LONG SHOT. A revolutionary mob surges outside the Chancellory. Its massive doors swing open ... the shouting dies ... a solitary figure leaves the protective stronghold, walking slowly through the parting crowd.

The long shot. The long shot similarly helps us to establish location and atmosphere, and to follow the pattern or purpose of movement. But now the figures that appeared impersonal and detached in the very long shot become characters and personalities. The effect of the setting and lighting will lessen, the closer we get; while the impact that people within the scene make upon us, will become progressively greater. Facial expression and gestures, too, will become correspondingly more forceful.

An early productional habit of beginning most scenes with an establishing long shot has died hard, for it has the merit of letting the viewer know where he is supposed to be, from the start. Sometimes it is preferable, however, to build up his impressions gradually, shot by shot. By satisfying his curiosity a little at a time, we can

223

encourage a continually growing interest, and a sense of expectation.

Badly handled, however, this delayed, piecemeal introduction of a location can prove confusing; especially when the locale is unfamiliar. Sufficient pointers must be provided for the viewer to be able to interpret correctly time, place and action.

Medium shots. Ranging from full-length to mid-shot, their function lies between the environmental effect of the long shot and the intimate scrutiny of the close shot.

From full-length to three-quarter-shot, large bodily gestures are effective, but below mid-shot such movements will have to be restricted, or go out of frame.

The close-up. The close-up concentrates the viewer's interest. Being restricted, it is an extremely powerful shot—but one with somewhat double-edged properties.

Carefully used, it can point detail that we might otherwise overlook, or have difficulty in seeing. Wrongly used, the close-up can be an embarrassment. Then we encounter over-enlarged faces spreading grotesquely across the screen, and unsharp subjects merging in a shallow-focused field. The closer the subject, the greater the care needed.

The close-up puts emphasis on what it shows but, in doing so, makes us lose sight of the whole subject and its setting.

As we have only a limited view, there is comparatively little for the attention to dwell upon, so the close-up rarely sustains our interest for long periods. When held overlong, a close-up fragment tends to become detached, and this can cause us to lose our orientation, sense of location, or forget the relationship of the portion to the whole.

If it is not to outstay its welcome, no shot should remain long enough to satisfy the viewer's curiosity completely. This is particularly true of close shots.

He should be induced to feel that he wants to look that close; not that he has been robbed of the wider view, with a suspicion that he is missing something more interesting, now out of shot.

He should not feel thrust disconcertingly close to the subject. Nor will he welcome detail being over-emphasized or pointed out, when he has already grasped the situation.

Close-ups present their operational snags, too, for the cameraman. The narrower the lens-angle, the greater they are. These will be restricted depth of field, precluding sharp overall focus, and more

THE CLOSE-UP. In close shots, action must be kept within a restricted area if it is not to go out of shot. Subjects should not be handled in and out of picture. Either will draw attention to the limitations of our viewpoint.

Fig. 14.4.

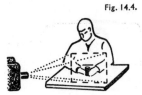

difficulty in maintaining composition—even slight subject or camera movement will appear exaggerated.

Furthermore, if we use a wide-angle lens, perspective distortion will become excessive.

The closer the shot, the slower, smoother, and more deliberate all movement should be. For the tensing of a facial muscle in a very-close shot can be as pronounced as a large bodily movement in a long shot.

Panning in a big close-up over a large, flat surface, presents problems, particularly when we are filling the screen with small details of maps or photographs. Limited depth of field makes sharp overall focus difficult. Keystoning (trapezium distortion) occurs whenever we look sideways-on at a flat picture.

By using a long-focus lens further away, we can avoid having to pan far off the perpendicular. But, unfortunately, this also leads to noticeably rougher camera movement.

Preferably, we should move the lens parallel to the subject's plane by trucking across it.

Varying the length of shot. We can vary our effective closeness to the subject in four ways; each of which has its characteristic features. (See Fig. 14.6.)

Depth of field—its artistic evaluation

Sensitive camera-tubes give us the opportunity to choose lens-apertures for the depth of field we want.

With a small aperture, most of the scene will be acceptably sharp-focused.

Large apertures give a shallower depth of field; at most working distances, only the main subject plane being dead sharp.

Do we normally want an extended, or restricted depth? Deep-focus techniques may help to achieve an illusion of spaciousness and depth, when scenic planes stretch from foreground into the far distance.

TVP—H

225

Fig. 14.5.

FRAMING. A shot that can accommodate one person in close-up, will frame two or three people at the same distance in the proportions shown (*top*). Similarly for a medium close-up (*centre*), and for a mid-shot (*bottom*).

They relieve the cameraman of the need for accurate focus-following, for there is little danger of important subjects becoming fuzzy and indecipherable. Moreover, we can more readily compose in depth; distributing subjects at different distances from the camera.

Against this, overall sharpness has its weaknesses. Where we have little camera movement, or few progressive planes of depth, the picture can become unattractively two-dimensional. Surfaces at quite different distances away can merge.

So, even where small lens-apertures are technically possible, it may be better to select a shallow depth of field. Excessive shallowness is disturbing, but carefully limited depth has its advantages.

Sharply-defined planes attract the eye more readily than defocused ones. So, by keeping our principal subject focused and allowing others to become soft, we can localize the viewer's attention. A well-composed picture will direct the eye, but this differential-focusing, too, will prevent its wandering.

It is possible by *pulling-focus* between subjects at different distances away, to move the viewer's concentration about. When co-ordinated with action, this device can be quite effective. Too often, though, it can misfire or become just another gimmick. (Fig. 14.7.)

226

Fig. 14.6. Pt.1

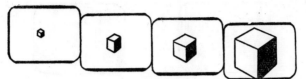

VARYING THE LENGTH OF SHOT. Tracking: Forward tracking can offer a gradual build-up of interest or tension, the importance of the subject growing progressively. Tracking back allows interest, tension, or importance to be released gradually. The change can be made imperceptibly or fairly rapidly, as we choose, the size and rate being controlled throughout. There is no editing interruption during the change, and only one camera is needed.

Fig. 14.6. Pt.2.

Cutting: Cutting shock-excites, thrusting the second viewpoint forward forcefully. Cutting to closer shots helps to emphasize momentarily the subject's strength and importance, while cutting to longer shots tends to cause sudden drops in tension and subject-importance. The principal value of the cut is to show instantly important detail, or relate immediately a part to the whole subject. Tracking would be laborious. However, where one shot is particularly satisfying to the viewer, cutting to a closer or more distant viewpoint may prove frustrating, although gradual tracking would have been acceptable.

Fig. 14.6. Pt.3.

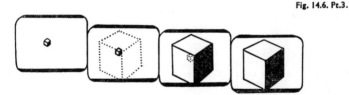

Mixing: Mixing has something of the transitional smoothness of tracking while taking less time. It enables the viewer to interrelate two viewpoints more readily than cutting, but the double-exposure of slow mixes is highly distracting.

Fig. 14.7.

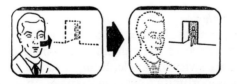

PULLING FOCUS. As the subject turns his head to welcome a newcomer, the camera refocuses.

227

Fig. 14.8. Pt.1.

FOCUS TECHNIQUES. Both deep and shallow focus techniques have their productional value. We need considerable depth of field for overall scenic detail (*top left*); to relate subjects at different distances (*top right*); when composing in depth (*bottom left*); when subjects are grouped (*bottom centre*), and for close-up detail (*bottom right*).

Fig. 14.8. Pt.2.

Restricted depth of field is required to soften-off scenic artificiality (*top left*); for spatial isolation from other people (*top right*); for spatial isolation from a distracting background (*bottom left*); for pictorial depth (aerial perspective) (*bottom centre*), and for ethereal effect (*bottom right*).

Differential-focusing lends aerial-perspective to studio-built exteriors, softening off hard lines of artificiality for backgrounds in general. It prevents our being distracted by more distant parts of the scene. Moreover, limited depth gives spatial isolation, so that the subject stands out from its surroundings. And, if correctly lit, there will be an illusion of solidity about it.

Variable lens-angles

Earlier we stressed the various distortions and perspective changes that can arise whenever the camera's lens-angle differs markedly from the audience's viewing angle. It might seem illogical, therefore, to provide the opportunity to vary lens-angles on the regular studio camera.

228

Fig. 14.9.

THE PROBLEM OF LIMITED DEPTH OF FIELD. Where the depth of field is too limited for our purpose, there are several routine solutions. The vertical broken line and triangle in the diagrams indicates focused plane. *Left:* Stopping down increases depth of field, but light levels may not permit this. *Right:* Focusing on the most important subject, allowing others to soften-off is convenient and does not affect picture quality, or require additional light. However, it may necessitate focus pulling between subjects, and some subjects may become further defocused.

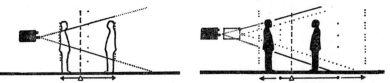

Left: Splitting focus (spreading the degree of defocusing evenly between important subjects) is tolerable where subject distances are not appreciably different, but if too split, it may be hard to see what *is* in focus. *Right:* Moving back for a longer shot. Depth of field increases with camera distance, but subject becomes smaller.

Left: Using a wider angle lens. Depth of field increases with wider lens-angle on a static camera, but subject becomes smaller, and perspective exaggerated. Tracking-in obtains the original subject-size, but re-introduces limited depth. *Right:* Adjusting the subjects' positions, bringing them more equidistant from the camera. For large close-ups of small subjects this may be an unavoidable solution, although subjects tend to become lined up across the lens axis.

Luckily, distortion can often be tolerated or overlooked. We may even introduce it deliberately for scenic effect.

Alternative lenses offer several productional opportunities.

(i)   For obtaining otherwise inaccessible shots:
With remotely-placed subjects (e.g. on a balcony; separated by uneven ground, etc.).
When the camera is isolated (e.g. situated on a rostrum; behind scenery; in confined space).
Where the camera cannot be pushed into the required position (e.g. in a crowded setting; where obstructions prevent dolly movement).

Fig. 14.10.

MULTI-LENS TREATMENT TO ADJUST EFFECTIVE PROPORTION. Where subject proportions are not those required, switching to a different lens-angle, and readjusting camera distance, may correct matters.

(ii) To adjust effective perspective:
Assisting the illusion of space (e.g. making a small setting appear spacious).

Compressing depth (e.g. to bunch together a straggling procession, giving it homogeneity).

Adjusting the proportionate sizes or distances of subjects for artistic effect (Fig. 14.10).

Including or excluding at will foreground subjects that might strengthen or weaken a shot; when moving the camera position would spoil the subject-proportions (Fig. 14.11).

(iii) When there is insufficient time to move cameras closer to or farther from the subject:
During fast intercutting between cameras on the same set.
When repositioning would involve complicated camera moves.

(iv) Permitting otherwise impracticable mechanics:
Avoiding a close shot camera coming into the corresponding long shot camera's picture.
Obtaining a wider field of view than a normal lens would provide at the furthermost possible camera position.

From being an operational convenience (sometimes a necessity), switching between different angles can become a productional habit. The persuasiveness of camera movement will be forfeited unnecessarily, and replaced by semi-static viewpoints. At worst, this leads to a series of intercut pot shots—an over-prevalent habit with zoom lenses.

Camera head movement

Many of our everyday attitudes and moods are accompanied by commonly-associated actions: we move closer to a subject to scrutinize it; we look up expectantly, and so on.

As many feelings have their associated movements so, conversely, many physical movements have their associated feelings. On a garden swing, we feel elated as we rise; apprehensive as we fall. It is

230

Fig. 14.11. Pt. 1.

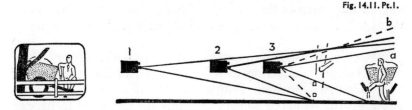

CHOOSING LENS ANGLE TO ADJUST COMPOSITION. *Left:* The effect required. *Right:* The treatment possible; 1, distant narrow-angle lens; 2, normal lens; 3, tracking in with (a) normal lens (b) wide-angle lens.

Fig. 14.11. Pt. 2.

The effect obtained shows that: (1) Distant narrow-angle shot gives the required subject and foreground proportions in this case. (2) From closer, with the normal lens, foreground proportions are correct but subject too small. (3a) Tracking-in with the normal lens produces correct subject size but foreground is out of shot. (3b) Wide-angle lens brings the foreground into shot but subject is small and subject-foreground distance increased.

no accident that we often use physical-description terms to refer to their associated emotions. A person looks up to his father. He may look down on an inferior. He feels depressed.

When we make similar kinds of movements with our cameras, in a great many cases we shall be able to evoke similar feelings through our pictures, so strong are these associations.

The pan. There is the *surveying-pan* in which we search the changing shot expectantly, looking at whatever happens to interest us. Done slowly, it is restful, anticipatory; done rapidly, it provides us with an exciting transition between two spaced points.

In the *following-pan*, we concentrate on a particular moving subject, while the background becomes an incidental environment. Here a visual interaction develops between the subject and its passing background. The interesting speculation this opens up we shall discuss in Chapter 15.

During a following-pan, we generally hold the subject around centre-frame throughout. There are some exceptions to this maxim, however, as we shall see later (Figure 14.20).

Panning shows us the spatial relationship between two subjects or areas. Cutting between viewpoints does not provide the same sense of continuity or extent. When panning over a wide arc, the

231

intermediary parts of the scene connect together in our minds, helping us to orientate ourselves, and creating an impression of spaciousness.

Whatever the pan's purpose, it should be smooth; neither jerking into action, nor juddering to a halt. Erratic or hesitant panning attracts irritated attention. With a correctly adjusted panning-head, and a properly-stanced cameraman, this is only likely to happen when a narrow-angle lens is used to follow close movement, or when the subject makes some quick, unpredicted, move.

The *interrupted-pan* is a long, smooth, movement which is suddenly stopped (occasionally reversed), to provide visual contrast.

Its commonest use is to link together a series of isolated subjects. In a dance-spectacle for example, the camera follows a solo dancer from one action-group to the next, pausing for a while at each group as it becomes a centre of interest.

The technique has its comic applications: the camera is following a bunch of sailors along a sidewalk. They pass a pretty girl. The camera stops with her. One sailor runs back, grimaces to the camera, beckoning it to hurry and catch them up.

A dramatic application: escaped convicts making their slow, painful way over treacherous marshland. One man falls, exhausted, but the camera stays with the rest. A moment or two later, it stops and looks back–to see that only the last traces of the straggler now remain.

PANNING SPEED. A prolonged pan can be full of opportunities or disappointments—according to how wisely we use it. Where we pan over subjects having a progressively increasing significance, building up to a climax, the movement will have continual interest.

A slow, surveying-pan over a scene which provides no reactional build up (e.g. just looking around a landscape) will invite falling attention, once the initial expectancy has subsided.

When a lengthy pan is made fairly quickly—but not rapidly enough to be accepted as a whip-pan—the intermediary scene merges annoyingly into a meaningless blur, and conveys little.

Fig. 14.12.

SLOW PANNING to provide a dramatic development. Panning slowly from the sleeping victim ... along the intruder's shadow ... then a rapid upward-tilt, revealing the intruder's identity.

The *whip pan* (swish or zip pan) turns so rapidly from one subject to the next that the intermediary scene appears as a streaking, indistinguishable blur. The momentary confused excitement the whip-pan arouses, depends for its success upon how the preceding and following shots are developed. Dragging our attention rapidly from one shot it gives—momentarily at least—an added importance to the next.

Used for this purpose, the whip-pan can provide a dynamic change —

Of viewpoint from one camera-angle to another. But the new angle must be related to the direction of the pan. A right-hand pan must take us camera-right of the earlier viewpoint of the same scene.

Of interest from one aspect of the subject to another (e.g. showing a different interpretation of the same subject).

Of attention-point from one area of concentration to another (e.g. to emphasize a different feature in the same subject).

From one subject or locale to another.

Comparing or contrasting situations (e.g. wealth with poverty).

The whip-pan implies spatial (not temporal) relationship between subjects. As a transition between two rapidly moving scenes, it continues the pace. It can provide a tempo-bridge between a slow and a fast scene. But, at all times, the whip-pan continues to be something of a stunt.

A live whip-pan also has its operational hazards. Unless the subjects are similar distances away, the second shot may begin defocused. The second picture's composition must be immediate; there is no opportunity to control the second picture's quality beforehand.

Any correction of focus, composition, or quality must, therefore, be carried out on the air, and cannot be covered up. One tolerable solution is to intercut momentarily a filmed whip-pan between the two static cameras' shots.

Tilting (Figure 3.1). Tilting, like panning, enables us to connect together visually subjects or areas that are spaced apart, and would otherwise require intercutting or a longer shot to encompass them.

We can also emphasize height or depth by tilting (e.g. from the eyes of a watcher on a roof-top down to his victim in the street below).

Basically, upward tilting becomes an act accompanied by rising interest and emotion, expectancy, hope, anticipation; whereas downward tilting becomes allied to lowering interest and emotion, disappointment, sadness.

Through appropriate conditioning beforehand (i.e. by dialogue, action, etc.), more complicated, modified responses arise. When the viewer has been primed to expect a particular conclusion to a tilting movement (justifiably or falsely), the downward tilt can become an act of enquiry; the upward tilt a gesture of despair. But downward tilting does not automatically evoke a feeling of enquiry, unless circumstances have led us to expect some sort of climax.

Again, when a subject no longer holds our interest (e.g. a speaker's face held overlong), the relief of a downward tilt brings an anticipatory response, even although the downward tilt is not itself anticipatory.

So we see how misleading any over-simplified analysis could be. We need to take into account not an isolated operation, but what is happening before, during, and after it (i.e. concurrent stimuli). Then we can arrive at quite useful, workable hypotheses.

Camera height

An audience's attitude towards the subject will be influenced by the camera's height. That is, whether we are looking down, up, or level with the subject. How we get that viewpoint is of little account. We might, for instance, get an overhead shot from a suspended camera, via a mirror, or even by setting the subject upright against the studio wall (e.g. for an aerial view of a model building).

Unusual camera-angles have an unhappy knack of drawing attention away from the subject, to the ingenuity or abnormality of the camera's position. This need not necessarily happen, though, if the presentation has been well arranged. Looking down from an upper storey window, or looking up from a seated position, we have, after all, a naturalistic viewpoint.

Even extreme angles can be introduced as naturally as any other productional device, providing they appear as a subjective effect. Shooting upwards through glass floors, or down through a ceiling, can look like an eccentric visual gimmick; but when such a viewpoint represents that of an eavesdropper—looking through from the room above, for example—the viewer will accept it without question.

Again, when analysing a picture one has to remember the circumstances in which the camera treatment is being used. Supposing we

234

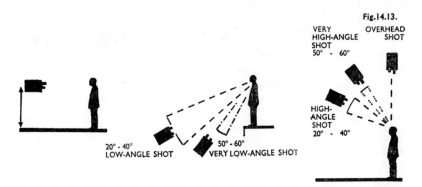

Fig.14.13.

VERY HIGH-ANGLE SHOT
50° - 60°

OVERHEAD SHOT

HIGH-ANGLE SHOT
20° - 40°

20° - 40°
LOW-ANGLE SHOT

50° - 60°
VERY LOW-ANGLE SHOT

CAMERA HEIGHT. *Left:* Level shots. A normal viewpoint, usually + 10° of chest height; carries no special significance. Camera height: 4-6 ft. for a standing person, 3 ft. 6 ins. for a sitting person.

Centre: Depressed shots. Low angle shots make most subjects acquire a stronger, more imposing aspect. Many appear overpowering, strange, ominous. A person can seem threatening, pompous, authoritative, determined, dignified, benevolent, according to his attitude and environment. Dialogue and movement become significant and dramatic. The closer we are, the stronger these impressions. In a very-low-angle shot, the subject takes on a strangely distorted, even mystical, appearance. At greater distances, it appears remote and unknown.

Right: Elevated shots. High angle shots give the audience a sense of strength or superiority, an air of tolerance, even condescension, towards the subject, increasing with distance. Therefore one can imply unimportance, inferiority, impotence, etc. Very high angle shots give an attitude of peering down, scrutinizing. From a height of 10-20 ft., surveillance gives way to complete detachment. Overhead shots emphasize pattern and movement in formation (as in ballet and dance spectacle). Reveal isolation or congestion.

see a high-angle shot of an old man trudging through rain-swept streets.

We do not seem to be experiencing any noticeable superiority or strength in relation to him. We pity him, or feel sorry for his loneliness. But, thinking about that reaction more closely, we see that it is really a psychological variation on the same theme, namely, of viewer-strength and subject-weakness.

Our consequent reactions to a particular viewpoint are not always immediately obvious. In a very high-angle shot we have an overall view; an impression of seeing much at a glance. More, in fact, than people on the spot. From such a basis, an underlying feeling of superiority may spring unconsciously that colours our attitude to what we are looking at.

If there are not to be unexpected side-effects, the more extreme camera viewpoints will need to be selected cautiously. Bird's-eye views, especially of mass-spectacle, can cause their subjects to lose their vigour and appeal, becoming transformed instead into interweaving patterns having no emotional association with lower level

Fig. 14.14.

VERTICALS IN OVERHEAD SHOTS. In overhead shots, the position of the camera relative to strong verticals in the scene can considerably affect the impact of the shot. Verticals at the bottom of the frame cause tension and instability in the shot (*left*), but this impression is absent where foreground verticals are at the top of the frame (*right*).

shots. Pattern predominates; depth and vastness are reduced to a minimum. We have the least dynamic interpretation of the subject.

Some subjects, such as balancing, juggling, and acrobatic acts, lose almost their entire appeal when shot from above. A low viewpoint underlining their difficulty shows them to best advantage.

There are also more intimate aspects to camera-height. Elevated shots (especially on wide-angle lenses) tend to emphasize baldness, plumpness in women and aggravate foreshortening and optical distortions.

Depressed angles can emphasize dilated nostrils, upturned or large noses, heavy jawlines, scrawny necks. They may show possessors of high foreheads or receding hair as entirely bald.

Movement of the camera mounting

Freedom of camera movement is determined largely by the type of mounting we use. (See Figure 3.6.) Well-chosen movement gives a picture vitality. The parallactic changes that take place provide an illusion of solidity and realness. But the movement has to be motivated and appropriate, or it will look like excessive restlessness.

Fig. 14.15.

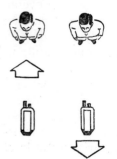

CAMERA TRACKING (dolly shot, travelling shot). Moving to or from the subject. *Left:* Tracking-in causes increased interest, build-up of tension, but the closer view may result in disappointment, disillusionment and, consequently, diminished interest. *Right:* Tracking back results in lowered interest, relaxed tension; unless hitherto unseen subjects are revealed, or when curiosity, expectation, or hope have been aroused, when we experience rising interest. Attention tends to be directed towards the edges of the picture. Fast tracking is visually exciting, but space and safe speed restrict fast moves.

236

Fig. 14.16.

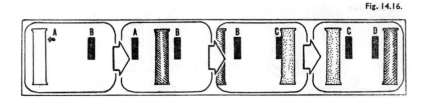

TRUCKING MOVEMENT (travel shot, crabbing). Moving across the scene, parallel with it. Trucking becomes associated with an attitude of inspection, critical observation, expectancy, intolerant appraisal. The lateral displacement of planes (parallactic movement) introduces a strong illusion of depth and solidity, but the restriction of the picture's frame becomes over-apparent if trucking stops abruptly.

Fig. 14.17. Pt.1.

CRANING MOVEMENT. Elevating or depressing the camera by raising or lowering the crane boom-arm (or pedestal height). Our reactions to camera-craning will be allied to those experienced at different camera heights, and on tracking. But the continuity of craning results in something more than just a combination of these effects. Whenever we vary the camera height, the scene's horizontal planes will seem to tilt. A rising camera will show the floor as tilting forward. With quick height changes, this illusion can be disturbing; particularly where the craning is not clearly motivated.

Left: Static mounting. When craning up from level, audience response is elation, superiority, eventual detachment as height increases; while the subject becomes less significant, even unimportant. *Centre left:* When craning down from level, the response is depression, inferiority, even a feeling of being dominated; and the subject becomes more important, impressive, significant. *Centre right:* Craning down from an elevated height with superior viewpoint will produce an effect of return to normality, with some sense of depression. *Right:* Craning up from low-level to level produces the same effect, but with a slight feeling of elation.

Fig. 14.17. Pt.2.

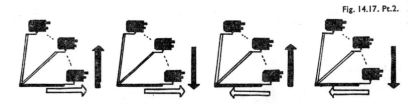

Left: Moving mounting. Tracking forwards and craning upwards creates an air of freedom, flight, lack of restriction, elevation, joy. *Centre left:* Tracking forwards and craning down, implies swooping, power, strength, importance relative to subject, when fast; but when done slowly suggests depression, return to normality. *Centre right:* Tracking back and craning upwards gives a feeling of complete release; detachment from the subject or action. *Right:* Tracking back and craning down is a recessive move, often saddening, disillusioning; a return to earth; no longer detached from subject or action.

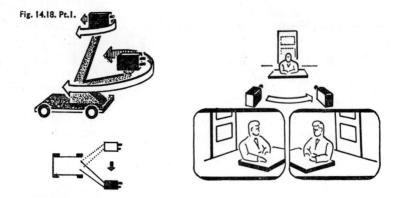

Fig. 14.18. Pt.1.

LATERAL CRANING (jibbing, slewing). Swinging the crane's boom laterally. *Top left:* The available slew will be greatest at lower camera heights. *Bottom left:* Slewing "on the air" will produce a limited trucking motion. *Right:* We must avoid any spurious side-slip effects.

Fig. 14.18. Pt.2.

Left: Slewing enables us to correct composition laterally, when trucking is not possible (e.g. where performers are out of position and masking one another). *Right:* It also provides flexibility of viewpoint, since the camera's viewpoint is normally tied about its tracking line (the path of the mounting around the studio floor).

Fig.14.18. Pt.3.

Additional diagonal and sideways movements can be provided at will by slewing the crane's boom on a straight or curved tracking line.

238

The prominence of camera movement

The greater the camera's movement, the stronger will be its viewer impact. But pronounced movement tends to draw attention to itself, and then one becomes more conscious of the mechanics than of the stimulus it is intended to impart. This would suggest that forceful manoeuvres are liable to frustrate their own purpose, by distracting attention from the subject. Fortunately, this dilemma is not insoluble.

How noticeable any camera movement is will vary with: its effective speed and extent, the amount of compositional change that occurs, and whether there is any movement in the scene at the same time—its effective direction, speed and prominence.

Camera movement will be most noticeable when shooting a static subject. Where the subject is moving, any camera moves are masked to a certain extent, for the viewer is preoccupied with the subject's action. The camera's movements may not even be observed consciously, but their effect will still remain influential.

In a highly dramatic situation, the double impacts of panning, then tracking-in, on the static subject, will provide stronger tension build-up and a faster pace.

For a less tense occasion, the combined subject and camera movement will have a more even continuity, and a suitably slower tempo.

So the old rule-of-thumb adage that camera movement should be made on subject movement, tells only half the story. It really depends upon the impact we require.

We can illustrate this point with two simple examples.

Fig. 14.19.

MASKING CAMERA MOVEMENT WITH SUBJECT MOVEMENT. Someone enters a room (in long shot) and crosses to a table (arriving in close-up). There are two methods of treatment, each providing a different audience impact. *Left:* The entrance and walk are taken in long shot. When the subject has stopped, the camera tracks in to close-up. *Right:* The camera pans and trucks throughout the walk, with subject and camera arriving simultaneously in a close-up.

239

That brings us to the emotional consequence of having concurrent subject and camera movement. As we would expect, their influences combine. The subject's movements may be more prominent than the camera's. Conversely, the camera's movements may predominate; the subject's creating a secondary effect. Where their mutual actions have similar associations, they will produce a cumulative influence. Strongly opposing associations can nullify or confuse.

So, for example, a camera movement promoting rising interest (e.g. tracking-in), accompanied by a recessive subject movement (e.g. his walking away), can frustrate the possibility of either move arousing a complete, definite, audience reaction.

Following the subject

When somebody moves from one place to another, there are six basic methods by which we can follow him. If we look at these closely, we shall see that each has its own kind of audience impact. Just as the camera's viewpoint influences our feeling towards the subject, so will the way in which we follow his movement also affect our ideas on his relationship to his background.

Sudden changes in viewpoint

The viewer's interpretation of the scene comes from the impression he gets through the camera's "eye". Whenever his virtual viewpoint is changed, he must readjust his sense of direction (balance, too, perhaps) to these new conditions. Providing this change is made skilfully, he reorientates himself immediately. Inappropriately altered, though, ambiguity arises.

In daily life, rapid and instantaneous changes in viewpoint are comparatively rare. The act of cutting is a filmic convention, and has accepted purposes.

We change our viewpoint of a subject to redirect attention to another aspect of it, to keep it in view when it would otherwise pass out of shot, to retain a particular viewpoint when the subject moves, or to show the subject's position relative to some other.

Where a quick change in viewpoint seems obtrusive, we shall usually find that it does not fulfil any of these purposes. It is a change that the viewer has no reason to want. An unmotivated change will not satisfy, even when it nominally meets these demands.

When cutting to closer shots, we must normally avoid too extreme changes, otherwise the viewer may become momentarily lost in relating the fragment to the complete scene. In general, the more concentrated the audience's attention upon a localized part of the

240

Fig. 14.20. Pt.1.

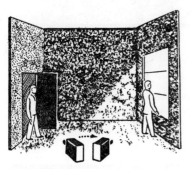 

FOLLOWING THE SUBJECT. Method 1: Following in long shot throughout. *Left:* Audience impact comes chiefly from the moving environment; subject importance being limited. The pace of action is relatively slow (Chapter 18). *Right:* By varying composition or tone during the panning, we can produce emotional changes—for instance, the emotional uplift of panning from low to high key. Sudden mood changes, however, are obtainable only by such devices as having the actor open window shutters in a darkened room, letting in daylight, or switch on room lights, during his walk.

Fig. 14.20. Pt.2.

Method 2: Following in close shot throughout. The subject dominates, its strength depending on camera height. The influence of the environment is limited, and pace varies with dynamic composition. Slightly off-centring the subject in the direction of movement creates a feeling of anticipation, expectancy.

Left: Profile—weak against a plain background, but an impression of speed and urgency against a detailed background. *Centre left:* Three-quarter-frontal—can be dramatically strong. By preventing us from seeing the subject's route or destination, marked curiosity can be encouraged. *Centre right:* Elevated frontal—although himself weakened, the subject dominates his environment. *Right:* Depressed frontal—the subject especially forceful, dominating his environment.

Fig. 14.20. Pt.3.

Rear shots introduce a subjective effect. *Left:* Three-quarter-rear—partially subjective; we move with the character, expectancy developing during movement. *Centre left:* High rear shot—an almost entirely subjective effect; produces even more pronounced anticipation than providing increasing tension; searching. In low shots there is a striking sense of depth. The subject is strongly linked to the setting and other people, yet remains separated from them.

(*Fig. 14.20 continued on page 242*)

241

Fig. 14.20. Pt.4.

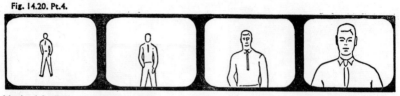

Method 3: Following from long shot to close-up. The subject walks up to the camera. Environment predominates at first, but the subject grows increasingly stronger as it approaches. The impact can be modified by tracking relative to the oncoming subject.

Fig. 14.20. Pt.5.

Method 4: Following from close-up to long shot. A weak, recessive movement, generally accompanied by lowering tension as the subject's strength falls and that of the environment increases. Used to depict pathos, anticlimax.

Fig. 14.20. Pt.6.

Method 5: Following the performer, arriving at his destination before him. *Left:* Beginning as 1, 2, 3, 4, increasing the panning speed to arrive before him. The quickening pace provides an exciting introduction to the new position, but this pace must be sustained to be effective. It shows spatial relationships of subject and destination, reduces intermediaries where irrelevant, and gives a short time for the viewer to assimilate the destination before the speaker arrives. *Right:* Beginning to move with the subject, cutting to another camera, awaiting his arrival. Often mechanically convenient, the cut giving the destination shot impact. Audience-superiority can arise through fore-knowledge of the subject's destination. Can provide pathos, bathos, stage irony, climactic effects. It avoids lengthy tracking, and can imply non-existent spatial relationships. Irrelevant intermediate movement is eliminated and pseudo filmic time effect achieved, but interest and tension can flag while awaiting the subject.

Fig. 14.20. Pt.7.

Method 6: Cutting from the static performer to his destination, awaiting his arrival. A simple but weak treatment, unless dialogue or gesture has indicated that he is going to move, and/or the destination-point. Spatial relationship or significance may not be clear to the viewer. Static action cuts of this kind can make one over-conscious of the time taken for the performer's move. On commenced-action cuts, as in Method (5), this is less pronounced.

242

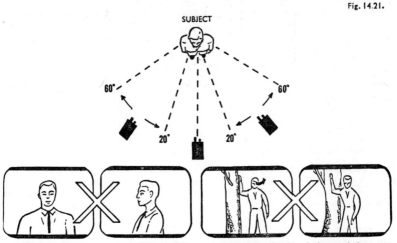

Fig. 14.21.

CUTTING. *Top:* Intercutting between cameras taking similar-sized shots of the same subject is rarely successful, especially for deviations of less than 20° or more than 60°. *Bottom left:* The subject is given a sudden twist on the cut. *Bottom right:* Where some conspicuous piece of the setting is common to both shots, a similar-sized shot may be accepted, providing the cut is motivated by the subject's action. Action that appears to be motivated by the cut (e.g. a singer turning to face the new camera) is seldom unobtrusive. Slow dissolves barely suffice to disguise such viewpoint changes.

long shot, the closer the viewpoint to which we can cut; but a change of over 2 : 1 in subject size on the cut, draws undue attention to itself. Where interposed cutaway shots are used (Chapter 18), cuts between more extreme subject sizes or angles can be accepted. Too rapid cutting between different sized shots, however, gives rise to a series of spasmodic lurches.

Unorthodox camera treatment

Occasionally, one finds it necessary to depart from straightforward camera usage, and to introduce less orthodox methods of approach, either for visual effect or mechanical convenience, or through sheer necessity.

The commonest instances of unorthodox camera treatment include:

Overhead views; some mention of these has already been made.

Under-views; the camera shooting up through a toughened glass sheet (via a mirror) which supports the subject. Most subjects are least recognizable from this angle; pattern predominates.

Inverted-views; achieved by reversing camera field and line scanning-circuits, or shooting via suitable mirrors or prisms. Among the applications are ceiling-walking and tank effects (see Fig. 20.2).

Fig. 14.22. Pt.1.

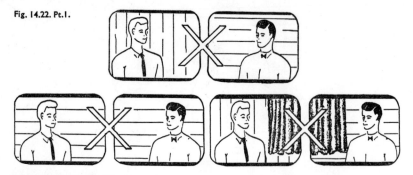

CUTTING AND VISUAL CONTINUITY. *Top:* The viewer can lose his sense of direction through inappropriate cutting where there are dissimilar backgrounds. *Bottom:* This is unlikely to happen where there is background continuity (*left*), or compositional links between pictures (*right*).

Fig. 14.22. Pt.2.

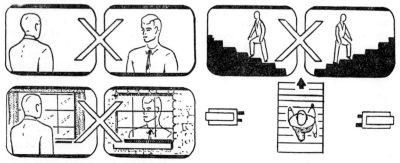

Top left: Location can be lost through reverse-angle cutting. *Bottom left:* Where a common scenic unit (e.g. a door or window) relates diverse viewpoints naturally, this re-orientation problem may be overcome. *Right:* But this is not always the case.

Fig. 14.22. Pt.3.

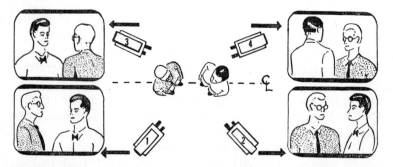

Only cameras on the same side of the centre-line joining two subjects will intercut well) e.g. between (1) and (2) or between (3) and (4). But not between (1) and (3) or (2) and (4).

Fig. 14.23. Pt.1.

THE CANTED SHOT. *Left:* Canting is most pronounced with subjects inherently associated with horizontal or vertical stability. *Right:* Canted shots of indeterminately angled subjects mean little.

Fig. 14.23. Pt.2.

Left: Slight canting is largely ineffectual, seeming accidentally lopsided. *Centre:* Compare a suitable degree of canting with too little or too much. *Right:* Too steep a cant looks like an eccentric overbalancing.

Fig. 14.23. Pt.3.

In canted low-angle shots (*left*), and high-angle shots (*right*) the direction of the cant causes the subject to lean into and out of the picture.

Mirroring producing a mirror image of the original shot, and achieved by reversing camera line scanning-circuits, or shooting via suitable mirrors or prisms. This can be used for correcting mirror shots, reversing mirror writing, or for compositional correction.

The canted shot

The usual purpose of off-vertical shooting is to suggest instability, uncertainty, abnormality, dynamism. Combined with an elevated viewpoint it may also suggest imminent action. Combined with a depressed viewpoint and right cant it suggests active forcefulness; with left cant—impotent force.

Imitative camera movement

The use of camera movement to simulate jogging vehicles, rolling ships, etc. is too familiar to need detailed description.

245

More subtle, though, is camerawork that visually echoes the subject's movement: staggering as it tracks after a drunk, copying the bounce of a fop's gait, swaying in waltz-time, when following a girl home from her first ball.

This near-subjective movement draws attention to itself quite openly; often as a tongue-in-cheek comment between the director and his audience, upon the subject they are seeing together. When it comes off, this device has a persuasive appeal.

Using the zoom lens

A zoom lens enables us to alter the effective lens-angle continually over a wide range. We can set it to any intermediary angle, or adjust it on the air to produce a zooming action.

Zooming-in narrows the lens-angle, blowing-up the centre of the shot to fill the whole screen—an all-too-overworked technique.

Zooming-out widens the lens-angle, giving a longer shot of a larger area.

The zoom lens has its attractions.

We can, for example, adjust its angle for the exact proportion and coverage we want. There are no lens changes or compromise lens selection. Zooming can be done quickly, where fast tracking would be impracticable. A shot-box enables pre-set angles to be button selected.

The zoom lens seems to possess enormous advantages. But there are pitfalls for the unwary. Lens angles may be used indiscriminately.

One must pre-check zoom-in shots to avoid defocusing as depth decreases with closer shots on a narrow lens angle.

There is a temptation to zoom instead of tracking. Tracking provides naturalistic perspective changes. The effect of zooming is entirely artificial and, on static cameras, can result in continual perspective changes.

Where zooming is slow, and accompanied by tilting or panning movements, these distortions may be disguised—especially when shooting large, open settings.

Zooming allows rapid, highly dramatized swoops, taking the eye from a wide field to concentrate on a central subject, or vice versa. But in so doing, it tends to fling the subject at the viewer. The effect is dramatic, or annoying, as the case may be.

In any event, zooming must be decisive. Slight quick changes in lens-angle show as disturbing jumps in image size to the viewer, and have no value, save as an irritant.

Fig. 14.24. Pt.1.

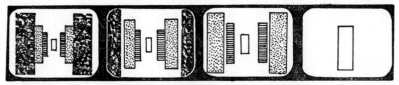

THE EFFECTS OF TRACKING AND ZOOMING COMPARED. When a camera moves towards a setting, the relative sizes of subjects in the latter will change proportionately, in a natural manner. There is continual parallactic movement as different planes become visible, and perspective remains constant.

Fig. 14.24. Pt.2.

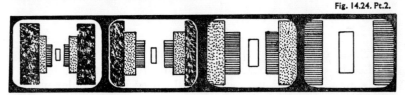

When using a zoom lens, the proportions of all subjects remain the same as the field of view is altered. There is no parallactic movement, simply an enlarging or diminishing action. Zooming-in produces visible compression of depth, and zooming-out produces visible depth stretching.

Single-camera techniques

The elaborations of multi-camera treatment have sometimes been avoided by shooting the entire production from a single camera crane. Well handled on suitable occasions, this technique undoubtedly has its benefits, but as a practice is double-edged.

Advantages. It precludes productional *faux pas* by its very nature.

We cannot have incorrect editing, inadvertent perspective variations, other cameras in shot, reorientation difficulties.

Performers cannot play to the wrong camera.

Disadvantages. The production's pace is slow, depending entirely upon camera and subject movement, and sound.

Visual variety relies entirely upon subject and camera movements, and upon repositioning.

Precise camerawork rests upon one cameraman—and this may prove unduly tiring.

On balance, however, true single-camera techniques are restrictive. Occasionally they meet all one's needs. More usually they needlessly abandon persuasive devices (e.g. reaction shots) without providing alternatives. A compromise solution, therefore, is to supplement one main camera with other auxiliaries wherever necessary.

247

15

PICTURE COMPOSITION

Is pictorial composition bunk? Intelligently studied and applied, no. Treated blindly as a magical panacea, it certainly can be.

We may not know why we are influenced in particular ways by certain visual arrangements, but their effects are regular enough to provide us with rational working principles. So we no longer need to distribute subjects around tentatively, hoping that they will produce the effect we want. We develop a background of understanding that helps us to arrange, correct, and improve camera shots in an organized fashion.

Purpose and aim

Composition helps us in two ways. It helps us to direct the viewer's attention to a selected subject, and to influence the viewer's feelings.

Composition is deliberate selection and arrangement for a specific effect within the picture frame. It is based upon an appreciation of the way in which man seeks subconsciously to form patterns, to relate what he sees to his past experience, and to think in terms of generalized symbols.

Some people store up this kind of knowledge unwittingly; others acquire it through study. Many are suspicious of its value, but most are appreciative of its effects.

Attractiveness is not enough; pictures must be appropriately meaningful. They must persuade the viewer in a particular way, suitable for our purpose.

A television production may be so complex as to prevent ideal composition throughout, but we can achieve a high pictorial standard nevertheless by arranging careful *key shots* for each action sequence.

Looking around a scene, our attention flits between subjects that

Fig. 15.1.

SUBJECTIVE LINES. One's mind tends to think in terms of pattern, seeing relationships even where none really exists. Thus we see the first group of symbols (*top*) not as odd individual marks, but combining or interacting to form patterns (*bottom*). These subjective effects form one type of dominant lines.

happen to attract our attention. But try peering through nearly closed eyes. Now we can begin to see broad effects more readily. Colour becomes subdued. We can assess monochrome values more easily. Areas now resolve themselves into tonal masses. We can see tonal balance between parts of the scene. Broad compositional lines are clearer. We begin to see essentials. And these essentials are the underlying influence of composition that become evident in a flat picture.

Dominant line

Looking at a picture, we are subtly influenced by various aspects of its arrangements. Many pictorial lines and patterns have come subconsciously to have symbolic meaning for us. So when we recognize them in a picture, our feelings towards that scene can be affected correspondingly. Some of these lines and shapes are those of the actual subjects themselves. Others are formed from real or imaginary lines joining parts of the scene, as our mind's pattern-making habit comes into play.

By deliberately arranging our scenes so that particular types of line predominate, we can endow them with an underlying mood or atmosphere.

Fig. 15.2.

REAL DOMINANT LINES. There are also the actual dominant lines in scenes. These will depend upon the subject, how we look at it, and how we arrange people and other subjects relative to it.

249

Fig. 15.3.

VARIATIONS OF DOMINANT LINE. This group of illustrations all show the same subject doing the same sort of thing, but each has a different atmosphere. Why? They contain different kinds of dominant line which, through their associatory influences, affect how we feel about the scene as a whole.

Fig. 15.4. Pt.1.

THE IMPLICATIONS OF SHAPE. The shapes, forms and patterns made by dominant line fall into a series of basic units, which have remarkably constant influences upon atmosphere and mood of the picture.

Straight lines suggest definiteness, directness, rigidity, masculinity, simplicity. *Left:* Vertical, they imply solemnity, dignity, importance, formality, strength, restriction, slenderness, height, dynamic stability. *Centre:* In perspective, hope, inspiration. *Right:* Horizontal, rest, inactivity, peace, breadth, openness, settled stability.

Fig. 15.4. Pt.2.

Diagonal lines suggest forceful action; the instability brings excitement and attracts attention. *Left:* Less than 45° off vertical implies forward movement, insecurity. *Centre:* Around 45° it becomes rapid forward movement, antagonism. *Right:* More than 45° indicates downward fall, instability, excessive speed.

Fig. 15.4. Pt.3.

Curved lines create impressions of rhythmical movement, change, beauty, elegance, grace, femininity. The sharper the curve, the greater the speed and sense of movement. *Left:* When the curve is downward the feeling is of dominant pressure, expression, restriction. *Right:* When upward support, freedom.

Fig. 15.4. Pt.4.

Left: Broken lines can suggest interruption, lack of organization, casualness, muddle, absence of formality or dignity. *Centre left:* Broken patterns, impermanence, confusion, dynamic clash, quick, irregular movement. *Centre right:* If the lines are coordinated, the impression is of unity, regularity, pattern, monotony. *Right:* If uncoordinated, confusion, disorder, individuality.

Fig. 15.4. Pt.5.

Derived shapes in plane and perspective forms have various effects. *Left:* The influences of the horizontal and vertical components tend to become nullified; the suggestion is opposition, contrast, unorthodoxy, informality, equivalence. *Centre left:* Concentration, tension, impending action. *Centre right:* Orthodox conformance, evenness of forces, symmetry, equal overall attention, masculinity. *Right:* Rectangles carry their corresponding horizontal or vertical influences.

Left: Balanced, powerful, forms at rest: stability, unity, balance, compactness, potential power, dominance, climax. *Centre left:* Even, non-stop movement, continuity of interest, order, formality, equalized forces, magnitude. *Centre:* Continuous movement, charm, femininity. *Centre right:* Smooth change of movement, beauty, charm, elegance, procedure, sequence. *Right:* Sharp change of movement, violent change, vigour, excitement.

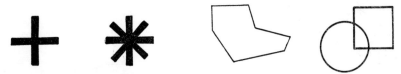

Left: Equally-opposed forms of additive strength: resistance, cohesion, converging interests, unity, sometimes sacrifice. *Centre left:* Forces focusing on a common point: Converging or radiating force, unity, dramatic concentration, expansion. *Centre right:* Broken, irregular shapes: indecision, instability, disorder. *Right:* Continuous, regular shapes: regularity, order, system, montonous continuity.

Tone

A picture's emotional key is largely set by its predominant tones. In general, we find that where predominant tones are light, the effect is cheerful, delicate, airy, open, gay, simple, trivial. Predominant dark tones create an effect that is heavy, dramatic, sombre, sordid, mysterious, forceful, dignified, significant.

Large, well-marked areas of tone give a picture strength, vigour, significance.

A dark tone, relieved by smaller, distinct light areas has grave, solemn, mysterious, dramatic qualities.

Fig. 15.5.

THE IMPLICATIONS OF TONE AND CONTRAST. Sharply-defined tonal contrast isolates and defines areas, suggesting crispness, hardness, attractiveness, vitality, dynamism, definiteness. *Right:* Tonal gradation blends areas together, and guides the attention from darker to lighter areas, suggesting vagueness, mystery, softness, beauty, restfulness, lack of vigour.

Fig. 15.6.

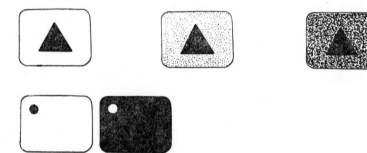

THE EFFECT OF CONTRAST. *Top:* Tonal contrast places emphasis on shape and mass. *Left:* A small dark area tends to become subjugated when placed within a large white area; whereas a small light area stands prominently from a dark surrounding tone.

Fig. 15.7.

SIMULTANEOUS CONTRAST (spatial induction). *Left:* Adjacent tones interact. A light tone tends to make its neighbour look darker, and a dark tone to make it lighter. This effect is known as simultaneous contrast, or spatial induction. *Right:* The sharper and greater the contrast, the stronger will be this effect.

The apparent chromatic values of colours can be affected similarly. Against a dark background a colour may look cool (bluish), against a light background it can appear warm (brownish)—again, an effect to be guarded against or utilized.

A light tone, relieved by smaller, distinct dark areas suggests brightness, liveliness, cheerfulness, delicacy.

An unrelieved light or dark tone, containing few contrasting areas, is dull and uninteresting.

Balance

Balance is stability. And although we may occasionally seek for the dynamic restlessness that unbalance can give, we shall normally want our picture to appear in equilibrium. But to achieve this, we must know something about the nature of pictorial balance.

Thought is largely a comparative process, so it is not surprising to find that we habitually carry over many of our ideas from one field of experience into another. Our assessment of the effects of gravity, proportion, weight, etc., have become applied illogically but firmly, to our judgment of pictorial effect. (A black box will look physically heavier to us than an identical white box, although tone and weight obviously have no true connection at all.)

Fig. 15.8. Pt.1.

BALANCE. A picture's centre forms the pivot about which the whole composition balances. A formal, precise balance of equal values suggests formality, solemnity, method, etc., but can become monotonous and uninteresting.

Fig. 15.8. Pt.2.

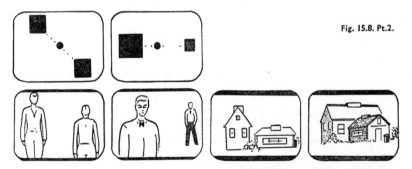

Left: Where there is no balance, the result is unharmonious or unstable. Centre left: With asymmetric balance, however, weight is distributed about the pivot to provide variety and interest. The larger the mass, the nearer to the centre must it be to preserve balance. Centre right: Straight-on views are usually dull. Right: The most pleasing balance is in both plane and perspective, angled viewpoints being preferable.

253

Fig. 15.9. Pt.1.

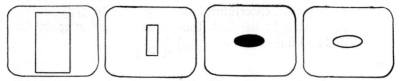

THE INHERENT BALANCE OF A PICTURE is affected by several factors simultaneously. The position of principal subjects in the frame. A subject's weight tends to increase when farther from centre-frame. Lower frame suggests heaviness, support, importance. The upper part of the frame suggests suspense, lightness, unimportance. Verticals near the upright edges imply firmness, balance, stability, upthrust. *Left:* Effective size. Large areas have greater weight.

Right: Predominant tones. Dark tones appear heavier, but light tones appear larger.

Fig. 15.9. Pt.2.

Left: Size and tone of areas between subjects. In the first example the interposed space is unobtrusive and weak; in the second it is stronger than the main subjects.

Right: Isolation—this gives weight to a subject.

Fig. 15.9. Pt.3.

Left: Dominant scenic lines. Balance depends more upon vertical than horizontal lines, although overall balance in a horizontal plane determines the final effect. Regularly shaped subjects are heavier than irregular ones.

Right: Close grouping increases the collective weight of individual subjects.

Fig. 15.9. Pt.4.

Balance in depth is affected by the proportion of foreground to middle distance we show, depending upon our viewpoint.

Balance is affected too by intrinsic interest—the subject of our attention is given greater weight; by associated stability (see Fig. 14.23) and by colour. Warm colours appear heavier than cool colours, and bright or saturated colours seem heavier than darker or desaturated ones.

Fig. 15.10. Pt.1.

BALANCE AND TONE. Tonal distribution affects balance; weight increasing with dark tones. The position of tonal masses can create a direction of thrust.

Fig. 15.10. Pt.2.

Left: Darkening the top of the picture produces a strong downward thrust, top-heaviness, invoking a depressed, closed-in feeling. *Right:* At the bottom, darker tones form a firm base for composition, lending it solidity.

Fig. 15.10. Pt.3.

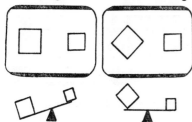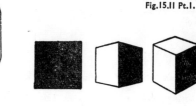

Both size and tone influence how far a subject's position affects overall balance. Smaller areas and lighter tones alter it less. Large area dark tones require careful positioning.

Fig.15.11 Pt.1.

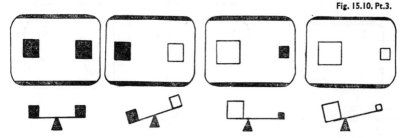

BALANCE AND SHAPE. *Left:* Shape, too, has its effect upon balance. *Right:* Shape will change with viewpoint.

Fig. 15.11. Pt.2.

Angles formed by implied subject-lines, or subjects and parts of the frame, affect balance. *Left:* Outward angles lead the eye out of the picture. *Right:* Re-entrant angles take it back into the scene.

255

Fig. 15.12. Pt.1.

METHODS OF ALTERING PICTORIAL BALANCE. Changing the sizes of subjects. *Left:* Varying camera-distance. *Right:* Varying lens-angle.

Fig. 15.12. Pt.2.

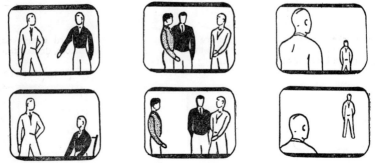

Readjusting the position of subjects in the frame. *Left:* Varying subject height. *Centre:* Repositioning subjects. *Right:* Changing camera height.

Fig.15.12. Pt.3.

Altering scenic or subject tones. *Left:* Lighting changes. *Right:* Changing our viewpoint.

Fig. 15.12. Pt.4.

But remember that in adjusting pictorial balance we can inadvertently focus attention on, or give prominence to, a subsidiary subject.

256

Fig.15.13.

UNITY AND BALANCE. *Left:* Unity is often confused with balance, but a picture can be balanced without being unified (*top*), or unified without being balanced (*bottom*). *Centre:* To achieve unity, generally speaking, we must relate the principal masses by real or imaginary compositional lines. Not isolation (*top*), but unification (*bottom*). *Right:* We should avoid lines or tonal changes that subdivide the screen (*top*) and aim at unification (*bottom*).

Changing the pictorial balance

We usually want to readjust a picture's balance in order to re-direct attention to another subject, or to alter the picture's significance, mood, etc.

Several methods lie open to us, which we can introduce at a speed suitable to the occasion, as in Figure 15.12.

Unity (order)

Unity is an essential of successful composition. It is arranging the picture as a whole, so that all its separate parts are co-related into a unity of mood, interest and attention, a unity of purpose and style.

Visual rhythm

As in music and poetry, the ear prefers to hear a recognizable, but not too elementary, rhythmical beat, so the eye is attracted to a variety of pattern.

Visual rhythm is not just an academic consideration. It is a pattern-term, embracing the combined effect of subject and setting, in a particular arrangement. In a well-composed picture, we shall find a rhythmical arrangement that suits its occasion and purpose (a slow-flowing visual rhythm for a solemn, quiet mood; a rapid, spiky staccato for an exciting one).

Fig. 15.14. Pt.1.

VISUAL RHYTHM. *Left:* A picture may be balanced, but dull and monotonous; we can imagine its aural counterpart—tum-tum-tum. *Centre:* With slight rearrangement, a more pleasing rhythm is obtained; more like a tum-ti-tum rhythm. *Right:* But an even greater improvement presents a more complex rhythm; perhaps ta-ti-tum.

Fig. 15.14. Pt.2.

  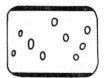

The variety of visual rhythm can become out of hand (*left*) or confused (*centre*) or too complex (*right*).

Fig. 15.14. Pt.3.

Rhythm can imply the repetition of an idea. Through repetition, an initial theme gains strength, providing there is a progressive development (*left*); rather than mere identical reiteration (*right*).

Fig. 15.14. Pt.4.

Visual rhythm comes from:—*Left:* Subjects—their size, shape, tone, texture, etc. *Centre left:* The combined shape they make with other subjects. *Centre right:* The space-pattern left around them. *Right:* The shapes of shadows on the subjects, and on nearby surfaces.

Proportion

Beauty defies analysis. But from ancient times man has tried to pin down exactly what constitutes harmonious proportions. He discovered a widely accepted principle in the so-called Golden Section (Golden Mean). If we examine the art of centuries, we can detect how innumerable great painters, sculptors, architects, have em-

258

ployed certain ratios of size, over and over again. Have they selected these divine proportions unwittingly when expressing their concept of beauty?

Draw a straight line. Now try dividing it into two unequal parts, giving the most pleasing relationship. Subdivide the parts similarly. With incredible regularity, we find the line divided into a constant ratio, so that the small part is to the larger as the larger is to the whole. If we then take this smaller portion and divide it too into unequal parts, similar proportions again appear. And so on for larger and smaller multiples.

This proportioning has quite far-reaching applications, for it gives us a guide to proportions that most readily please the eye. Be suspicious of such formulae, but see how often they seem to validate themselves. Through their use, we can develop our own personal ability to select satisfying proportions instinctively.

Fig. 15.15.

THE GOLDEN SECTION. Dividing for ideal proportions tends to produce YZ: XY in the same proportions as XY: XZ

Making several such divisions, a series appears, and this ratio of division, the Golden Section, is about .617 to .383 or, more conveniently, 1.618 to 1.

Framing—the subject's position in the frame

Does it matter how and where scenic-lines divide the frame or where we place subjects within it? Our thoughts on the Golden Section suggest that it does. The whole business of subject-placement, we shall find, has many unexpected ramifications.

As Figure 15.9 has already suggested, all parts of the frame do not have equal pictorial value. Experience has shown that we can expect fairly regular effects when using certain regions.

According to the parts of the screen we use, and the subject's closeness to the frame, we shall experience an implied effect upon the subject itself.

However, we cannot be dogmatic about how much space to leave around it. The headroom between the top of the principal subject and the upper frame-edge will need readjusting to suit the shot length and the compositional balance.

Fig. 15.16. Pt.1.

POSITIONING THE SUBJECT IN THE FRAME. *Left:* An equally divided frame allows only formal balance—usually dull and monotonous. *Right:* Thirds can lead to quickly recognized mechanical proportions. Dividing the screen in a 2 : 3, or a 3 : 4 ratio achieves a far more pleasing balance.

Fig. 15.16. Pt.2.

The centre of the screen is generally its weakest concentrational area. Unless strong compositional lines concentrate upon it, continual or sustained use of picture-centre for concentrating attention leads to monotony and lack of interest.

Fig. 15.16. Pt.3.

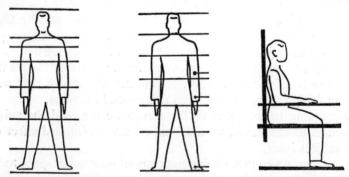

Left: Positions near the edges seldom give good balance, especially if vertical or horizontal scenic lines run parallel to them. *Centre:* Frame corners tend to exert an outward pull upon subjects placed there. *Right:* Leaving about 1/10th border within the screen edge allows for receiver masking. Strength and importance of the subject tend to be greater higher in the frame, and towards the right.

Fig. 15.17.

FRAMING PEOPLE. *Left:* When framing people, it is best to cut the figure at intermediate points. *Centre:* Cutting at natural joints is not advisable. *Right:* We should also avoid them touching the frame edge (standing, sitting, leaning on it).

Fig. 15.18. Pt.1.

VERTICAL FRAMING. Headroom is the distance between the tops of heads and the frame. *Left:* If people are placed too high in the frame (too little headroom), the picture looks cramped. *Centre:* If too low in the frame (too much headroom), the picture looks bottom-heavy. *Right:* Correct headroom provides good balance.

Fig. 15.18. Pt.2.

Headroom that would be excessive with a plain background may provide a suitable balance when there are strong compositional masses at the top of the picture (*centre*). Headroom can even modify the meaning of the picture, as where the foliage (*right*) can be either an incidental border, or an overhanging, oppressive, mass, perhaps with some story-significance.

Fig. 15.19.

HORIZONTAL FRAMING is also important. *Left:* Centring may result in an unattractive balance. *Centre:* Off-centring slightly may improve matters. *Right:* Further off-centring can appear lopsided, but may be advantageous where we want to turn attention towards the empty side of the frame, or imply isolation.

Fig. 15.20.

REFRAMING. *Left:* As subjects leave or enter the frame, the cameraman normally re-adjusts his framing unobtrusively for the new situation. Very close (tight) shots can necessitate continual reframing. Recomposing then looks fidgety and obtrusive, emphasizing the confines of the frame. *Right:* Occasionally it is better dramatically *not* to reframe, but to leave the picture unbalanced. The compositional tension this creates emphasizes the departure **or absence** of the second subject.

Fig. 15.21. Pt.1.

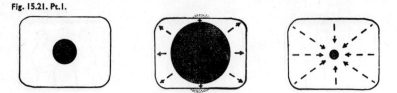

ILLUSORY ATTRACTION. A disc in the centre of the frame appears at rest, surrounded by empty areas of equal tension. A change in subject size alters this reaction between the subject and its frame. An over-large subject bulges the frame, squeezing-out the surrounding space, while a too-small subject becomes compressed by the large area of empty space surrounding it.

Fig. 15.21. Pt.2.

This compression effect gives way to a forceful thrusting as the intermediate space lessens.

Fig. 15. 21. Pt.3.

Tension is built up between subjects in a picture. The two subjects on the left convey a feeling of mutual interlinking—illusory attraction. In a more complex example, the isolated subject appears to be attracted towards the more stable, supported group.

Illusory attraction

Earlier, we talked of our mind's pattern-seeking habits, and how we tend to carry over ideas from other fields of experience. Another subjective effect springs from these same roots: that of illusory attraction.

By putting a frame around any scene, we appear to set up a subjective tension, both between subjects, and with the frame itself. A relationship becomes implied between lines and tones, that affects our impressions of adhesion and unity throughout the picture. The whole process is enormously complicated. But we can demonstrate the basic idea readily enough with simple diagrams, as we see in Figure 15.21.

Fig. 15.22.

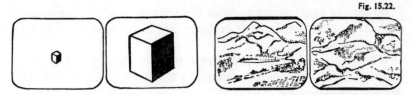

SCALE. *Left:* We judge scale subjectively, by the amount of screen the subject occupies. *Right:* But too close a view of our subject can obscure our reference-point, so that the portion becomes abstracted from the whole. While a long shot of a mountain range can convey its size, a closer shot becomes meaningless.

Fig. 15.23.

FOREGROUND AND SCALE. Foreground has a strong influence upon the scale suggested by the picture. *Left:* We get little feeling of depth from an isolated, remote subject without scale; but the foreground provides a comparative scale, tying together foreground and distance. *Centre:* Absence of scale and foreground can be corrected by perspective clues to distance. *Right:* This can be overdone. Without foreground there are vague clues to scale, but no unity in depth, whereas an over-emphasized foreground makes the scale relationship artificial.

Scale

Scale refers to the apparent size relationships that a picture conveys to us. We interpret scale in several ways: by knowing the subject's real-life size, by comparing this with adjacent subjects of known size, by relating it, subjectively, to the proportions of the picture frame filled, and by relating it to perspective clues.

Adjusting picture proportions

Where the subject and background are too large, but their relative sizes are satisfactory, change to a wider lens-angle, keeping the camera still.

Where the subject and background are too small, but their relative sizes are satisfactory, change to a narrower lens-angle, keeping the camera still.

263

Where the background appears too distant, although foreground subject size is satisfactory, change to a narrower lens-angle, increasing camera distance.

Where the background appears too near, although foreground subject size is satisfactory, change to a wider lens-angle, reducing camera distance.

Where subject appears too large, although background size is satisfactory, change to a narrower lens-angle, increasing camera distance.

Subject importance

How important or insignificant a subject appears to be will largely depend on how we present it. The effect of elevated and depressed camera viewpoints is widely known, but subject strength rests upon more than camera-height alone.

Fig. 15.24. Pt.I.

SUBJECT IMPORTANCE. The importance of a subject is affected by several factors. *Left:* Elevation. Lower viewpoints impart strength while elevated viewpoints suggest weakness. *Right:* The amount of frame filled by the subject. The larger the image, the greater its importance, regardless of the influence of elevation.

Fig. 15.24. Pt.2.

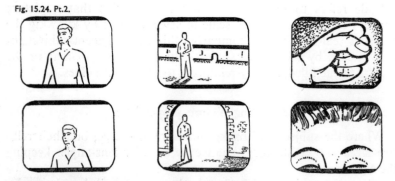

Left: Position in the frame—set higher, the subject gains strength. *Centre:* Size of a subject, relative to others of known size. *Right:* Associative significance of the feature shown. The clenching fist, grinding boot, snarling lips are obviously more powerful than less significant features.

264

Fig. 15.24. Pt.3.

The subject's surroundings further influence its pictorial importance. *Left:* A supported subject looks less forceful than an unsupported one. *Right:* Complete isolation provides more strength than when dominated by dynamic scenic lines.

Fig. 15.24. Pt.4.

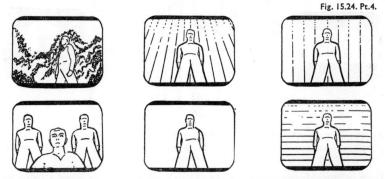

Left: The subject's strength can be reduced by weak background associations, and diminished considerably by backgrounds associated with dominance. *Centre:* Dominant lines of the background can modify the strength of the subject considerably. Backed by converging lines. the subject appears more powerful. *Right:* Strong verticals give it a less forceful appearance. Marked horizontals reduce its strength further.

Fig. 15.24. Pt.5.

Where a lot of background is seen, it can even dominate the subject. So its lines can accentuate, or nullify the strength of the subject. The first example shows the subject strong, but with background dominating; the second, the same subject dominating its surroundings; in the third, the subject's strength is emphasized by the background lines.

Fig. 15.24. Pt.6.

Tonal contrast will affect subject strength as these comparisons demonstrate.

265

The associative attitude of the subject itself can affect its apparent strength; whether it looks forceful, fierce, cowed, submissive.

Its general posture is significant—there are, for example, weak attitudes—lying down, bowed, looking down, stooping, clasped hands, slow movement; and strong attitudes—in a front view, uptilted head, hands clenched, stamping feet, fast movement.

Fig. 15.25.

ALTERING EFFECTIVE PICTURE SHAPE. *Left:* The crudest method is by cutting off parts of the picture with optical vignettes or electronic blanking. Introduced surreptitiously (during zooming or tracking in), the change need not draw attention to itself.

More persuasively, we can change the effective picture shape by scenic arrangement. *Centre left:* Through a scenic opening. *Centre right:* Using a scenic border. *Right:* By lighting.

Fig. 15.26.

VISUAL DETAIL. Scenic detail may be excessive, distracting; a lot of irregular shapes encourage a fussy, confused effect. *Centre:* Where shape and direction are co-ordinated, the effect can be stimulating and arresting. *Right:* If we exclude all but essentials, the effect may seem empty, meagre, cheap; although, suitably applied, it can lead to an open, expansive atmosphere.

Fig. 15.27.

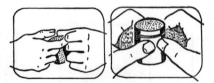

CLARITY OF VIEWPOINT. A good camera viewpoint can mean the difference between obscurity (*left*) and clear demonstration (*right*).

266

Picture shape

A picture's shape influences our feelings towards the scene it contains. A horizontal format can suggest stability, restfulness, extent. ... A vertical format can imply height, balance, hope. ...

Motion pictures were faced with the problem of providing a standard-shaped format that would accommodate a wide range of situations, and 4 : 3 proportions were chosen as the best compromise. Later, television adopted this same shape.

In television, the picture's format is unalterably fixed by the system's standards. But there are means by which the director can modify its effective shape, should he wish to, as we see in Fig. 15.25.

Clarity of composition

If a picture is to get its message over quickly and clearly, we must be able to see its relevant details easily, clearly, and without distraction. Visual confusion can take many forms.

Fig. 15.28. Pt.1.

AMBIGUITY. In a flat picture, visual ambiguity can arise whenever lines or tones in one plane merge into those of another; as when adjacent patterns or adjacent tones are similar or where scenic lines continue.

Fig. 15.28. Pt.2.

Parts of the setting may convey the illusion of being extensions of the performer.

Obscurity

A poor camera viewpoint encourages visual confusion. An inapt camera-angle can make the most common object seem unfamiliar. Very occasionally, we may aim to obscure action for dramatic or altruistic motives. But, more often, we want to show what is going on, clearly and unmistakably. This may even mean resorting to a contrived arrangement, to obtain a result that looks naturalistic. A demonstrator, for example, may need to handle or place things in an unaccustomed fashion, so that we see them better on-camera.

Poor lighting treatment can similarly add confusion, by hiding true contours, merging planes, casting misleading shadows, or false drawing.

Distraction

Amongst the most familiar forms of visual distraction we have: hotspots; flares on shiny surfaces; excessive contrast in background; lack of homogeneity in background line or tone; unmotivated shadows—especially from unseen objects.

Slightly defocused detail can prove distracting, too, particularly when we have to scrutinize carefully to detect what it is (e.g. lettering we cannot quite read; small background patterns whose tones vary with focus).

A background with strongly-marked detail sometimes inveigles the cameraman into focusing harder there than upon a foreground subject near to it. Faces can appear unsharp in comparison, even when focused. A close inspection of the eyes, necktie, jewellery, etc., will show us if this is happening.

Focusing visual attention

So far, we have dwelt upon how far composition affects the picture's emotional values. But as well as influencing mood, we usually want to concentrate the viewer's attention upon a particular focal point in that shot. There are obvious ways of doing this, e.g. leaving everything else unlit. Fortunately, we have more subtle and persuasive methods at our disposal.

Holding and localizing the audience's attention requires a calculated effort on the director's part. We should not overrate the amount of trouble the viewer is willing to take to evaluate our picture. It is certainly less for television than in the cinema. And, as always, the individual's attitude will affect his perceptiveness.

In practice, therefore, this means that in television we must continually redirect the viewer's attention, if we are to hold his interest along the lines of our choice. Redirection implies change; change through movement of subject, viewpoint, or scenery. Indiscriminate change, however, can lead to confusion or irritation. It must seem to be motivated; allowing the viewer to readjust himself easily.

Fig. 15.29.

FOCUSING ATTENTION. All these pictures demonstrate an essential fact ... that we can attract the attention of the viewer to a chosen centre of interest by the way we arrange and present our subject.

Fig. 15.30.

UNIFYING THE CENTRE OF INTEREST. Where a picture is not completely integrated, the fragmentary sections will destroy its unity.

Left: We should, therefore, avoid split centres of interest. *Right:* One aims for a concentrated, unified centre of interest.

Fig. 15.31.

CHANGING THE CENTRE OF INTEREST. *Top:* Where there is no transition or continuity (e.g. no co-relating lines between successive focal points), the picture lacks unity and appears disjointed. *Bottom:* Arranging compositional lines to guide the eye to the new centre of interest improves the situation.

269

Fig. 15.32.

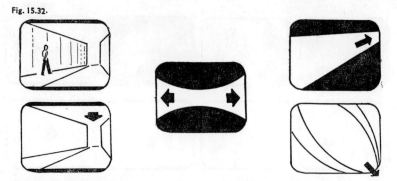

MOVING INTEREST ABOUT THE PICTURE. In moving interest about the picture, we have to guard against the eye being led to an accidental focal point, where there is no real centre of interest when it arrives. *Centre:* The eye may have an ambiguous choice of directions. *Right:* Attention may even be led out-of-frame altogether.

Fig. 15.33. Pt.1.

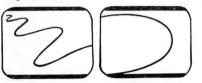

SPEED OF INTEREST MOVEMENT. A pattern or direction-of-line over which the eye lingers is termed "slow". Curved lines usually predominate, suggesting leisure, beauty, deliberation.

Fig. 15.33. Pt.2.

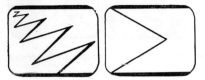

"Fast" lines are usually straight or angular, creating an impression of speed and vitality.

Fig. 15.33. Pt.3.

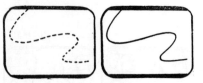

Unbroken lines are faster than broken ones.

Fig. 15.33. Pt.4.

Direction, too, will affect "speed": Lines pointing in the directions illustrated on the left are generally faster than those on the right.

270

Television pictures fall into three broad categories:

Those localizing our attention upon a specific centre of interest.
Those encouraging eye-movement along ordered, detectable lines.
Those encouraging free eye-wandering.

Each of these has its practical purposes, according to whether we want to encourage respectively: concentration, or transition, or ruminative inspection.

Although we can hardly expect to direct an entire audience's attention with any great precision, we can do much to lead it along lines we select. The more he looks at the right things, the greater the likelihood of the viewer's thinking in the right direction also. It is up to us to persuade him as strongly as possible to concentrate on those aspects.

We shall normally arrange a picture's composition so that the eye can find satisfaction within the confines of the frame; concentrating on certain features while the rest serve as a subordinate background. It is generally agreed that we should try to convey only one idea at a time. Three-ring circus techniques, where several different things are happening simultaneously, breed divided attention, ambiguity, confusion. One picture—one focal point.

Unifying the centre of interest does not mean riveting attention to one spot continuously. It can be moved around freely. This continual flow that links together parts of the picture, is often termed transition or continuity. The eye moves naturally from one area to the next.

This brings us to the speed of compositional lines. The eye scans some forms at a leisurely rate, while others are quickly appreciated. So through suitable selection we can adjust the picture's vitality and the speed with which the eye examines its contents. Faster lines assist in moving the attention quickly between areas, e.g. to allow rapid intercutting of shots.

Simple lines and shapes direct the eye more readily than more complex or disjointed ones. Interest is strongly attracted by clear-cut, geometrical shapes, especially where they seem to arise accidentally. An excess of simple lines soon palls, however, and is likely to lead eventually to a somewhat stark, unsympathetic atmosphere.

How can we focus visual attention?
There are numerous ways of focusing attention on to a particular area. Some are so obvious that we may overlook them; others are so unobtrusive that we may not realize their value. By summarizing

them in an easy-to-find list, we can see the limits of persuasion that lie open to us, at a glance.

By subject attitude. Having performers use strong movements (i.e. diagonal or upwards; stand-up; play to camera), move in front of other scenic elements.

By movement. Changing direction during a movement, rather than carrying it through. Interrupting and continuing movement, rather than maintaining sustained action.

According to how it is employed, movement can attract attention: to itself (e.g. a moving hand), to the subject that is moving (the person whose hand it is), or to the purpose of the movement (the subject at which the hand is pointing).

By contrasting the subject with its surroundings. Through differences in tone, position in frame, relative size, shape, proportions (scale), type of line, movement, association (e.g. elaborate with plain) disposition (e.g. seated with standing), etc. The centre of interest may be arranged, perhaps, to contain the lightest tones or the maximum contrast in the picture.

By camera control. Concentrating interest by differential focusing; avoiding weak subject viewpoints (e.g. side or back views), or weakening camera-angles (e.g. high shots, long shots).

Synchronizing a movement with dialogue, music, effects, etc. (preferably the subject's own), gives it strength and draws attention to it (e.g. having the subject move on a line of his own dialogue emphasizes both action and speech).

Shifting visual interest

It is quite as important for us to be able to shift the viewer's concentration to another aspect of the picture as it was to localize his attention originally. We can do this, of course, by readjusting any of the influences we have just listed.

A few examples demonstrate how smoothly this can be arranged:

Have a person stand up within a seated group.

Change the camera's elevation, length of shot, etc.

Give the original subject weaker movements and strengthen those of the new subject.

Transfer the original emphasis on contrast or similarity to the new subject.

Fig. 15.34.

HIDDEN CENTRES OF INTEREST. Where an implied centre of interest falls outside the frame or is obscured, we can arouse feelings ranging from expectant anxiety to frustration and annoyance. Carefully introduced, such tactics make the viewer welcome a change to a new viewpoint; over-prolonged, we shall only antagonize him.

Fig.15.35. Pt.1.

HOW CAN WE FOCUS VISUAL ATTENTION? *Left:* Avoiding spurious centres of interest, e.g. prominent lettering, tonal contrast, etc., in supporting subjects. *Centre left:* Compositional lines (real or implied). *Centre right:* Tonal gradation. The eye naturally follows gradation from dark to light areas. *Right:* By deliberate unbalance (tonal or linear), with the subject in the heaviest position.

Fig.15.35. Pt.2.

By isolating the main subject. *Left:* In depth. *Centre:* Horizontally. *Right:* By body position.

Fig.15.35. Pt.3.

Left: By height. *Centre:* Using a stronger part of the frame. *Right:* With lighting.

Fig. 15.35. Pt.4.

By similarity between the subject and its surroundings (*left*), although too great a similarity can pall or lead to confusion (*right*).

273

Use linking action; e.g. the original subject looks over to camera-right... cut to the new subject appearing on screen-right.

Weaken the original subject and have it join the new, stronger subject.

Pull focus from the old subject to the new.

Change the apparent sound-source—i.e. the new performer now speaks instead.

Continuity of centres of interest

At any given moment, our interest is localized to something like 1/12th of the screen area. The position of this area is continually moving as our eye follows compositional lines, moving subjects, or turns to a new centre of interest.

Where successive centres of interest are widely-spaced about the frame, the viewer's concentration must be disturbed to some extent while he seeks each new spot. Fast interlinking compositional lines help here.

Our principal problem arises whenever we cut from one shot to another; for a new shot presents a new centre of interest, and unless these successive centres are within about 1/8th of a frame of each other, this change can be very distracting.

Poorly matched shots are probably the clearest sign of bad editing. A succession of unmatched cuts between short duration shots, and the viewer's mild resentment can build quickly into a marked hostility, overwhelming all else.

Fig. 15.36.

MATCHED CUTTING. On cutting to a new shot, a new centre of interest presents itself. *Left:* If the centres are too dissimilar, we become over-aware of the change. *Right.* But where reasonably matched, the transition may be almost unnoticed.

Fig. 15.37.

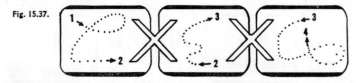

MOVEMENT OF INTEREST BETWEEN MATCHED SHOTS. In matched cuts, we do not have to keep action localized throughout the scene. We have unlimited freedom of interest-movement providing successive shots are matched at the transition.

Fig. 15.38.

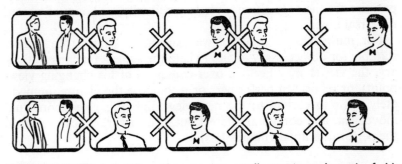

INTERCUT SHOTS. *Left:* Where there are two equally prominent alternative foci in intercut shots, the viewer will usually accept spaced centres of interest (*top*), rather than the confusion that true matching would give, when similar subjects occupy the same part of the frame (*bottom*).

Continuity of composition in multi-camera production

Composition that looks good in one set-up can prove quite unsatisfactory when shot from a different distance or viewpoint. Many settings that are attractive as a whole prove disappointing when dissected in multi-camera shooting.

The setting may disintegrate; lacking cohesion, unity, balance. It may remain well-composed but inapt; laying wrong emphasis, concentrating attention on the wrong subject, and so on. Sometimes even our method of shot-selection interferes with picture continuity.

Where the compositional set-up is elementary or haphazard, this problem will not be manifest. But the more painstaking and elaborate the visual treatment, the greater the difficulty in maintaining good continuity of composition. In motion pictures one can rearrange people, set-dressing, lighting, for the best effect from each new position. Sometimes this cheating is so extensive that even the most gullible viewer becomes aware of it. However, live transmission

Fig. 15.39

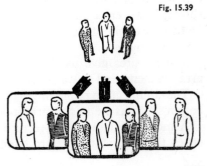

MISMATCHED CUTS. We can see how camera 1's shot intercuts with camera 2 or camera 3; but the shots of cameras 2 and 3 do not intercut, because of the apparent lack of continuity.

275

precludes such devices, and careful planning and judicious compromise are the only alternatives to mediocrity, or hours of wasted rehearsal time.

For multi-camera shooting we can either arrange the subject suitably for all necessary angles (where the subject remains stationary, the viewer may become over-conscious of his changing viewpoint), or reposition performers or recompose the picture, immediately prior to cutting to the new viewpoint (this gives more fluid movement, and continual interest but, if overworked, can result in the performers appearing restless).

The choice of compositional treatment

In composing a shot, our aim, we have said, is not merely to present an attractive picture. Its composition has to be meaningful; its influence appropriate to the occasion.

By studying a couple of examples (see Fig. 15.40) in some detail, we can get a fuller idea of the striking variation that even a change in viewpoint provides. More important, we see how extensively we can modify the viewer's feelings about the subjects, their environment, and their relationship to it, simply through our choice of compositional treatment.

Try taking each picture in turn (see Fig. 15.42) and noticing how the comparative strengths of the individuals change, how the relative strengths of different parts of the setting alter, how the dominant line varies, and how the environment becomes alternatively closed-in, open, dynamic.

Dynamic composition

Distinctions between still and moving pictures

So far, we have been examining principles long accepted in the composition of still pictures. Where our shots are static, these ideas can be applied usefully to television production. Most television shows, however, are a mixture of still (or semi-static) and moving picture throughout, so the question arises how far these principles continue to apply.

Still pictures give us time to deliberate and study, both as we arrange them, and as we enjoy the resultant effect.

The composition of moving pictures, on the other hand—dynamic composition as we might term it—is a fluid, evanescent thing.

A moving (or movable) picture possesses certain valuable properties that the static picture cannot have. It enables us to show

Fig. 15.40. Pt.1.

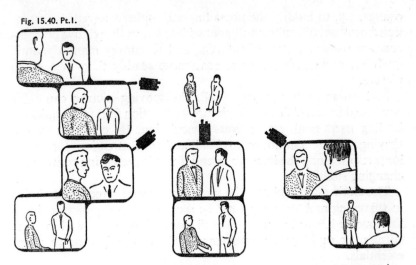

MULTI-CAMERA CONTINUITY OF COMPOSITION. Subjects can be arranged to provide appropriate composition from several viewpoints. *Above:* In a two-shot.

Fig. 15.40. Pt.2.

Continuity of subject interest in a three-shot from various camera positions.

Fig. 15.40. Pt.3.

Compositional continuity by moving a performer to recompose the shot before changing to a new viewpoint.

change, e.g. to modify the prevailing atmosphere, to redirect attention from one subject to another at will, to adjust the relative prominence, importance, etc., of subjects, and to convey many dynamic qualities, movement patterns, etc., more readily than in a static picture.

Perhaps most important of all, the moving picture can offer continued interest. The static picture's attraction falls fairly quickly; lending itself readily to private speculation. By comparison, the moving picture continues to hold audience attention for much longer, and along the lines we intend, for it is presenting continually changing stimuli.

The static picture allows the viewer prolonged freedom of choice in surveying and assessing what he sees. The moving picture limits the time he has to linger over detail, particularly when we have fast cutting, or rapid camera-movement. There is only time to grasp essentials.

The moving picture has its weaknesses.

Picture movement can introduce inaccuracy or uncertainty.

The impact of a frozen-action still can be stronger and more sustained than if we see the actual movement (e.g. showing a boxer falling from a knock-out punch). This indefinable feeling of "suspended animation about to recommence" is the underlying appeal of many great paintings and sculptures. A compositional balance that appears forceful in a static shot may be ineffectual, or impossible to maintain, in a moving picture. Certain action that is intriguing in a fragmentary glimpse can look commonplace when carried to its conclusion (hence the productional techniques of cutting between snippets of uncompleted action, or fading before a sequence has logically concluded). Further, a well-composed static shot can become unpleasing when its component parts change their relative positions through movement. And, finally, a picture can rely more for its effect upon a subject's significance than upon its movement impact, and so be equally powerful in static or moving shots.

These various distinctions between the static and moving picture are, we have seen, quite substantial. They mean that we cannot simply treat a moving picture as a static shot that just happens to change. The moving picture merits study in its own right.

With the advent of video discs, it has become possible to extract a brief segment (e.g. 30 seconds) from a normal-speed programme, and to reproduce this at any speed from freeze-frame to double speed, forward or backward, so permitting closer scrutiny of action, repeat action, speed changes and animation.

Fig. 15.41.

CARE IN GROUPING. *Top left:* People must normally work closer together than in everyday life. Otherwise shots will be too wide. *Top right:* Close spacing may look ridiculous in long shots, however, and widely-spaced subjects may have to be shot from carefully selected viewpoints.

Bottom right: People tend to converse face-to-face, but this allows only a limited range of shots. Slight angling is standard TV practice, giving an improved selection of shots and better visual continuity.

Fig. 15.42.

THE INFLUENCE OF VIEWPOINT. Our feelings about the subjects, their environment and their relationship to it, can be modified according to the viewpoint we select for our camera.

279

Fig. 15.43.

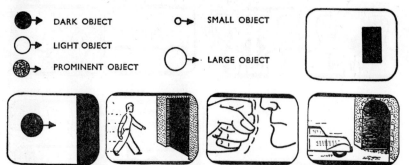

REPRESENTING MOVEMENT. Symbols can represent a whole number of different situations, each of which has the same basic dynamic composition. Our subjects can have their direction and path indicated; while the TV frame and any static compositional elements it contains, can be shown equally simply (*top right*).

A theory of dynamic composition

Representing movement. Although we cannot show movement on the printed page, we can use representative symbols—and our imagination. Symbols, moreover, help us to think in general, rather than particular, terms.

Our assessment of movement. Our chief impressions of subject movement spring from several factors:

Speed—i.e. its rate of progress, usually relevant to its surroundings.

Tempo—i.e. its rhythm, rate of acceleration, etc.

Direction—of its travel.

Pattern—the shape of its travel-path.

We may see this movement directly (e.g. the subject moving within a static scene) or as an implied action (e.g. a moving background behind a static subject; or moving light and shadow passing over it). But these are the more evident physical signs of movement. Several strongly-associative ideas will colour our interpretation of the picture's dynamics. These are:

Those due to our subconsciously relating the position and movement of the subject to gravity.

Those due to our tendency to judge a movement's strength according to apparent size, power, or force involved.

Those due to our association of relative movement with the forces of compression (impact, collision); expansion (stretching strain).

Those due to our comparing the apparent relative speeds of subjects in the same picture.

Those depending upon where we happen to fix our attention (when looking skywards, we see moving clouds and static buildings, or static clouds and toppling buildings, according to which we fixate).

The amount of border surrounding the subject within the frame. A medium shot of a travelling subject appears faster-moving than in a long shot.

Interpretations influenced by the accompanying sounds and their associations.

The influence of gravity. When we examined pictorial balance earlier, we saw how one invariably arranges pictures with reference to gravity, even when there is no real justification. How strongly this gravitational reference affects us seems to change with the subject's height in the frame (the higher, the stronger the effect), its apparent size, weight, energy, speed and direction in the picture, and our associated ideas about the subject's properties.

When we see these preconceptions refuted, the result normally strikes us as being astounding or highly amusing. Cartoon films take advantage of this situation when they show us a featherweight elephant, or bubbles that are cannon-ball heavy.

Fig. 15.44.

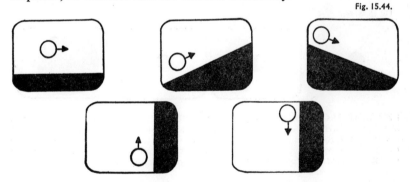

THE PSYCHOLOGICAL EFFECT OF GRAVITY. *Top left:* Horizontal movement will usually suggest only normal, progressive advancement. We tend to interpret movement having an upward or downward bias as an effort being made relative to gravitational forces. *Top centre:* Forceful progress, uphill, climbing. *Top right:* Ordered descent, dynamic progress, vigour, downhill, slipping. *Bottom left:* overcoming obstacles, rising, scaling. *Bottom right:* Forceful progress (when fast), lack of control, falling.

The influence of surroundings upon effective movement

As we shall see, both the subject's position in the frame and the nature of its surroundings will influence our impressions of movement.

281

Fig. 15.45. Pt.1.

OUR IMPRESSION OF MOVEMENT. Imagine a shot of an aircraft flying in a cloudless sky. *Left:* Where the subject is held firmly-framed against a plain background, there is little sense of movement. *Centre:* Where its position in the frame vaccilates, as the cameraman tries to follow it, we infer that the movement must be swift. *Right:* Where background clouds race past, we sense movement and speed more strongly.

Fig. 15.45. Pt.2.

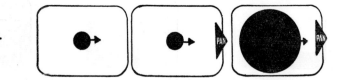

To this we can add the four general axioms. *Left:* Movement across the screen tends to appear exaggeratedly fast. *Centre left:* Movement to or from the lens seems slower for narrow-angle, faster for wide-angle lenses. *Centre-right:* Camera movement in the direction of subject movement reduces its speed. *Right:* Camera and subject moving in opposite directions increase their mutual speed and impact.

Fig. 15.46. Pt.1.

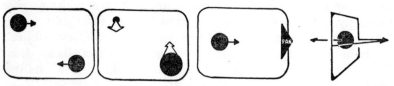

RESTRICTION OF MOVEMENT BY THE FRAME. *Left:* We interpret a movement in free space as having unrestricted freedom. *Centre left:* But once a frame is placed around the subject, freedom is relative to these limits. There is potential restriction. *Centre right:* Where we hold the moving subject near centre-frame, this restriction is not excessive. *Right:* But for close shots must always be cramped.

Fig. 15.46. Pt.2.

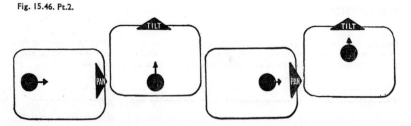

Left: Restriction becomes less when we keep the subject near the lagging edge of the frame. *Right:* It is emphasized when kept near the leading edge.

Fig. 15.47. Pt.1.

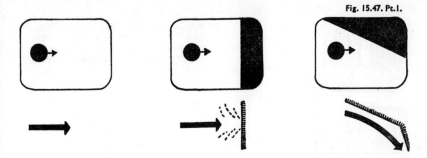

RESTRICTION BY MASSES. *Left:* This alone is an unprepossessing action, free relative to the frame. *Centre:* Add in its path a large mass, and the movement is given new significance, in this case suggesting imminent impact or potential arrest. *Right:* Progressive restriction.

Fig. 15.47. Pt.2.

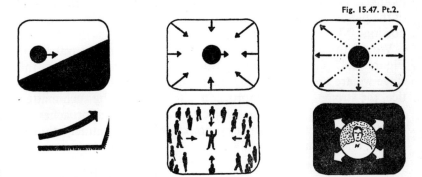

Left: Potential ascent and freedom. *Centre:* Converging, encircling, crushing, as when a crowd presses in on a central figure. *Right:* Expansion, new freedom, isolation, as when using an iris-out wipe.

Fig. 15.47. Pt.3.

Even more complicated situations can be arranged to emerge quite naturally. *Left:* Imminent escape from annihilation (lowering skies). *Centre left:* Freedom barred. *Centre right:* Potential annihilation (panning left with the pursuer; he and the wall move inwards on to the victim). *Right:* New freedom, expansion (tracking-in).

Fig. 15.48.

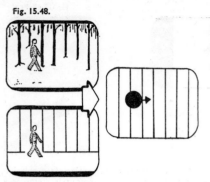

BACKGROUND PATTERN DERIVA-
TION. The background pattern illustrated may be derived from natural forms
(e.g. trees, standing people), or artificial
ones (e.g. railings, decorative line).

So we arrive at the concept that a movement's effect can be modified by the environment in which we see it. A mass in one part of the picture can influence how we feel about movement in another portion.

We see this underlying implication in each of the illustrations in Figure 15.47. Granted, the suggestion may be very weak, but at this stage we are only considering the most elementary trends. In practice, the strength of this impression will be decided by such matters as the speed and size of the subject; how closely it approaches the mass; subject associations; any pre-knowledge or anticipation of events; and so on.

Progress against restriction. The shape and position of the basic masses in the picture can have several unexpected effects, altering our impressions of the subject's implied direction, affecting the apparent difficulty it has in getting there, and introducing an underlying feeling about the action.

So strong are these concepts that we even have them on seeing neutral patterns moving under similar conditions.

Fig. 15.49.

BACKGROUND AND SUBJECT MOVEMENT. Where movement is in the direction of background pattern it will appear unimpeded if slightly restricted. The line emphasizes direction. Where the subject moves across (against) the pattern, effort is emphasized; movement appears stronger. Subject shape may modify these effects.

284

Background pattern. Quite often the picture does not contain large masses. Instead, we can distinguish a general background pattern.

Whichever it is, we shall find that owing to its repetitive nature this pattern exerts a strong, continual influence upon any subject moving in front of it.

Fig. 15.50.

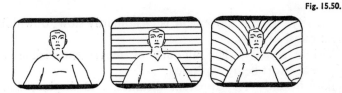

BACKGROUND AND STRENGTH. A moving subject's importance can be affected by its passing background. Against a neutral background, the subject's strength remains constant, whether still or moving. In front of a horizontal background, the vertical subject is strong. When he moves towards us, with the camera tracking back to hold the same shot, downward moving horizontals pass his vertical form, giving it added strength. In the last example, the downward movement of the background pattern as the subject comes forward emphasizes the depressed camera angle, thrusting the subject upwards and strengthening it further.

Directional trends. Through the combined effects of the subject's movement and the prevailing lie of the background pattern, we may feel an implied undertow of movement that is not present in either.

Fig. 15.51. Pt.1.

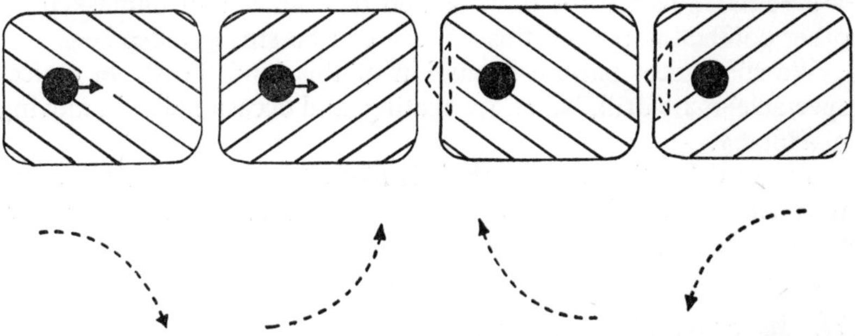

BACKGROUND AND DIRECTION. *Left:* When the subject moves across static backgrounds, we feel an undertow of movement. *Right:* Where the same background passes the static subject, this undertow takes a different form.

Fig. 15.51. Pt.2.

Taking this a stage further, in a sequence of shots the upward-sweeping background seems to exert an upward force on the subject. The subject continues on its path despite this influence, and so becomes noticeably stronger, especially as it crosses the border.

285

Effective speed. How quickly a subject seems to be moving will be affected by various prejudices:

Subject associations—i.e. the speed we normally associate with that kind of subject.

Mass—We tend to think of large subjects as moving slowly, and vice versa.

Gravity—Uphill movement is customarily slow; downhill movement relatively fast.

Visible activity—The amount of movement visible, and the apparent effort being expended.

Background—The apparent distance, and displacement of the background.

Shape—See Figure 15.52.

Intrinsic rhythm. Where the subject moves across a regularly changing background, the eye seeks to recognize rhythm and regularity in what it sees,—rhythm in shape, in line, in tone.

Patterns of movement

When a subject moves, we tend to retain a mental note of its path, especially when the paths forms a simple, well defined shape. Where the movement is obvious and repetitous, and unmasked by other conflicting action, this pattern can be strong indeed.

We shall meet pattern-formation of this kind mostly in dance-spectacle, mass drill, ballet, ice skating, and occasionally, in moving machinery.

Where the action is smooth, it is likely to suggest control, evenness, simplicity; while jerky, irregular, motion suggests the clumsy, the erratic, or the uncontrolled.

The subjective effect of tonal values

The tone of an area, you will remember, alters our assessment of its size and "weight". The lighter the tone, the larger it appears. But the less its "weight" and "strength". Its dynamic properties will naturally vary, too.

To give a practical example, let us take one situation of Figure 15.47 and repeat it with tonal effects in mind (page 288, Fig. 15.56).

We see here how far-reaching the influence of tones can be. The tones of costume, staging and lighting treatment can affect the overall potency of the picture. Shades of meaning can come about purely through tonal relationships.

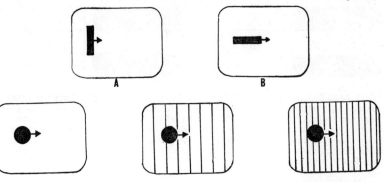

Fig. 15.52.

SPEED: The subject's shape can affect its apparent speed. *Top:* Although moving at the same speeds, A appears to be moving faster than B. *Bottom:* The apparent speed is also increased as the background becomes more detailed.

Fig. 15.53.

RHYTHM. Visual rhythm from the passing background will seldom be as regular as these examples, although quite strong rhythmical patterns can be built up naturally from passing flags, trees, railings, etc.

Fig. 15.54. Pt.1.

PATTERNS OF MOVEMENT. Some movement patterns have widely accepted associative feelings. *Left:* Continuity. *Centre left:* Vigour, excitement, indecision. *Centre:* Beauty, charm. *Centre right:* Expansion. *Right:* Contraction, collapse.

Fig. 15.54. Pt.2.

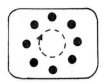 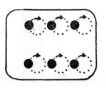

Left: In massed spectacle we find simple pattern formation. *Centre:* Also patterns built up from individual movements. *Right:* And from a combination of individual and group movements.

Fig. 15.55.

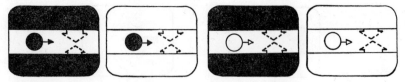

TONE AND STRENGTH OF MOVEMENT. The effective strength of static and moving masses varies with tone. *Left:* When a dark subject moves through converging dark areas, the mutual impact is powerful. *Centre left:* The "crushing" effect is lightweight against a heavy, substantial subject. *Centre right:* The apparently larger but lightweight subject suggests less resistance to approaching forces. There is a hint of destruction. *Right:* Both subject and approaching forces are lightweight, so the effect is less dynamic than in the first example.

Changes in visual significance

An overwhelming advantage the moving picture offers, is the opportunity to change its subject's relative significance. We can do so by moving the subject to a new position (actually or apparently), by moving it to a different environment (i.e. with a different line, tone, etc.) or by moving the camera to a new viewpoint.

Dynamic composition between pictures

When we make a transition from one shot to another, the viewer becomes aware simultaneously of several things:

A retained memory of the previous shot (its composition, associations, etc.).
The nature of the transition (cut, wipe, etc.).
The new shot's initial impact (its composition, associations, etc.).
A comparison between the two shots.

This must always happen to some extent, whether we intend it or not (although the viewer himself will hardly analyse the fleeting impression, he will react none the less). And these reactions we can put to good use.

Fig. 15.56.

THE DYNAMIC EFFECT OF TONE. *Left:* Dynamic pursuit. *Centre left:* Ineffectual pursuit of strong subject. *Centre right:* Annihilation of vulnerable subject. *Right:* Undramatic. Following rather than pursuing.

Fig. 15.57. Pt.1.

CHANGING VISUAL SIGNIFICANCE. Typical ways of altering a picture's visual significance. The movement to the higher position takes attention from the right to the left.

Fig. 15.57. Pt.2.

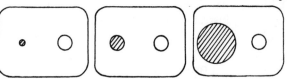

The striped object is approaching. At first it is insignificant, but with increasing size, its importance and interest exceed that of the static subject.

Fig.15.57. Pt.3.

As the subject passes from the dark background to the light one, there is an emotional uplift.

Fig. 15.57. Pt.4.

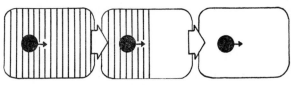

Passing from a region of restrictive verticals to a plain background produces a sense of freedom, and an effective speed change.

Fig. 15.57. Pt.5.

Left: Canting the viewpoint changes the pictorial balance. The picture becomes unstable.
Right: Changing the viewpoint can alter compositional line. Tilting transforms verticals into converging diagonals.

Where we find two pictures that do not intercut well (despite matched interest-areas), the answer can lie in the lack of visual continuity. The eye seeks relationships in what it sees. Where two successive pictures have a false, unintended relationship through their juxtaposition, the transition between them usually proves unacceptable.

Cutting to another shot can at times produce quite marked secondary effects implying movement. We may use these deliberately, to get over a particular idea. They may creep up on us unsuspected and prove highly disturbing, like the sideways-jump of Figure 14.18. Again, diagrams illustrate this situation quite well (Figure 15.59).

The last of these would be very suitable during a description of a city being destroyed by bombing, but hardly appropriate in a sequence describing post-war reconstruction; yet such things can happen during fast montage shots.

Accepted maxims

Although the composition of moving pictures is seldom studied as such, we see its principles applied daily in the best television and film productions. Many directors work from rough rules-of-thumb learned from experience, rather than from an understanding of fundamentals, but some excellent working axioms have thus emerged that are widely accepted.

Slope of movement. Like vertical lines, vertical movement is stronger than horizontal movement.

A rising action is stronger than a downward one. Thus, an upward lift of the head, a raised hand, a rise from a seated position, will have greater powers of attraction than their downward counterparts. An upward movement, moreover, tends to appear faster than a horizontal one.

Horizontal lines are the weakest compositionally, owing to their associations. Horizontal movement is therefore the least arresting. A left-right move is stronger than a right-left.

Diagonal movement, like diagonal lines, is the most dynamic.

Movement towards the camera being more striking than movement away from it, we find that any forward gesture or movement is more powerful than a recessive action (e.g. a glance, a turned head, a pointed hand). A shot approaching a subject arouses greater

290

Fig. 15.58. Pt.1.

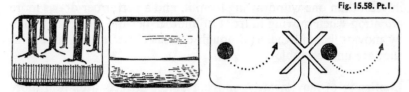

MUTUAL CONTRASTING emphasizes the composition of a new picture more than when the pictures are viewed separately. *Left:* Cutting from a vertical composition to a horizontal one. *Right:* Similarity emphasizes the aspect repeated, e.g. repeating the shape or direction of an action.

Fig. 15.58. Pt.2.

Cutting between a series of shots having an upward thrust: in this case providing an underlying of uplift, ascension, vigour, etc.

Fig. 15.59. Pt.1.

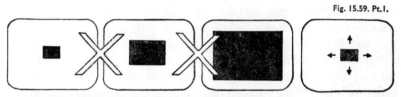

EXAMPLES OF SECONDARY EFFECTS WHEN CUTTING. Quick cuts between shots produce the overall effect, in this case, of a series of expansive jumps.

Fig. 15.59. Pt.2.

An implied cave-in, resulting from the comparative effects of successive shots.

Fig. 15.59.Pt. 3.

A diagonal slide, resulting from the comparative effects of successive shots.

interest than one withdrawing from it, and a performer draws more attention when moving in front of another person, or scenery, than by moving behind them (although crosses closer than about mid-shot are unacceptable).

Continuity of movement. A moving object attracts attention more readily than a static one, although constant-speed continuous movement cannot maintain maximum interest. Action that is momentarily interrupted, or changes direction, will be more striking than one carried through to the end. Converging movements are more forceful than expanding ones.

Right-handedness in composition

In the Western world, it is customary for us to read from left to right and from top to bottom of the page. Possibly as a carry-over habit, we tend to find this approach applied to other activities too. When devising or looking at a picture, for example, there seems to be reasonable evidence that we have this right-handed bias.

Collect together a number of photographs. Look at each direct, and then in a mirror. You will invariably find the two versions have a slightly different appeal. These differences may be great or small, but they are generally there. So consistent are the results of reversing a picture laterally that we can list them, with a certainty that one or more will be present.

These findings have very practical applications. They demonstrate that good composition alone is not enough. It is often advisable to bear in mind that audience impact can vary according to which way round we arrange our picture. (see Fig. 15.64).

Basically, we find changes occurring in balance, stability, proportions, subject strength, areas of interest, concentration of attention, picture significance, visual interpretation and dynamism.

(i) **Pictorial balance.** (a) The picture may look balanced or lop-sided, depending on whether the predominant mass comes on picture left or right. The right of the picture appears to lend subjects a greater weight, with the result that large masses placed there are liable to unbalance the overall picture. Conversely, we find that the left can support more weight (i.e. density or mass).

(b) A secondary effect is that of apparently excessive space between subject and frame.

(ii) **Pictorial stability.** (a) The picture may look more stable one way round than the other.

(b) Dormant in one version—but dynamic when reversed.

(iii) **Apparent proportions.** Apparent proportions can vary. For example, when using large, dark, foreground-areas on the right of the picture, the scene may feel crowded, shut-in, or heavy, while the laterally-reversed version has an open atmosphere.

(iv) **Subject strength.** A compositional element that seems unimportant when composed on the left can become obtrusive on the right (although the reverse seldom holds).

Moreover, a subject we may overlook altogether on the left side may become compositionally or dramatically significant, placed on the right.

(v) **Area of interest.** The eye's response to diagonal lines can vary with the direction of their slope.

So too, our centre of interest is likely to change; particularly where there is marked perspective. This situation enables us to direct the eye in depth within the scene.

Fig. 15.60.

AREA OF INTEREST. Sloping to the right, the eye tends to favour the distance, and so run out of the frame more easily than with the reverse slope, where the eye tends to favour the foreground.

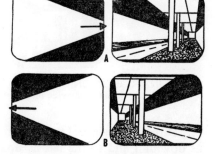

(vi) **Concentration of interest.** Our right-handed bias seems to influence too the ease with which one can concentrate attention within the frame. Broadly speaking, the eye tends to wander over to whatever is on the right of the picture. Even with a highly-dominant left-hand subject, this still seems liable to happen.

But the situation is not reversible; for subjects are placed on picture-right, anything left of them may go almost unregarded.

Fig. 15.61.

CONCENTRATION OF INTEREST.
Top: Placed on the left, the window is just part of the scene; the person dominates. When on the right, we anticipate important action through the window; the person becomes less important. *Bottom:* We are very conscious of the watcher on the right, but only half-aware of his victims. Placed on the left, he is nearly overlooked, attention focusing on his victims.

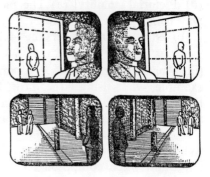

(vii) **Picture significance.** What the picture appears to mean can vary initially according to where its subjects are placed. We may find, for example, that a quite nondescript part of the setting in one version of a picture, will take on some implied significance in a mirrored version.

(viii) **Visual interpretation.** In more extreme cases, we may even have difficulty in recognizing, in one version, what the picture is. This usually arises where tonal gradation alone gives us our sole visual clues to light sources.

Fig. 15.62.

VISUAL INTERPRETATION. *Left:* The highlights are interpreted as obvious window openings. *Right:* Here they may be interpreted as patterns of light on the wall.

(ix) **Direction of slope.** One tends to interpret down slope to the right as downhill and to the left as uphill. And when something is seen moving along such a slope, we tend to have our feelings about its effort and energy coloured accordingly. (15.63).

Somehow, it looks easier for someone to move "downhill" in version (A) and harder in (B), and this in turn will modify how forcible the action appears to be. The whole business is entirely subjective. It can vary with tonal balance, our predilections, and so on, but the bias remains substantially true for many situations.

Fig. 15.63.

DIRECTION OF SLOPE. Direction of the slope can alter a picture's attractiveness, there being a tendency for one version (*left*) to seem more dynamic than another (*right*). In a still picture of a falling tree or chimney, for instance, this effect can be most marked.

Fig. 15.64.

(i—a), (iv—b), (vi)

(ii—b), (vii)

(i—a), (iii)

(i—a), (v), (vi)

(i—b), (ii), (iii)

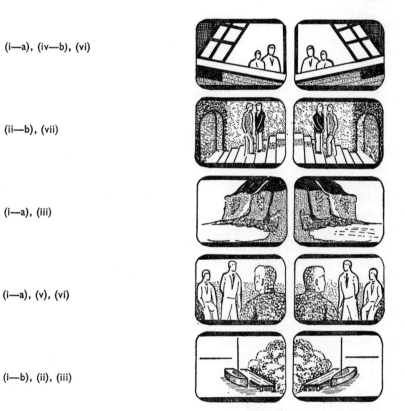

EXAMPLES OF COMBINED EFFECTS. The above illustrations demonstrate how several aspects of right-handedness can appear simultaneously.

16

EDITING

BASICALLY, we can say that the editing process is chiefly concerned with questions of:

When and how we are going to change from a shot on one camera, to a shot on another.

The order and duration of shots.

The maintenance of good visual continuity.

But editing proper has more far-reaching consequences, as we shall see.

The potentialities of editing techniques are seen best in motion picture production. As television uses both live and filmed material, it is important from the outset to show the distinction between their respective editing problems. Their relative audience impact is identical, but their nature is quite different.

The *film editor* is presented with numerous coded lengths of film, containing most of the material shot during production. Scenes will have been retaken, often from several viewpoints, usually out of sequence. His task is the selecting and blending of these into a meaningful whole. He must avoid unintentional discontinuity, omissions, or repeats; but beyond these mechanics, he must make subtle aesthetic judgments. For through editing techniques one can evoke emotions, imply, condense, or expand time, as we shall consider later. Having assembled his film, the editor can examine and re-examine his choice, until he is satisfied with his use of the material he has been given.

The *television director* is his own editor. He may himself carry out the mechanical operation of the *video switching console* (vision mixing desk), or have a technical director, or a switcher (vision mixer), follow his instructions.

The director may decide his editing arrangements during pro-

296

duction-planning, during camera-rehearsal and occasionally on the air. He sits before his preview picture monitors in the production control room (four to eight monitors are typical). These give him continuous pictures from his various sources. The shots each source offers in turn will be those he has pre-arranged (the exception is in situations where the cameramen find shots for themselves, the director modifying and choosing from these).

From these sources he can select at will by controls on the video console. The push-buttons and variable-faders here allow most normal editing techniques to be achieved immediately. (This is unlike film, where cuts only are easily made—by mechanical jointing—while all other transitions must be made by the processing laboratories from a cue-marked copy). Herein lies the two-edged character of television editing. Transitions are easy. But on that account, the editing process can degenerate into mere mechanics. Its audience impact can become overlooked or ignored. One's transmitted decisions are irrevocable, unless video-tape editing is to follow (see page 133).

Remember also the general circumstances. The monitors will be showing us not only what each source is going to transmit; they will be showing cameras getting their next shots, moving position, film-sequences running-on to their next cue-point; and so on. Concurrently, the director will be guiding, instructing, selecting, and generally co-ordinating the production team's work. This in addition to keeping on schedule, coping with contingencies, and the rest.

If the finer points of editing are lost in the process, it is understandable. But we are providing the viewer with stimuli—not making excuses to him—and our choice of production treatment will have its effect, whatever the behind-the-scenes hardships.

Types of editing approach

Editing techniques range from the simplicity of *continuity-cutting* to the more sophisticated, perceptive, forms of *dynamic-cutting*. Present-day television uses the former almost exclusively, employing dynamic-cutting only occasionally, in dramatic productions where accurately-timed action is practicable. In the better motion pictures we shall usually find a blend of the two styles.

Continuity-cutting
This relies primarily upon the continuity of dialogue and action to tell the story. Here is editing at its simplest. At its worst, purely

functional; at its best, it gets on with telling a story in a clear, straightforward, manner. It makes appropriate emotional use of transitions, but avoids editing devices that might distract from the programme proper. We cut to a close-up simply to see more clearly.

Dynamic-cutting

The process was conceived by early film experimentalists, who realized that continuity-cutting was forfeiting important aesthetic opportunities. Editing now became a persuasive art, not merely a linking process.

Here, the film-maker's aim was not so much to tell a story directly through continuous dialogue and action, for vital visual relationships were discovered. It was found that *intercut shots* do not necessarily retain their original interpretation. New ideas can arise that may not be implied in the component shots themselves. Moods, abstract ideas, can be conveyed, which could only be half-expressed, if at all, in more direct terms.

Dynamic-cutting relies for its impact upon the effects that inter-related shots can evoke.

At best, dynamic-cutting is highly stimulating, for it exercises one's imagination. We are not allowed to see and hear the obvious. At worst, this style degenerates into bewildering or ambiguous symbolism.

The basic units of editing

The cut

The cut is the simplest transition—an immediate change from one shot to the next.

Sudden change has a more powerful audience impact than a gradual one, and here lies the cut's strength.

The cut is dynamic, instantly associating two sets of circumstances. Thrusting the new upon us, while the memory of the old shot is fresh. It is usually introduced to emphasize, or to transfer attention.

Fig. 16.1

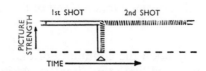

THE CUT. The nature of the sudden cut can be shown graphically in this form.

298

Cutting, like all production treatment, should be purposeful. An unmotivated cut not only interrupts continuity, but can create false relationships between shots.

It is sometimes mistakenly thought that cutting is akin to repositioning the eyes as we glance around a scene. But we move our eyes with a full knowledge of our surroundings. We realize how we have moved them. So we always remain correctly orientated.

Cutting, on the other hand, jumps one's attention between new viewpoints or locations, with the result that each transition means that the viewer has to find out what the new picture is about, and where he is. This is a situation that is most likely to arise when intercutting between action in entirely different settings.

Where there is some common point of reference between the conjoined shots, readjustment is easy. Continuity of thought can come through dialogue (introducing or implying the next shot's substance), action (establishing a cause/effect relationship), the showing of subjects common to both shots, or aural continuity.

Occasionally, for dramatic or comic effect, we may deliberately mislead the viewer's continuity, but very rarely.

The moment for the cut. Because of its strength, the cut requires caution both in how and when we introduce it. Cuts should be made preferably on an action or reaction (e.g. a turning head; gasp of astonishment; a rise). Exactly when we should cut is controversial.

Some people maintain that the cutting moment should be just before or just after an action (the action necessarily being the centre of attention at that instant). A movement interrupted by a transition is disturbing, they maintain. It breaks the continuity of the subject's action, creating a spurious cutting rhythm.

Others contend that if we cut during an action (e.g. while opening a door, lifting a glass), we prevent the illusion of its being jerked into action, or to a halt. This can sometimes occur on cutting from a shot showing the movement, to another of the subject at rest. Cutting is less of a visual shock when hidden within action, for we are preoccupied with the movement itself. This can help us to reduce the cut's impact when it is not clearly motivated.

The delayed-cut. A late cut is frustrating. Badly-timed, it has missed the optimum moment for change. But the delayed-cut has exciting possibilities. It deliberately withholds cutting to the new shot until after the expected moment, to gain audience suspense, interest, and anticipation.

299

Cutting on a reaction:
  Knock at door. Writer looks up—

CUT Door opens. Woman enters. "May I come in?"

  Delaying the cut:
  Knock at door. Writer looks up. Door heard opening.
  Woman's voice, "May I come in?"—
CUT To woman at door, who enters.

Accentuation by cutting. Cutting accentuates whatever dialogue, sounds, or movements are happening at that moment, and thus enables us to lay emphasis. However, we must guard against giving false emphasis. Even an innocent remark—"Drink this. It won't kill you!"—may suggest different undercurrents of meaning, according to the moment at which the cut is made.

Cutting between moving pictures. By this means we can even imply spatial relationships where none exist—trains seemingly roaring towards each other, for example, or the hero and heroine departing for ever. This illusion arises easily—too easily. Associative ideas can arise when we do not want them. To avoid this, an intermediary shot, mid-way between the two extreme directions, will normally do the trick.

   Where we are dealing with a single moving subject, its direction should normally seem continuous. Where its direction changes, we must be able to see or infer this easily. Otherwise, distracting direction jumps can arise, or we lose orientation (see Fig. 14.22).

Cutting between static and moving pictures. Cutting between static and moving scenes brings its particular opportunities and hazards.

   Cutting from a static scene to a moving one will suddenly accelerate the viewer's interest. There is an effect of springing into action, the speed or violence of the new scene's movement being accentuated by its sudden appearance.

Fig. 16.2.

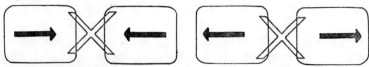

CUTTING BETWEEN MOVING PICTURES can produce strong subjective effects. Between movement in the same direction, suggests continuance of action or direction. Cutting between movement in opposite directions suggests (*left*) converging forces (impending meeting, collision), or (*right*) diverging forces (parting, expansion).

Fig. 16.3.

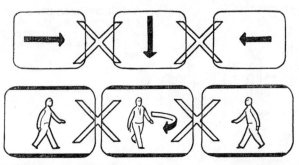

AVOIDING DIRECTION REVERSAL. Where two successive scenes show movement in opposite directions, accidental subjective effects can be avoided by using an intermediate buffer-shot containing movement mid-way between these extremes (*top*), or an actual change in direction (*bottom*).

Changing from a moving scene to a static one, invariably causes a sudden collapse in tension. A momentary anticipation which, if not satisfied, can lead to a rapid fall-off in interest.

As always, our use of these devices must be justifiable. The sudden interruption of movement by a cut to a static shot can be highly dramatic, e.g.—

Panning with a searcher in an empty room ...
we cut—to a static shot of someone watching his every move.

The borderline, however, between exciting scenic-transition and distracting gimmick, is slender.

Fig. 16.4. Pt.1.

MISMATCHING. *Left:* Mis-matched elevation (on intercutting between slightly different lens heights). *Right:* Mis-matched headroom.

Fig. 16.4. Pt.2.

Left: Too closely matched shots can look like Jekyll and Hyde transformations. *Right:* Cutting between identical shots at different distances can look like instantaneous growth or shrinkage.

301

Fig. 16.5.

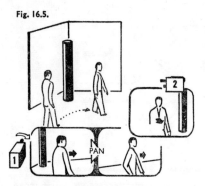

REDISCOVERING ON A CUT: Camera (1) pans, following the subject past the pillar. Cutting to camera (2), we rediscover the pillar which has just passed out of shot.

Matched cuts. We have already talked of the need for matching areas of interest when making transitions. Mis-matching can take other forms, too (see Fig. 16.4).

The cut as a concluding or introductory action. Both of these editing devices need to be used sparingly and with great caution.

THE CUT-IN. As we would expect, the cut-in provides shock treatment. A cut-in to a field of thistles would be a dynamic introduction to a programme on "The Weed Menace", but hardly right for a placid occasion.

Fig. 16.6.

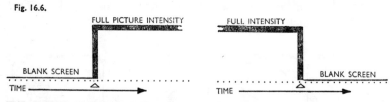

THE CUT-IN AND THE CUT-OUT. *Left:* The Cut-in. *Right:* The Cut-out.

THE CUT-OUT. The finality that comes from concluding a scene with a cut-out to blank screen, is definite and severe. Occasionally it may be used as a gesture of completeness. But it has the appearance too, of a technical fault ... or a cut by the censor!

The Fade

The fade as a concluding or introductory action. The fade-in heralds the start of action; the raising of a curtain. Again, its speed is variable, but it will normally have the air of quiet introduction; the forming of an idea. A fast fade-in is nearly as striking as a cut, although it has rather less vitality and shock value.

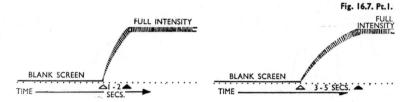

Fig. 16.7. Pt.1.

THE FADE-IN AND FADE-OUT. *Left:* Fast fade-in. *Right:* Slow fade-in.

Fig. 16.7. Pt.2.

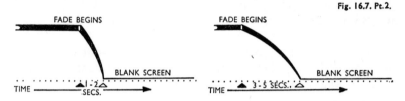

Left: Fast fade-out. *Right:* Slow fade-out. The fade-out and fade-in may be combined in permutation (slow-out, fast-in etc.) to link two shots producing a fade out-in.

We can fade-out a picture from its full intensity to a black screen at any rate we choose, taking several seconds to die away, or vanishing almost instantly. Regardless of its speed, it always preserves something of its integral character of gradual change.

A quick fade-out has rather less finality and suspense than the cut-out.

A slow fade-out is a smooth, peaceful, dying-away to darkness; a fall to rest; a cessation of action.

Used to link two sequences, the fade out-in introduces a pause into the flow of action. The mood and pace evoked vary with their relative speeds, and the pause-time between them. We find this transition used mostly to connect slow-tempo sequences, where a temporal or spatial change is required; while between two fast-moving scenes, it acts as a fragmentary pause, emphasizing the renewed activity of the second shot.

Fig. 16.8.

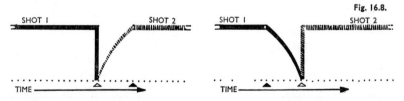

COMBINATIONS OF FADE AND CUT. *Left:* The cut/fade-in results in a surge, a visual crescendo, strongly introducing the new shot. *Right:* The fade-out/cut-in produces a visual punch; used between a succession of static shots (e.g. of captions, paintings), provides a dynamic introduction to each.

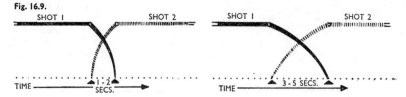

Fig. 16.9.

THE MIX, which is usually comparative in purpose. *Left:* Fast mix. *Right:* Slow mix.

The cut-out/cut-in with a pause between shots is valueless, for it always looks like a lost shot, or an operational error.

The mix (dissolve; lap-dissolve). To *mix* we fade-out one picture while fading-in the next. The two images are momentarily superimposed; the first gradually disappears, being replaced by the second.

Mixing between shots provides a smooth, restful, transition with a minimum interruption of the visual flow.

A quick-mix usually suggests that their action is concurrent (parallel action).

A slow-mix can suggest a spatial change, or a temporal one.

It is generally comparative: pointing the scenes' similarity or difference, comparing time (especially showing time passing), or comparing space (e.g. in a series of shots progressively higher up a mountain).

The mix is probably television's most misused transition. It is often introduced as a "softened-off" cut, to hide the absence of motivation in moving to a new viewpoint. This is regrettable, for, by allowing ambiguous interpretation, it misapplies an established piece of filmic grammar, and might conceivably destroy the conventional meaning of the mix. But it does permit a gentle transition for unhurried occasions, and covers a multitude of sins in slovenly editing.

Mixes are fraught with potential problems, especially where movements, size-changes, or repositioned attention occur at the same time. Their sustained double images can make very slow mixes tedious and muddling; while mixes between similar-sized shots of the same subject, taken from different viewpoints, produce disturbing twin-images of little or no value. However, the mix can help to relate areas visually, when we want to intersperse shots of part and whole of the same subject.

A *matched-dissolve* between identical shots of different, but similar, subjects, is fine for transformations (man changing into

304

boy); but nothing else—except, perhaps, to convey confusion or disorganization.

Mixes during movement are usually only completely satisfactory when their relative directions are similar. Mixing opposite directions of movement can sometimes arouse feelings of expansion or impact, without necessarily suggesting that these subjects are involved. But, again, this is an occasion for caution.

Defocus-dissolve and ripple-dissolve. (See Chapter 20.)

The wipe. (See Chapter 20.) The wipe is a visual stunt: a transition that still smacks of the slick "schmalz" presentation of motion picture trailers. It has limited use in lighter types of entertainment, but is rarely found in more serious production.

The wipe is, therefore, either an uncovering, revelatory action, or a concealing, or fragmentation, according to how it is applied. In all forms, the wipe draws attention to the flat nature of the screen, destroying any three-dimensional illusion the picture might have.

According to the shape and speed of the wipe, its emotional impact will vary, following the principles we discussed in dynamic

Fig. 16.10. Pt.1.

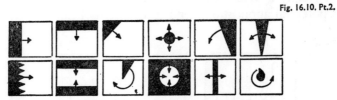

THE WIPE. While looking at one picture, a portion of another begins to break through. This break-through grows in size, eventually obliterating entirely the first shot.

Fig. 16.10. Pt.2.

Over 100 shapes of wipe are used in motion picture making. In television, a dozen meet most requirements. Obtainable by electronic-switching units or inlay masks.

Fig. 16.10. Pt.3.

More elaborate wipes usually require specially prepared film, for example: paper tear; peel-off; explosions; sparkle break-through; slats; push-over.

305

composition. Its direction can aid or oppose the subject's movement, and gain in vigour accordingly. The wipe may be hard-edged or soft-edged—the latter being much less distracting. Broad soft-edged versions can have the unobtrusiveness of a mix.

Wipes have been used tastefully, as a transitional link between a full shot of an orchestra, and a single instrument (iris out), and when reverting to the full shot again (iris in). But like all gimmicks, they can be easily overdone.

Visual transitions in tempo and mood

All these transitional devices possess their own special impact-value. They are not only links, they give the new shot a certain "flavour". So, as we might expect, some transitions are more suitable to conjoin shots of particular mood or tempo than others.

The split-screen

This is an artifice occasionally used to divide the screen into two or more sections, each showing a separate scene.

It is obtained:—electrically (by a switching-generator or electronic-insertion), by vignetting, or by mirrors.

Despite its artificiality, the split-screen can help us to show simultaneously two or more concurrent events, or the interaction of events in separate locations, or to compare the appearance, behaviour, etc., of two or more subjects (e.g. a relief map of terrain, with aerial photographs of the same area).

Virtually a visual cliché for two-way telephone conversations, its potential applications tend to be overlooked.

 Fig. 16.11.

SPLIT SCREEN. The screen is split into two or more segments, showing respective portions of component pictures.

306

Fig. 16.12. Pt.1.

VISUAL TRANSITIONS IN TEMPO AND MOOD. To link shots where tempo or, the mood is slow to slow, we can use a slow fade, a slow mix or where appropriates slow, soft-edged wipes, ripple-dissolve, defocus-dissolve.

Fig. 16.12. Pt.2.

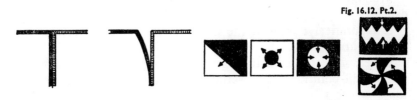

Where the change is fast to fast we normally cut or have a fast fade/cut-in. Alternative, if practicable, are fast, hard-edged wipes, or broken (e.g. zig-zag, slatted) or patterned wipes (spokes, diamonds).

Fig. 16.12. Pt.3.

The three most customary transitions from fast to slow are cut-out/fade-in, fast slow fade or fast-slow mix, with the possible alternatives, soft-edged wipes ripple-dissolve or defocus-dissolve.

Fig. 16.12. Pt.4.

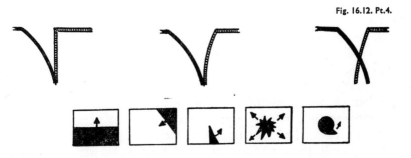

To change from slow to fast we can have a slow fade/cut-in, a slow-fast fade or a slow-fast mix. Any of the wipes illustrated here may also be used in suitable circumstances.

Superimpositions (half-lap dissolve). (See Chapter 20.)

In film, superimposition necessitates exposing each frame twice or more (double-exposure) in the camera, or in the laboratory's optical printer. In television, we merely transmit the two or more pictures simultaneously.

Superimposition has a number of useful applications, both artistic and operational.

Spatial montage: suggesting that two or more events are occurring concurrently.

Bringing together action, events, locale, having parallel significance.

Comparatively: showing the similarity or difference between the subjects juxtaposed.

Development: demonstrating stages in a process's development (e.g. a half-built machine, overlaid with an image of the completed job).

Relationship: showing the relationship of our subject to another (e.g. a cosy fireside, overlaid with the image of miners digging coal, or an internal part of a machine solidly superimposed upon a ghost of the entire apparatus, to show its position relative to the whole).

Thoughts: suggesting a character's thoughts, by superimposing their subject upon a close-up of the thinker.

To obtain transparent-looking images (e.g. fairies, ghosts).

To obtain larger or smaller proportions (e.g. giants, dwarfs, by "solid" superimposition).

To juxtapose subjects that are spatially remote from each other.

To permit appearances and disappearances.

To combine written or drawn material with another picture (e.g. titles, sub-titles, etc.).

To insert people (solid or transparent) into a separate scenic background (e.g. a photograph).

To provide surface textures to pictures (Chapter 20).

The tones of superimposed pictures are additive, so the light areas of any picture inevitably break through the darker tones of other pictures with which it is superimposed. The combined superimposition easily becomes muddled, therefore, unless we take precautions in arranging their relative tonal distribution. To prevent mutual degrading of final picture quality (which is difficult when superimposing multi-toned pictures), use dark tones in background areas. Coloured graphics superimposed on multi-hue backgrounds similarly need care.

Fig. 16.13.

SUPERIMPOSITION. Two or more pictures faded up at the same time will be superimposed, their relative strengths being controllable by the respective faders on the video switching console.

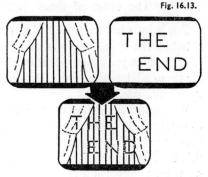

Also, unless we want unorthodox effects such as growth, shrinkage, or side-slip, we have to avoid all movement of superimposed cameras.

The elaboration possible in superimpositions must depend on the number of picture-sources at our disposal, and on the flexibility of the video switching console.

Montage

Montage is a term that has ambiguously acquired quite diverse meanings. It is applied sometimes to a rapid succession of brief shots having a common theme, these individual fragments combining to imply some overall idea (usually an abstract one, such as "Progress", "Gaiety", "Age" ... that is not readily conveyed by other means).

Nowadays, montage can refer to the practice of showing several images on the screen at the same time. They may be superimposed or juxtaposed, for any of the artistic reasons we listed earlier.

Fig. 16.14.

MONTAGE can mean a rapid succession of connected shots, juxtaposed images or superimposed images.

The order of shots—and their associative effects

Our minds are for ever trying to detect meaningful relationships: between the things we see or between the sounds we hear. These relationships may be evident or they may be obscure. We may even see them where none exist. But a relationship of some sort we shall eventually discover.

Whenever we see a succession of pictures, we try unconsciously to establish a connection between them, particularly when cutting between short-duration shots. (Very slow mixes or fade-outs reduce this tendency considerably, hence the use of such transitions when we want to suggest a completely fresh train of ideas.)

We can interpret successive shots in several ways:

Shot A's may give rise to (or lead to) Shot B's situation ...

$$A \longrightarrow B$$

Shot A may be explained by Shot B ...

$$A \longleftarrow B$$

Shots A and B together may give rise to Shot C ...

$$A + B \longrightarrow C$$

Shots A and B together may be explained by Shot C ...

$$A + B \longleftarrow C$$

Shot A, when juxtaposed with Shot B, may imply an idea "X", that is not implicit in either A or B alone ...

$$A + B = X$$

Now this simple analysis reminds us that when we show a succession of events, the viewer's ideas about what is happening will depend upon what they appear to mean to him. How he thinks they are related. And that can vary according to the order in which we show him these happenings. Let us take an example:

We have three brief shots of: a burning building—a violent explosion—two people running to a stationary car.

Changing the order of these shots will alter or modify what seems to be going on:

fire—car—explosion
   People fleeing from a fire were killed while trying to escape.
fire—explosion—car
   Fleeing from a fire, people escaped just in time, despite an explosion.
car—explosion—fire
   That fleeing people had caused an explosion, leading to a fire outbreak.

By implication rather than direct statement, we can stimulate the audience's imagination to the full, at the same time overcoming a great many practical difficulties. Take a close-up of a boy looking upwards, and join it to a shot showing a tree falling towards the camera.

Showing these as ...

Boy—cut to—tree falling,
suggests that he is merely watching a tree being felled.
Reversed as ...
Tree falling—cut to—Boy looking up,
can imply that the tree is falling on to the boy who, sensing danger, looks up.

The pictures themselves may originally have been quite un-related.

Sometimes the pictures will convey practically the same idea, whichever way they are combined, e.g.

woman screaming — lion leaping
lion leaping   — woman screaming

But we can usually trace a nuance of difference between the arrangements, especially when any cause/effect relationship is suggestible.

A cause/effect or effect/cause relationship is one of the principal things we can show by cutting between two shots. Somebody turns his head. We cut to see the reason. The viewer has become accustomed to this idea. Occasionally we may take advantage of this, and deliberately show an unexpected outcome. e.g. A bore and his victim are walking along a street.

Cut to—A close-up of the bore, who eventually turns his head towards his companion;
Cut to—A shot showing his victim far behind, window-gazing.

But this surprise can arise inadvertently; then the viewer may find himself mystified or frustrated, having jumped to wrong conclusions. One such situation crops up regularly.

e.g. Speaking on world trade, a lecturer stands, and moves towards a wall-map.

Cut to—A close-up of the map;
Cut to—The speaker in an entirely different setting.

Instead of finding him near the map, as we should expect, we have to re-orientate ourselves.

311

An even more disturbing situation is not exactly unknown:

Hearing a knock at the door, the actor turns.
Cut to—A shot of a train speeding through the night.

Here, there is no visual continuity whatever; no effect/cause relationship, although the cut has implied one. Even where we are prepared for a scene change (viz., the speaker has referred to his forthcoming trip), the cut is still an unsuitable transition.

Duration of shots

What is the shortest time a shot can last and yet convey its information? How long can a shot be maintained and still remain effective?

These questions are continually being asked in one form or another, especially when comparing television and motion picture techniques. Some people maintain that the small televised picture requires greater concentration and, hence, longer assimilation time. Others point out that when standard feature films are televised, we can follow their faster pace easily enough.

What is often taken for tolerance of long-held shots in television turns out, in fact, to be a case of the viewer's watching instead of looking, or hearing instead of listening, after he has become satisfied. Certainly, more shots in live television outstay their welcome than under-stay it.

We may, perhaps, summarize the whole situation as follows:

If a shot is of too short duration, there will be insufficient time to understand its full intended import.

Held too long, attention will wander. Having assimilated the visual information, thoughts dwell upon the sound. This eventually gives way to rumination—or channel switching. The limit for most subjects is around 15-30 seconds. For a static shot, much less. And for a mute shot, perhaps only 5-10 seconds. The claim that a close shot can sustain interest longer is a very arbitrary one.

A shot of correct duration retains interest in the aspects the director intends.

Many factors influence how long we can usefully hold a shot:

The amount of information we want the viewer to assimilate (i.e. a general impression, or minute detail).

312

The clarity of this information (i.e. how obvious; how easily discerned).

How familiar the subject is to the viewer (i.e. its appearance association, viewpoint, etc.).

How interesting the subject is to the viewer.

The amount of action, change, or movement seen.

The picture quality (marked contrast, detail, strong composition, hold most interest).

From even this résumé, it is obvious that we cannot hope to devise hard and fast working principles here.

The shortest shot-duration possible is one picture period (in television, 1/30th or 1/25th of a second; in the cinema, 1/24th sec.). Even in that short time, a simple image can register with us. A long documentary film, comprised of shots of 2-7 secs. duration, was televised several times, and received no adverse criticism. Its intended messages were imparted and understood. Although extreme, such cases have demonstrated that very short-duration shots can convey adequate information for limited purposes; but this sort of pace is scarcely practicable, especially in live television production.

The audience's attention is normally keyed to the speed of the respective production. A short flash of information during a slow-tempo sequence may pass unappreciated, even overlooked, while in a fast-moving sequence it would have been fully comprehended.

Cutting rate

The cut is dynamic. But how dynamic depends upon the matching, or contrast, of the shots it conjoins, and upon whether the viewer is expecting the transition. Skilfully matched and timed, the viewer may not notice it consciously at all. But we may not want an unobtrusive cut. We may wish to shock or surprise the viewer; to accentuate a beat, to emphasize the new shot. Then we have to be able to distinguish the borderline between pleasurable shock and distracting annoyance.

Fast cutting is usually impracticable in live production, for unless we are intercutting between almost similar shots, we soon exhaust those shots available, or we catch up unawares with a camera during a lens-change, or moving to a new shot.

Situations can be devised that do allow for rapid intercutting, providing good visual continuity without undue technical problems.

313

Fig. 16.15. Pt.1.

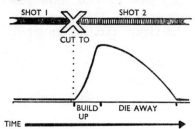

RESPONSE TO CUTS. Our response to a cut is not instantaneous. When the cut is made, the audience-impact is slightly delayed; building to a peak and dying away gradually.

Fig. 16.15. Pt.2.

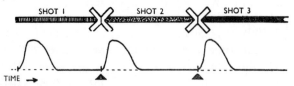

A series of cuts may achieve individual impacts; there being sufficient time for each reaction to subside.

Fig. 16.15. Pt.3.

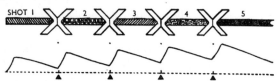

Faster cutting may produce a cumulative build-up.

Fig. 16.15 Pt.4.

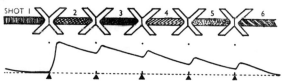

Or, through repetition, surprise may diminish with each cut, resulting in declining tension.

Fig. 16.16.

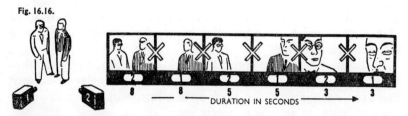

SEQUENTIAL INTER-CUTTING. Each subject accuses the other of treachery, tension is rising. The camera tracks closer and closer, shots become larger and larger, and of increasingly shorter duration.

314

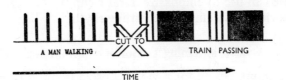

Fig. 16.17.

CUT TO
A MAN WALKING · TRAIN PASSING
TIME

TEMPO-CHANGE ON TRANSITIONS. A comparative rhythm set up between the two pictures.

Figure 16.16 shows one. But they are difficult to sustain for any length of time.

A fast average cutting rate imposes considerable strain on the entire production crew in the studio. Often we find operational standards deteriorating badly in consequence. Typical live telecasts have contained anything from less than 20 to over 250 shot changes in a half-hour show. But statistics alone tell us little, for they cannot convey the amount of operational complexity entailed.

Cutting rhythm

When we look at a sequence of pictures, we often become aware of certain rhythmical effects.

There is the intrinsic rhythm we find in each shot itself. This may be derived from:

Subject-movement—e.g. rhythm of a dancer's leg action.
Compositional change—e.g. scenic backgrounds moving past.
Superimposed movement—e.g. a regularly flashing light from a street sign.

Then, when we take two such pictures (each with a strong intrinsic-rhythm) and intercut them, we immediately provide a comparative effect.

Finally, we have the cutting rhythm that comes from the frequency (cutting rate) and the pattern in which we cut between a succession of pictures.

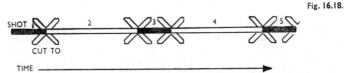

Fig. 16.18.

SHOT 1   2   3   4   5
CUT TO
TIME

CUTTING RHYTHM. The time-pattern created during fast cutting sequences.

This we can use to emphasize aural rhythm. Conversely, we can use aural rhythm to emphasize cutting rhythm.

The video switching console (vision mixing desk)

The optimum design for a video console is still a highly controversial matter. Several basic arrangements are widely used, each having its advantages and its limitations. The most advanced demand the operational dexterity of a cool-headed octopus. Whichever model we have available will naturally affect the ease with which we can carry out various editing operations. Several transitional effects necessitate careful pre-setting of controls before they can be made.

What do we want in a video switching console? In a large studio set-up, we should be able at any moment, at any speed, in any combination of strengths. ...

To cut; fade, up or down; mix; superimpose (and, possibly, control wipe pattern and movement, adjust electronic insertion); to add to, or subtract from, superimposed cameras already on the air; by cutting or mixing; to cut or mix to or from any given combination of cameras, to any others, in any proportions. At the same time, we need preview monitors to show all single and combined sources' pictures.

Fig. 16.19.

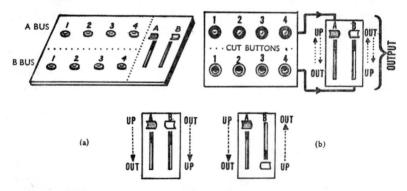

A TYPICAL BASIC VIDEO CONSOLE. (Only four sources are shown, for simplification.) *Top:* All the pictures sources are controlled by buttons, being duplicated on the A bus and on the B bus. To cut, we press the respective button on the bank selected by the main faders. *Bottom:* The main faders can be used linked together (a), moving in the same direction, to provide cross-fading (mixing) between A bus and B bus selections. Or they can be split (b) and used individually, to permit fading to blank screen, or to superimpose two or more cameras.

Fig. 16.20.

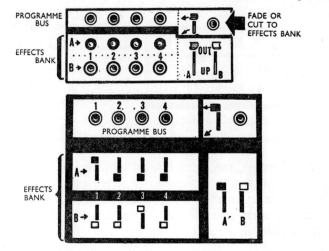

THE EFFECTS BANK. A variation on this theme is the effects bank, using individually controlled channel faders to permit simplified mixing and switching to blended superimpositions.

To get all these facilities without over-complicated manipulations is extremely difficult, especially when the show may involve some 6 to 12 picture sources.

On the console, we have fundamentally *push-buttons*, to switch (punch-up) any camera on to the main transmission channel, and a *fader* to fade each in or out, or adjust a picture's intensity relative to others, during a superimposition (Fig. 16.19).

Selecting any camera will also switch on several cue-lights, the tally-light on the camera, and an indicator inside the camera view-finder. In the production control room, a channel-number panel is illuminated above the master (transmission) monitor, and another above its respective preview monitor. The camera control unit, too, will have its tally-light.

A development is the 3-bus switching panel (Fig. 16.20).

Now, the A and B buses are renamed effects bank, and a further switching-bank added. This line or programme bus is now used for all straight cutting. An effects-button, or fader, enables us to switch or fade from this over to the effects bank.

On the effects bank we can set-up picture combinations in advance, using the A, B, or A plus B banks.

This enables us to arrange pre-set superimposed groups and to fade or mix selected sources within that combination.

Fig. 16.21.

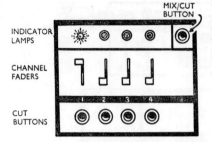

INDICATOR LAMPS

CHANNEL FADERS

CUT BUTTONS

MIX/CUT BUTTON

THE MIX/CUT BUTTON. Switching is by cut-buttons (faders being left full up for cutting sequences). Intermixing is by faders. A mix-cut button enables us to cut from any camera selected by cut-buttons to any combination of faded-up cameras.

From here on, the many-headed monsters of video consoles are derived, using two or more sets of effects banks.

Finally, in an attempt to reduce the number of controls, technical ingenuity has enabled one bank to provide two sets of operations, as in the version illustrated in Fig. 16.21.

According to design arrangements, remote-control cue buttons may be included at the video switching console. These enable one to have for example, remote control of the telecine (indicators denote the footage or sequence numbers). Similarly, one can change slides or film-strip frames in a caption scanner that is providing titles or caption illustrations.

Wipes are readily achieved by electronic circuitry, and advanced video-switching consoles include facilities for a range of plug-in pattern-generators. Controls adjust speed and direction. Stopped at intermediate positions, the wipes provide split-screen, vignettes and aspect-ratio adjustment.

Controls may be included too for *chroma-key* selection, so that the picture on a chosen subject camera(s) can be inserted into selected background picture source(s). Video adjustments may be remoted to the video operator's console.

A further facility, the *colour synthesiser* (page 367) selects lettering, provides separate push-button selection of caption lettering and background colours.

Edge-generator switching enables one to introduce black (or white) *edges* to caption lettering to improve clarity during super-impositions (subtitles, name plates, etc.).

17

AURAL COMPOSITION

IN the same way that we can find rational principles underlying our responses to pictures, so we can trace principles concerning one's choice and arrangement of sounds. This might be termed "aural composition".

Here again, we are all quite familiar with the effects, but we rarely think of analysing their nature. We recognize certain sounds as gay, exciting, melancholy, martial ... and so on. Music is the most obvious type of sound-form having strong emotional persuasion, but all sounds have this potential power. Many have such clear-cut subject-associations that we recognize them, and link them immediately with the circumstances of that type of sound, a thunderclap for example, or a crying baby.

Other sounds have a more abstract character. Their emotional influence stems partly from their subconsciously reminding us of particular subject/mood associations, partly from their psycho-physiological influences.

Aural composition is to the ear, what pictorial composition is to the eye. Our impressions of the complete programme derive from both.

If we realize something of the nature of these influences, we can avoid ineffectual treatment, and have a wider appreciation of the opportunities that lie before us.

An analysis of sounds and their effects

If we examine a series of sounds that all convey a particular mood, we shall find, naturally enough, that they have many common features. Furthermore, if we analyse them closely, we find that we can detect many characteristics which will be found in any sound conveying that mood. So now, instead of hopefully seeking sounds to fit a certain mood-situation, it becomes possible to appreciate more exactly precisely what features we are looking for.

When we get down to detailed analysis (Table 17.1), we discover some interesting connections between sound characteristics and their emotional impact.

For example, a sound containing the features—
1 . 4 . 9 . 10 . 13 . 15 . 19 . 20 . 23 . 24 . 26 . 27 . 31 . 33a . 35 . 39 . 42
can only be an exciting, vigorous, effect.
While one containing—
2 . 5 . 8 . 11 . 14 . 16 . 18 . 21 . 22 . 28 . 32 . 34b . 36 . 41
must be a sad, peaceful, sound.

The effect of combining sounds
Balance (i.e. the relative loudness, complexity, texture, speed, etc., of concurrent sounds).

Overall harmony—Balance, completeness, beauty.
Overall discord—Unbalance, uncertainty, incompletion, unrest, ugliness, irritation.
Marked differences in relative volume, rhythm, etc.—Variety, complication, breadth of effect, individual emphasis.
Marked similarity—Sameness, homogeneity, mass, weight of effect.

Overall movement (i.e. changes in the relative volumes, rhythms, pitches, etc., of concurrent sounds.) See Figure 17.1.

Focusing aural attention

We cannot concentrate upon both picture and sound simultaneously, nor will we find our attention equally divided between them. One invariably dominates.

Fig. 17.1.

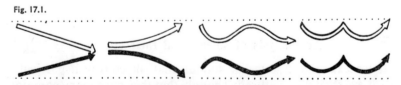

JOINT MOVEMENT BETWEEN TWO SOUNDS. *Left:* Towards a common focal point, suggesting conflict, concentration. *Centre left:* Away from a common focal point, suggesting divergence, broadening. *Centre right:* Contrasting sound movement, suggesting diversity, variety, interdependence. *Right:* Parallel sound movement, suggesting similarity, unanimity.

TABLE 17.1

ANALYSIS OF SOUNDS AND THEIR EFFECT

Sound characteristics	*Associated with—*						
VOLUME							
1 LOUD SOUNDS	Big	Strong	Assertive	Powerful	Energetic	Rousing	Earnest
Size, Force, Emphasis, Energy.	9.14.16.	5.14.16.	4.10.12.18.	5.10.14.	4.10.13.	4.10.13.	9.11.12.
[1]Especially when combined with	18.28.42.	18.22.24.	20.22.24.28.	22.24.28.	15.19.20.	19.20.22.	14.16.18.
		28.33B.	31.33B.	31.33B.	26.29.31.	27.31.42.	28.
				42.	42.		

	Small	Soothing	Peaceful	Gentle	Subdued	Delicate	Little energy
2 SOFT SOUNDS		5.11					
[1]Especially when combined with	8.15.	8.14.16.18.	8.14.16.	4. 8.14.	5. 7. 9.	4. 8.16	5.11.14.16.
	16.41	22.28.34B.	18.28.34B.	16.18.28.	16.28.34B.		18.25.36.
		36.41.		36.	36.		41.

	alerting	persuasive
[1]Against a quiet background ...	6.10.13.15.	4.7.8.22.
	20.38.	

PITCH 3 Pitch often suggests *physical height.*
Height generally increasing with 2.4.8.31.37.42.
Depth generally increasing with 2.5.11.14.28.38.42.

4 High-pitched sounds:—Exciting, light, brittle, stirring, invigorating, elating, attractive, distinct, sprightly, weak.

5 Low-pitched sounds:— Powerful, heavy, deep, solemn, sinister, undercurrent, depression.

[1] This example of fuller analysis shows how each idea is strengthened when several allied features combine.

TABLE 17.1—continued

	Sound characteristics	Associated with—
KEY	6 Major	Vigour, brightness.
	7 Minor	Melancholy, wistful, apprehensive.
TONAL QUALITY	8 Pure, thin (e.g. flutes, pure string-tone)	Purity, weakness, simplicity, sweetness, ethereal, daintiness, forthright, persuasiveness.
	9 Rich (possessing strong overtones; harmonics)	Richness, grandeur, fullness, complexity, confusion, boisterous, worldly, vitality, strength.
	10 Edgy, brassy, metallic	Cold, shrill, gay, bitter, snarling, vicious, forceful, hard, martial.
	11 Full, round tone (e.g. horn, saxophone, bowed basses)	Warm, rich, mellow.
	12 Reedy (e.g. oboe, clarinet)	Sweetness, nostalgic, delicate, melancholy, wistful.
	13 Sharp transients (a) High-pitched (e.g. xylophone, breaking glass)	Thrilling, exciting, horrifying.
	(b) Low-pitched (e.g. tympani, thunder)	Dramatic, powerful, significant.
SPEED and RHYTHM	14 Slow	Serious, important, dignified, deliberate, ponderous, stately, sombre, mournful.
	15 Fast	Exciting, gay, hopeful, fierce, trivial, agile.
	16 Simple	Uncomplicated, deliberate, regulation, dignity.
	17 Complex	Complication, excitement, elaboration.
	18 Constant	Uniformity, forceful, monotonous, depressing.
	19 Changing	Vigorous, erratic, uncertainty, elation, wild.

TABLE 17.1—continued

	Sound characteristics	Associated with—
SPEED and RHYTHM (continued)	20 Increasing (Accelerando)	Increasing vigour, excitement, energy or force; progressive development.
	21 Decreasing (Rallentando)	Decreasing vigour, excitement, energy or force; concluding development.
PHRASING	Repetition of sets of sounds—	
	22 Regular repetition	Pleasurable recognition, insistence, monotony, regulation, co-ordination.
	23 Irregular repetition	Distinctiveness, personality, disorder.
	24 Strongly-marked accents	Strong, forceful, emphatic, rhythmical.
	25 Un-accentuated sounds	Continuity, lack of vitality.
	26 Interrupted rhythm (Syncopation)	Character, vigour, uncertainty, unexpectedness.
DURATION	27 Brief, fragmentary	Awakening interest, excitement, forceful, dissatisfaction.
	28 Sustained	Persistence, monotony, stability, tiredness.
	29 Staccato	Nervous vitality, excitement.
MOVEMENT	30 Movement pattern	Movement pattern of sound suggests corresponding physical movement, e.g. upward—downward—upward glissando pitch changes suggesting swinging movement.
	31 Upwards	Elation, rising importance, expectation, awakening interest, anticipation, doubt, forceful, powerful.
	32 Downwards	Decline, falling interest, decision, conclusion, imminence, climactic movement.

TABLE 17.1—continued

	Sound characteristics	Associated with—
PITCH CHANGES	33 Sudden (a) rise	Increasing interest, excitement, uplift.
	(b) fall	Force, strength, decision, momentary unbalance.
	34 Slow (a) rise	Increasing tension, aspiration, rising motion.
	(b) fall	Saddening, depression, falling motion, reduced tension.
	35 Well-defined pitch-changes	Decision, effort, brightness, vitality.
	36 Indefinite pitch-changes (e.g. slurs, glissando)	Lack of energy, indecision, sadness.
	37 Vibrato	Instability, unsteadiness, ornamentation.
VOLUME CHANGES	38 Tremolo	Uncertainty, timidity, imminent action.
	39 Crescendo	Increasing force, power, nearness, etc.
	40 Diminuendo	Decreasing force, power, nearness, etc.
REVERBERATION	41 Dead acoustics	Restriction, intimacy, closeness, confinement, compression.
	42 Live acoustics	Openness, liveliness, spacious, magnitude, distance, uncertainty, the infinite.

One may be used to draw attention to the other, for example, a movement made immediately before a remark will emphasize that remark. Dialogue immediately preceding a movement gives emphasis to the move instead. So we can pass emphasis continually between picture and sound, to allow a regular distribution of attention.

We can list the characteristics that draw a listener's attention to a sound-source:

Loud sounds, increasing volume.
High-pitched sounds (around 1000 to 4000 c/s).
Sounds rich in overtones (harmonics); edgy, metallic sounds; transients.
Fast sounds; increasing speed or rhythm.
Complex rhythms.
Briefly-repeated phrases; syncopation; strong accents.
Short-duration, staccato, sounds.
Aural movement—especially increases in volume, pitch, etc.
        —clear-cut, unexpected, violent, changes.
        —interruption, vibrato, tremolo.
Reverberant acoustics.
Balance—Discord.
      Marked contrast,
      (a) between the principal and background sources;
      (b) between the sound and the picture (i.e. their associations, composition, etc.).
      Marked similarity,
      (a) between sounds—e.g. one source echoing another;
      (b) between sound and picture
         (e.g. simultaneous upward movements in both).

We can transfer aural attention to another subject by:—

Giving the original subject's sound-pattern (rhythm, movement, etc.,) to the new source.
Weakening the original subject's attraction and strengthening the new source.
Linking action—e.g. having the pattern of the original sound change to that of the new subject, before stopping it.
Transferring aural-movement through—e.g. by carrying over a solo sound, while changing its background.
Cutting to a shot of the new source alone.
Changing the original compositional lines—e.g. whereas upward

sounds lead attention towards high notes, downward sounds lead attention towards lower notes.

Dialogue attracting attention either to its source, or to its subject.

The selective use of sound

Like the eye, the ear is selective, and what one personally selects, will determine one's impressions of a scene. In a studio re-creation, we shall not attempt to imitate the original scene directly but, instead, provide whatever features convey the particular "flavour" we want. We may have to omit environmental noises one would normally expect to find, to avoid distraction. Alternatively, we may have to introduce atmospheric background sounds that are not strictly legitimate, to build up a convincing sense of location.

Our complete interpretation of a scene can be affected, according to how we select and blend a set of sounds. Imagine, for example,

TABLE 17.11

METHODS OF SOUND TREATMENT

Form of selection	Specimen treatment
Scene:—After a long hopeless day seeking work, a cripple returns through emptying streets.	
(1) All sounds audible—of subject and background.	His footsteps sound amidst traffic and crowd noise.
(2) The subject alone is heard.	His stumbling footsteps echo through quiet streets.
(3) The subject plus selected background sounds.	His slow tread contrasts with the brisk steps of passers-by.
(4) General background sounds alone.	Traffic noises. Passers-by.
(5) Significant background sounds alone.	The laughter and gaiety from groups he passes; contrasting with his abject misery.
(6) Interpretative sounds, not directly originating from the scene.	His echoing footsteps become increasingly louder and distorted. By progressively filtering out the higher audio-frequencies, his laboured tread becomes emphasized.
(7) Significant selected sounds from another scene (providing explanation or comment).	Voices of people refusing him work, echo in his brain.

the slow, even, toll of a cathedral bell, accompanied by the rapid footsteps of approaching churchgoers.

We could reproduce these sounds simply as they happen to arise, or we could deliberately draw attention to particular aspects, in order to give the sounds a certain significance:

The quiet, insignificant bell, for example, against loud footsteps.
The bell's slow dignity, contrasted with restless footsteps.
The bell's lingering echoes, contrasted with the staccato impatience of footsteps.
Its boom, overwhelming all other sounds.

So the sounds may merely set location; suggest hope, dignity, community, domination, simply through selective presentation. In this example, we adjusted only the quality and loudness of the two sources but, clearly, our selection can take much broader forms.

Aural-visual relationships

When discussing editing, we saw that there can be several relationships between successive pictures. So, too, we find that sound and picture can be related in several distinct ways:

The picture's impact due to its accompanying sound S———→P, e.g. A man crosses a busy thoroughfare to cheerful music: suggesting that he is in lighthearted mood; or to the accompaniment of car horns and squealing tyres: suggesting he is jaywalking dangerously.

The sound's impact can be due to the picture: S←———P, e.g. The sound of a carriage's wheels bumping over a rough road with a long shot: the sound becomes a normal background effect; but with a close shot of one wheel, every jolt may suggest impending breakdown.

The effect of sound and picture may be cumulative: S + P = E, e.g. A wave breaks, to a crashing crescendo.

Sound and picture together may imply a further idea: S + P = X, e.g. Wind-blown daffodils, accompanied by birdsong and lambs bleating, can suggest Spring.

18

TELEVISION PRODUCTION METHODS

So far, we have been preoccupied with studying the underlying principles of television production. But to be realistic, we must recall the circumstances in which these principles have to be applied.

The background of production techniques

From the preliminary scheduling to transmission date may be a matter of months, but is more likely to be weeks, days, or even hours. The director's life throughout that time will be taken up with the vicissitudes of scripting, casting, costing, policy considerations; informing, consulting, and co-ordinating the machinery of production; filming, recording, rehearsing the cast ... and so on. In short, with a multitude of activities that leave very little time for meditation upon the medium's aesthetics.

Once in the studio, camera-rehearsal time is fairly limited. The camera and sound teams are meeting his brainchild for the first time, and will need to be guided into his interpretation. Snags arise —particularly in elaborate treatments—that cannot be anticipated completely beforehand. In such circumstances, there is much to encourage the director to use safe routines rather than creative techniques.

But safe routines are not inevitable. The greater the director's appreciation of his craft, the freer he is to choose, and to modify his choice wisely, as occasions arise. And therein lie the benefits of a soundly-based understanding of the medium.

A hazard of any entertainment medium lies in the familiarity that comes from working within it. One becomes so close to a production that it is not easy to appraise its first-time audience impact. One can lose a sense of the pace, timing and tension that a viewer will have. We have only to watch any film-length through several times to see how the stresses, emphasis, speed, change substantially with each showing. A word or gesture overlooked at first viewing may later appear significant; action that is heavily pointed at first, subsequently looks hackneyed or mannered.

Opinions differ as to how closely a production can be planned in advance on paper. Some directors rely heavily upon on-the-spot inspiration; others tabulate to the last detail. The creative artist must be allowed to choose his own approach, but in a complex medium such as television, the strategist has the advantage every time, for television demands tightly co-ordinated teamwork, and that comes most readily from precise imaginative planning. Too great attention paid to mechanics can breed a smooth-flowing but soulless production; but an artistically sensitive interpretation marred by operational errors has little to recommend it.

Types of production approach

Programme treatment tends to emerge in one of two ways: first, by starting with one's ideal interpretation, and modifying it to fit the budget and facilities available, or secondly, by deciding upon a practical layout with workable mechanics and then, with these particular facilities in mind, building up an interpretation of the programme.

Both methods have mixed merits. Over-insistence on a theoretical ideal may prove abortive when translated into reality. The busy street scene all too easily becomes a handful of walk-on extras by the time it reaches the studio.

The second scheme is more down-to-earth. Appreciating practicable limitations, space shortage, economy, and further hard facts of life, the director devises a working framework for his production. Knowing the order and relative importance of scenes, he contrives, with his technical collaborators, an operational campaign. A suitable number of cameras are allocated to each scene and general moves worked out. With this set-up, the director then builds up his treatment. At worst, this materialistic approach may never rise artistically above a competent routine; at best, it will use facilities to the fullest extent possible, providing all the artistic opportunity we might wish.

Most production treatment follows one of several detectable trends. We might term these:

(i) Mechanical treatment—in which consideration of the mechanics predominates.

(ii) Theatrical treatment—in which consideration of the action predominates.

(iii) Filmic treatment—in which consideration of the visual and aural arrangement of shots predominates.

(iv) Modified filmic treatment.

(i) *Mechanical treatment*

The primary purpose of this approach is to obtain smooth picture and sound continuity, and avoid operational problems. Treatment is almost entirely functional. Artistic motivation is comparatively incidental. The performer and what he has to say or do are heavily relied upon to influence the audience. Operations such as camera movement and editing are used for convenience—to change the viewpoint for visual variety; to get a closer or more distant shot; to release a camera for another scene, etc.

So emotional incompatibilities can arise between the programme's treatment and its intent. For example, a camera tracks back to include another person, although the track's audience impact may not be appropriate, or required. A performer is given meaningless business to do, simply to allow time for a camera to move.

Let us examine some mechanical production treatment more closely (Figure 18.1).

Seemingly, this is straightforward enough treatment. But there has been no mention of story-line. The visitor may be a casual enquirer, or a revengeful killer. Each situation needs quite individual interpretation. With mechanical treatment, it is likely to get much the same, whatever the occasion. It is only when we see such treatment repeatedly used, ignoring persuasive opportunities, that we see the situation for what it is. Mechanical treatment has a superficial air of competence, but little artistic integrity.

(ii) *Theatrical treatment*

Found mostly in drama programmes, this treatment is detectable by the way the director approaches production-mechanics. Action is conceived as a series of speeches, moves and business within a setting. As each sequence is acted out, the camera is used to emphasize and interpret these happenings. The production is built around the complete enactment of each scene. Camera and sound treatment is added. We seldom find productional imagery.

In slightly more advanced usage, the camera is used persuasively, and elementary editing principles explored.

330

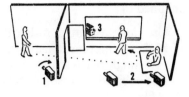

Fig. 18.1.

PRODUCTION TREATMENT. A visitor enters the office from the hall, walks to the desk, talks, moves to the window. Camera (1) Pan visitor down hallway to door. Cut to Camera (2) as he enters; track back, and pan him to the desk. He goes to the window, the camera tracking in with him. Cut to Camera (3) as he looks out of the window.

(iii) *Filmic treatment*

Whereas theatrical treatment plans the action first, in filmic treatment, the director thinks in terms of a succession of shots (a shot being the visual sequence between two transitions). Filmically, it is the picture treatment that largely determines the performer's action. The story-script having been broken down into brief shot-sequences, the camera set-up, movement, etc., for each is arranged to suit that part of the story-line. Picture and sound are used interpretatively; the cameras commenting upon, rather than simply displaying, the scene before them.

Many television directors draw upon their considerable theatrical experience, and interpret theatrically-contrived action in filmic terms, so that the final result is an amalgam of both. Where stage plays are only slightly adapted for television, instead of being re-conceived in filmic terms, this may be inevitable.

(iv) *Modified filmic treatment*

When appraising techniques, we have to bear in mind the characteristics of the medium in which they are being applied. Those that are suitable for pure cinema do not necessarily use television's particular properties.

In film, the emphasis is on the physical; upon freedom and variety of locations, elaboration of effect, spectacle. The very size of the screen makes it an extroverted, impressive, overpowering presentation.

Television's strength lies in its intimacy; its personal introverted appeal. The television screen lends itself more readily to psychological exploitation; to revealing human character and emotion. Spectacle loses much of its impact through smaller screen size and, in any case, can be prohibitively expensive.

The chief difference lies, then, in the scale and individuality of the two media. A television production approach that explores the aesthetics of filmcraft, while exercising the special properties of the television medium, we have called modified filmic treatment.

331

This is fundamentally similar to the filmic approach, but more inclined to:

Use camera-control rather than editing to achieve an effect.

Make a more extensive use of close-ups to reveal character through reactions than in film.

Use subjective personal treatment, than the objective methods of film.

Practice the art of concealment; suggesting a situation rather than trying to show it entirely.

Use partial and incomplete sets extensively.

Use representational parts and symbolic associations to convey ideas.

Objective and subjective approaches

The usual audience role is an objective one, in which we remain essentially onlookers watching the action. Our position may be that of an invited audience at a vantage point, eavesdroppers, passers-by who happen to witness an event.

Occasionally, by personal identification, we may subconsciously associate ourselves with a character in the programme. Our reactions thus range from those of privileged or unsuspected observers, to those of a casual bystander.

The alternative approach is subjective. The camera moves amongst action, often participating in it. The lens becomes our eyes —or those of a character in the programme. Instead of being shown things, we go in and have a look for ourselves. At its simplest, subjectivity occurs whenever the camera moves as we ourselves would have done, e.g. tracking-in to see a book being held open towards us.

In its most forceful form, subjective treatment links us very intimately with the action. Wandering around as a disembodied extension of ourselves, the camera meets our fellow performers and behaves in a humanized fashion. In the process we may find ourselves squirted, kissed, punched, blown-up, fired at, even drowned. But, at this extreme, subjectivity degenerates into distracting stunting; for the absence of bodily sensations, and our inability to respond to, or direct, the camera's moves, reveal the artificiality of the situation.

Providing the viewer realizes his intended relationship to the scene, there is nothing to prevent his moving between subjective and

332

objective viewpoints. Sometimes the difference is indistinguishable. A camera watching a crowd may represent either. Should it move down amongst them, treatment becomes definitely subjective.

Another subjective approach is frequently found in television. Here, a speaker addresses us directly, using the screen as a communicating opening into our home. Talking straight at us, subjects are demonstrated or displayed for our perusal, while maintaining the convention that he is speaking from a study, a living-room, and so on.

Finally, there is an approach that may be either subjective or objective. Artifice is thrown aside. We are obviously in a television studio. Cameras, equipment and staff are all visible as he talks. In some quarters this idea is thought to bind together performer and audience into a frank, unconventional relationship. Others feel it smacks of exhibitionism, distractingly revealing, and emphasizing the means of communication, to little purpose.

Audience interest

Interest or concentration patterns

A fundamental aim of all programmes is to arouse in our audience certain definite ideas and emotions, to stimulate their thoughts along particular, prearranged lines.

Throughout the show, however, the audience's interest will fluctuate. Concentration mounts and subsides. Tension builds up and relaxes. Excitement rises and falls. We cannot maintain them at full pitch, even should we wish to. One of the skills of good direction is to engineer these variations so that they come when we need them most. A judiciously arranged lull will make a subsequent peak appear all the greater.

Although our audience is comprised of individuals, we shall find nevertheless that there will be strong common reactions amongst them. We can show their varying concentration or interest as a rough graph (Figure 18.2).

The shape of this pattern will be influenced by:
The script's construction (i.e. the order and duration of events; their dramatic strength; plot outline; etc.).
Interpretation (i.e. the manner of the performance; its delivery, etc.).
Production treatment (i.e. the staging; camera-control; editing; sound treatment, etc.).

Of course, most writers and directors rely upon experience and intuition rather than analysis, when judging how they will introduce these stresses and lulls. But a graphical outline helps to remind us that these audience-interest or concentration patterns have to be contrived to be completely successful. They do not come automatically—at least, not in a form that makes the best use of the programme material.

In Figure 18.2, we have used a rough breakdown of a well-worn story-line, to show how this works. We see where we hope to build tension, where interest will be allowed to fall. We see the attempt to grip audience-interest at the start of the programme, the minor variations, the build-up to an encouraging peak just before the interval (or commercial). The second half of the show begins with a bid to recapture interest, fluctuating with the story-outline, giving a build-up to the finale.

We could go further still, and analyse a single scene in some detail, to demonstrate the way various aspects of the production contribute to the eventual viewer impact. We see how an emotional peak can be heightened by combining several allied influences, and how we vary the particular methods of persuasion used.

A typical example of such an analysis has been detailed in Figure 18.3.

Both strengthening and relaxing tension require a certain amount of care. It is as well, for instance, not to attempt to reach a high emotional peak in a short time—especially after a long period of calm, otherwise the peak becomes obtrusive; the viewer is over-aware of his own sudden reaction. It may even pass him by completely, because he has not been preconditioned to receive it. It is normally better to build up interest, concentration, or tension in well-defined stages, each peak surmounting the last. Too slow a build-up allows interest to flag.

Relaxation of tension after a peak brings its problems, too. Interest can fall away to become indifference. A sudden tension-release and the audience can be precipitated into laughter, after the nervous strain of concentration. Extremely effective when we want this to happen, like a peak during an uneventful period, it must not happen when unwanted.

Where action or story-line run close to a recognizable pattern, the viewer may have grasped a point long before we have had time to make it. Then, unless we can disguise the obvious, through careful presentation, or by-passing all superfluous detail (as with filmic space and time), a complete exposition can prove a painful process.

334

Fig. 18.2.

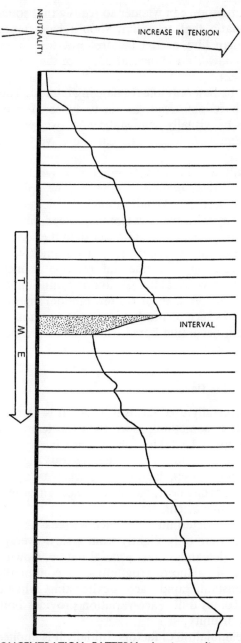

NEUTRALITY

INCREASE IN TENSION

TIME

OPENING TITLES

CRIME COMMITTED

CRIMINAL'S ESCAPE

HERO APPEARS AT SCENE OF CRIME

VILLAIN HATCHES PLOT TO INCRIMINATE HERO

VILLAIN'S FALSE ALIBI ESTABLISHED

EVIDENCE ACCUMULATES PROVING HERO GUILTY

HERO CANNOT ESTABLISH ALIBI

HERO WRONGLY ACCUSED

INTERVAL

(COMMERCIALS)

EVIDENCE ACCUMULATES AGAINST HERO

HEROINE DESERTS HERO

AND JOINS VILLAIN

FURTHER EVIDENCE PRESENTED AGAINST HERO

FLAW FOUND IN EVIDENCE

SUSPICION THROWN ON VILLAIN

HERO ESCAPES

SEARCH FOR VILLAIN

VILLAIN FOUND
HEROINE HAS GONE

HERO ARRIVES AT HEROINE'S APARTMENT

HEROINE MISSING

HERO FINDS HEROINE

THEY ARE RECONCILED

VILLAIN JAILED
HERO PARDONED

END TITLES

CONCENTRATION PATTERN, showing audience concentration throughout programme.

335

Duration of attention

The television viewer's attention can wander so readily that some directors believe it axiomatic that television should have more frequent climactic peaks than film or the theatre. But too many climactic peaks eventually mean no climax at all, since the viewer becomes accustomed to them.

Mobility of attention is the best antidote for flagging interest. Too little variation leads to boredom. And, as many types of programme material tend to be visually semi-static and over-wordy (e.g. talks, interviews, demonstrations), this can be a very real problem.

Sometimes attention will wander despite great visual activity. When quickening pace and high mobility follow a slow episode (e.g. a film-sequence after slower-paced studio shots), the viewer will not necessarily be whisked up with the new tempo. Instead, he will find himself inspecting the new environment, rather than the scene's significance. Fortunately, this distraction usually passes quickly.

As to the dangers of introducing overmuch information at once, the viewer will rarely suffer from visual indigestion or confusion, as is sometimes thought. He more often absorbs a few random aspects, and then relaxes when he has seen as much as he wants to, not, as some film-makers maintain, becoming bored when he has grasped its full import.

Pace

Pace is one of those universally-used terms to which it is hard to pin a precise definition. It is something more than the tempo or speed at which events seem to be taking place. Pace has strong associative links for most of us. Thus a slow pace suggests dignity, contemplation, deep emotion, sleep, labour, etc., whereas a fast pace conveys speed, vigour, excitement, confusion, mechanization, modernity, youth, ease of achievement, etc.

The nature of the occasion determines the pace we seek. A religious programme would deviate only slightly from a slow average pace, while a variety-show demands a fast general tempo, slowing only occasionally. Any well-balanced programme is continually readjusting its pace. A constant, rapid pace is exhausting, while a slow, sustained tempo becomes dull. Pace-variation provides both changes and dramatic opportunity.

Pace comes from an accumulation of factors. Initially it is implied in the written script: the lengths of scenes, of speeches, of phrasing

336

FACTORS CONTRIBUTING TO THE EFFECT OF A SINGLE SCENE.

Fig. 18.3.

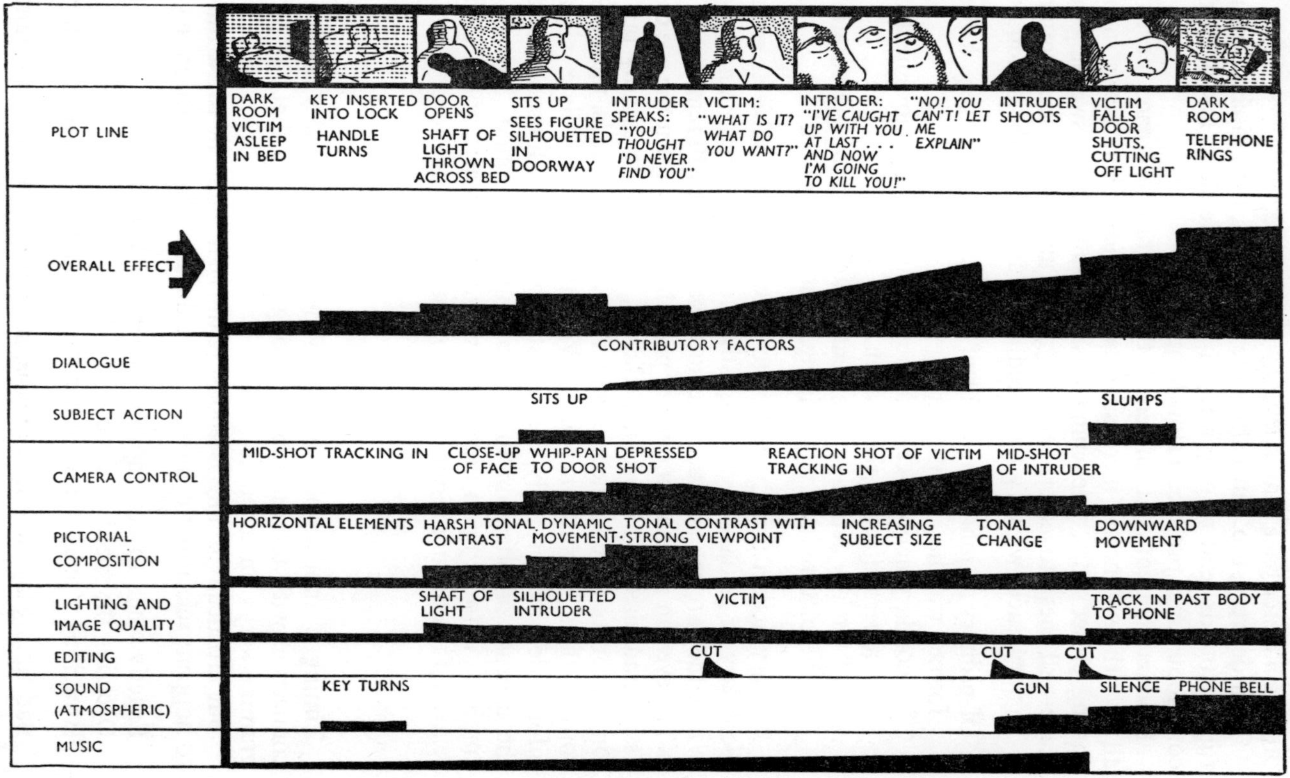

PLOT LINE

- DARK ROOM VICTIM ASLEEP IN BED
- KEY INSERTED INTO LOCK / HANDLE TURNS
- DOOR OPENS / SHAFT OF LIGHT THROWN ACROSS BED
- SITS UP / SEES FIGURE SILHOUETTED IN DOORWAY
- INTRUDER SPEAKS: "YOU THOUGHT I'D NEVER FIND YOU."
- VICTIM: "WHAT IS IT? WHAT DO YOU WANT?"
- INTRUDER: "I'VE CAUGHT UP WITH YOU AT LAST! AND NOW I'M GOING TO KILL YOU."
- "NO!, YOU CAN'T! LET ME EXPLAIN"
- INTRUDER SHOOTS
- VICTIM FALLS / DOOR SHUTS, CUTTING OFF LIGHT
- DARK ROOM / TELEPHONE RINGS

OVERALL EFFECT

CONTRIBUTORY FACTORS

DIALOGUE

SUBJECT ACTION
- SITS UP
- SLUMPS

CAMERA CONTROL
- MID-SHOT TRACKING IN
- CLOSE-UP WHIP-PAN TO DOOR SHOT
- REACTION SHOT OF VICTIM TRACKING IN
- MID-SHOT OF INTRUDER
- DOWNWARD MOVEMENT
- TRACK IN PAST BODY TO PHONE

PICTORIAL COMPOSITION
- HORIZONTAL ELEMENTS
- HARSH TONAL CONTRAST
- DYNAMIC MOVEMENT; STRONG VIEWPOINT
- TONAL CONTRAST WITH DEPRESSED FACE
- INCREASING SUBJECT SIZE
- TONAL CHANGE

LIGHTING AND IMAGE QUALITY
- SHAFT OF LIGHT
- SILHOUETTED INTRUDER
- VICTIM

EDITING
- CUT
- CUT
- CUT
- CUT

SOUND (ATMOSPHERIC)
- KEY TURNS
- GUN
- SILENCE
- PHONE BELL

MUSIC

337

of words. Sharp, snappy, exchanges between several people will provide a faster pace than a lengthy monologue.

Their delivery, too, must affect the pace. Fast, high-pitched, interrupted sounds tend to give a rapid pace, while slow, low-pitched, continuous ones produce a slow tempo.

Production treatment influences the programme's pace most of all: whirling dancers, increasing camera-movement, faster intercutting, closer shots, quickening music, pulsating lights sweeping the scene—all these together, cause a greater and greater momentum.

To vary pace, we vary the speed and duration of the action:

Of primary action—movement of the picture subject (e.g. a performer walking, lighting a cigarette).

Of secondary action—camera movement (i.e. any movement of the camera affecting the image).

And of tertiary action—movement by adjusting the presentation of the picture itself (e.g. editing, lighting changes).

Primary action usually dominates, but for static subjects either secondary or tertiary may be equally effective in determining pace. Where they are too subdued, the aural pace alone will remain.

The eye can maintain a quicker pace than the ear, for the eye can assess and classify almost immediately. The ear necessarily takes longer, for we have to piece together consecutive sounds, until they convey their overall meaning. We therefore assimilate visual information more rapidly, but a fast visual pace will usually be at the expense of attention to the accompanying sound. So, when emphasis is to be upon the sound, visual pace generally has to be relaxed if the picture is not to claim most of our attention.

Timing

Timing is another of those multi-meaning terms. We use it in connection with the duration-checking that ensures that the programme keeps on schedule.

But artistically timing is choosing the right moment for a thing to happen; selecting the most effective duration for an action, e.g.:

Exactly when to cut, move a camera, or a performer.
Deciding the speed of a mix.
How long a pause between a comment and an answering retort.

Bad timing lays wrong emphasis, so that the relative importance of events is distorted, and their continuity disrupted.

Filmic space and filmic time

Motion pictures have used the conventions of filmic space and time for so long that to many viewers these are now natural methods of story-telling. Both are techniques enabling us to omit irrelevant intermediary detail; to make rapid transitions in space and time, showing only the key-action. Through them we can sharpen pace, and follow parallel action in different locations.

In filmic space we show in juxtaposition action that is going on simultaneously at different places (e.g. a cut from a soldier, to his wife back home).

In filmic time two events occurring at different times are shown happening in immediate succession. Cutting from a shot showing a car drawing up, to the driver entering his apartment, will be readily accepted these days. Indeed, if we had shown all his intermediary walk without purpose, the viewer would resent this affront to his perspicacity, or lose interest.

In film, time and space can be rearranged at will on the editor's cutting-bench. Whenever we film television productions, or tele-record them they can be similarly edited. But for live-action shows, or for continuous shooting, people and equipment must have time to move into position; so, wherever filmic space or filmic time require anyone to move in an instant to another scene, we must find some means of making the impossible practicable.

There are several standard solutions here, to give performers or cameras time to move to another position.

Filming all or part of one of the scenes (especially to allow costume or make-up changes).

Having bridging or cut-away shots between scenes.

Providing covering-action or business to fill-out the start or finish of a scene during the move.

Where conversation is involved, having the performer who remains continue talking (in close-up), as if the other were still there, any replies by the absentee being provided by recordings.

Occasionally we may duplicate the performers themselves, by using twins or doubles. But where the performance is obviously live, the viewer is liable to be left wondering, instead of attending to the programme.

By conjoining the successive sets—regardless of their supposed locale—we can apparently achieve filmic time as the performer passes from one to the other.

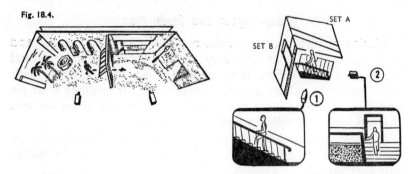

Fig. 18.4.

SET A

SET B

FILMIC TIME. *Left:* Filmic time can be achieved by conjoining the sets of two consecutive scenes (although spatially unrelated in the story-line). *Right:* Here two adjacent backings (e.g. back-projected) are intercut. Shot 1 shows subject about to climb the staircase; in shot 2 he has arrived at the stair-head.

Most important, the actor concerned needs to be fully aware of the production's mechanics throughout, so that he always knows exactly where he is supposed to be, and, at the same time, on which camera.

Time lapses

How can we indicate the passage of time in a dramatic presentation? Captions or spoken announcements are unambiguous and direct, but they are on the other hand unimaginatively primitive.

Numerous universally-appreciated conventions have evolved, via motion pictures, that will cover most occasions. Some of these are now established "grammar"—others have become somewhat threadbare.

Short time lapses (seconds, minutes, hours)

A slow fade-out to blank screen, slowly fading-in the new shot.

If we cut away from a scene and return to it later, one assumes that time has elapsed. The intermediate scene will usually imply how long—unless dialogue or pictorial changes suggest otherwise.

A time-indicator (e.g. a clock, burning candle, hour-glass), either showing its transition in quick motion, or at progressive stages.

Lighting changing with passing time (e.g. a room that earlier was sunlit is now dark).

Occasionally, a director arranges to have a setting's lighting-atmosphere changes before our eyes, as he holds a long shot of the scene. Daylight quickly fades ... the moon rises. A blatant artifice, but one which we manage to accept as such; providing he does not over-stretch credulity by having fires suddenly light themselves, room lights switch themselves on, and similar improbabilities.

Showing a subject before and after use (e.g. dissolving from a freshly-laid fireplace to the fire's spent ashes).

By transition between sounds that have time associations, e.g. from night noises (owls, frogs) to morning sounds (bird song, roosters).

The value of other devices like the ripple-dissolve, defocus-dissolve, or motionless-shot, to suggest time passing, is open to question. The last may be successful, when accompanied by suitable sound (e.g. clocks chiming). But the dissolve more usually indicates flashbacks in time.

Longer time lapses

A time indicator (e.g. a calendar losing its leaves, or changing date).

Seasonal changes (e.g. from spring flowers to winter's snow).

Showing changes in personal appearance (e.g. differences in wardrobe, beard-growth, wrinkles, etc.).

Age symbols—dissolving from a symbol of one age to that of another (e.g. from a boy's catapult ... to his man's rifle).

New to old (e.g. dissolving from a freshly printed newspaper, to its crumpled, discarded, self).

Flashbacks (reversed time)

The idea of going back in time, to see events that happened before our present action, is now a familiar narrative device. Flashbacks are mostly introduced by reversing normal time-lapse conventions, or through dialogue followed by "mist", ripple-dissolves, or defocused-dissolves.

Continuity problems

When filming, one photographs the planned shots in whatever order is most convenient and economical, later fitting them into their respective places during editing. This poses continuity problems, to ensure that intercut shots match in every detail.

341

Continuity difficulties in television can take several forms:

Studio shots must match film inserts, back-projection, etc., in lighting, costume, atmosphere, where continuity requires.

Make-up and wardrobe need to develop with the story-line (i.e. ageing, costume-changes, etc.).

Any visible indications of time passing have to be watched for continuity (e.g. burning candles, clocks, seasonal flowers). The same bouquet has remained untouched for centuries in a play before now! Furniture, drapes, and properties, too, have to change with time, to convince the observant viewer.

Rain or snow pose regular continuity hazards. Artificial snow on travellers' clothing does not disperse indoors. Furthermore, an exterior setting that requires authentic-looking rain or snow for one sequence, and parching sunshine immediately afterwards, will stretch staging ingenuity pretty far.

Sound from a crowd in one set can inadvertently hang over into the next scene—unless cueing is precise.

Where established characters leave or enter a scene unknown to the viewer, he may be disconcerted at losing or discovering them in later shots. All exits and entrances should be positive therefore, unless for a specific effect.

Cut-away shots (intercut shots)

The ill-advised adage that "the camera must show what the audience wants to see" is widely held. But it is sadly short-sighted, overlooking many productional opportunities. It robs the viewer of the satisfaction offered by tantalizing, indirect methods.

Far better to show him what he has been persuaded to want, to satisfy wants that we ourselves have created, and give greater enjoyment into the bargain.

During a football game, cameras turn from the field to watch the crowd's reactions for a moment. We can hardly say that the audience want this cut-away shot. It is a visual break that keeps their anxious enthusiasm. An attention-gambit that heightens audience interest.

Providing the viewer does not feel that he is missing vital information, cut-away shots from the main action offer some useful functions:

Permitting shots that are unmatched in continuity or action to be intercut, inaccuracies being disguised by disruption.

Effecting a time lapse, time having passed during cut-away.

Suggesting the environment of the main action.

Providing suspense; withholding knowledge of the main action builds tension.

Showing reactions. Reinforcing or guiding our response to the main action.

Preventing interest waning during a lull in the main action.

Allowing time for alterations of staging, mechanics, make-up.

Permitting intercut "nod shots" (post-interview "reactions").

Reaction, partial and cut-in shots

It is the art of concealing, rather than the revealing, that primes the viewer's imagination and arouses his curiosity. We shall often get greater impact by concentrating on an oblique aspect of an action, or its effect, than by showing the happening itself—A door opens ... the victim looks up ... horror spreads over his face, as he sees the intruder.

To strengthen tension, we might never see the intruder; but hold this reaction-shot of his victim.

Alternatively, we could cut to a partial shot showing the intruder's hand flicking open a knife, while the victim's frightened voice cries out to him ... and then is silent.

Again, we could take a cut-in shot of the victim's cat unconcernedly drinking milk, while the victim's frantic pleading goes on in the background. The victim falls—and his hand upsets the dish.

This business of showing by implication rather than direct reportage, provides enormous opportunities; often allowing the maximum effect with minimum facilities. But its purpose is to intrigue and tantalize. All too easily it frustrates and annoys, if ineptly introduced.

Creating atmosphere

Some pictures do little more than impart straightforward information. Others convey strong atmospheric qualities. Much of this atmosphere comes to us through associative ideas. Ideas rooted in symbol, convention, legend and so on, that may bear little or no relationship at all to reality.

Think of the "witching strokes of midnight": the hour of the uncanny, when evil and terror creep abroad. Hooting owls, thunderclaps, forked lightning, have all become familiar symbols of nocturnal "dirty work". Yet, in life, the basest of crimes are perpetrated in broad daylight. But courageous—or foolhardy—is the director who defies convention, and has sinister conspiracy under a blazing sun, or comedy on a nocturnal, storm-swept moorland. He may succeed, but the odds are against him, for such surroundings have strong atmospheric associations.

Radio has proved conclusively the potentialities of sound for creating atmospheric effects. The contention that the ear is more imaginative than the eye has firm foundations. Many associatory audio effects—such as the screeching seagull, depicting seascapes—have been so over-used as to become clichés.

Still, even clichés have their value, for they set a situation rapidly and unerringly, especially in a fast-moving production, where we want to convey location quickly, without drawing undue attention to it. Similarly a stock character has the advantage of merging into the background, avoiding distraction from the principal players. But all too easily, clichés become a substitute for imagination.

Atmosphere springs from cumulative details. But, as a picture can become fussy and overcluttered, so can we overdo these details. The country scene, accompanied by sounds of birds, wind, rustling leaves, babbling brooks *ad infinitum*, may exercise the effects man, but diffuses the audience's attention.

Lighting and atmosphere

We cannot discuss atmospheric effects for long without recalling how completely lighting influences the appearance and mood of the scene. The persuasive techniques of mood-lighting stir our imagination by:

Concentrating our attention—inducing us to look at particular aspects of the scene.

Revealing facts—letting us see the form, texture, surface-design, etc., of certain subjects.

Concealing facts—preventing us seeing other detail; subduing form, texture, etc.

Creating associations of light, by its direction, intensity, and distribution recalling certain luminants and their environmental moods (e.g. moonlight, candlelight, firelight, etc.).

Creating associations of shade—shadow formations recalling certain subjects, their environment and associative moods (e.g. of people, tree-branches, window-shutters, ornamental ironwork).

Mood-lighting is not engendered, as we should expect, by directly imitating natural effects. Copying the position and direction of a scene's customary light sources with studio lamp counterparts may simulate reality, but will not stir the imagination, for the spurious and ugly effects that a real environment can produce are less acceptable under the camera's scrutiny, especially in multi-camera production. The performers and the setting have to be lit selectively, emphasizing the aspects that are suitable to the mood or situation we are seeking, while suppressing others.

Perhaps even more interesting than the impressions conveyed by environmental lighting is the flexibility that lighting affords. Flexibility that enables us, during a scene, to change the prevailing mood, alter the subject's appearance, reposition one's attention, modify visual pace (by lighting fluctuations).

A scene opening in a gay atmosphere takes on a sinister aspect and finally becomes horrific. Lighting surreptitiously changes. Unobtrusively, the high-key treatment becomes more contrasted; shadows creep, until eventually underlighting dominates.

There may be no true environmental justification for these changes, although we might usefully help matters by providing one wherever possible (e.g. as daylight fails, the flickering firelight becomes prominent). But even where the entire readjustment is bogus, it will still remain convincing if properly handled.

Through animated lighting we can introduce life and movement into a scene where, in fact, little exists. It may serve as a background to other happenings, or dominate them, as we choose. For example, rain running down a skylight throws streaking shadows into the room below; fluctuating light and shade pattern the interior of a moving automobile; a room is lit only by the rhythmical glare from a flashing street sign.

In an instant, we can change the entire mood or significance of a scene:

We may reveal information that has hitherto been hidden: a nocturnal prowler is suddenly illuminated by a passing automobile's headlamps. We see that he is not the person we thought.

We can conceal information that has hitherto been visible: in a brawl, the room light is broken. The fight continues in darkness. We no longer know who is winning.

And, finally, we can transform its appearance, e.g. by room-light changes, introducing luminants, opening window blinds, etc.

345

Fig. 18.5. Pt.1.

CHEATED SUBSTITUTES. This sequence of illustrations shows a fugitive climbing a cliff-face, seeking escape from his pursuers. The fugitive looks up at the cliff-face. Hearing his pursuers, he decides to climb. We have a shot of the possible route, followed by one of running feet.

Fig. 18.5. Pt.2.

The fugitive seeks a handhold then, turning to see if his pursuers have seen him, he climbs. They stop at the cliff-base, his foot dislodges a stone, and they look up ...

Fig. 18.5. Pt.3.

And the studio set-up for this sequence. *Left:* A gravel-strewn floor. *Centre:* A photo-caption of a cliff-face. *Right:* And a surface-contoured flat, over which the fugitive crawls. The shot is canted to suggest vertical climbing.

Spectacle

As early motion pictures became ambitious, bigger and ever better spectacles filled the screen. Volcanic eruptions, battle scenes; all the pageantry and might of nature and man were drawn upon for these ends. Gradually, ingenious methods evolved that enabled the director to achieve such effects without the expense or danger of direct imitation. These methods were applied in turn to other, less imposing occasions, to reduce the time, expense or effort of normal setting construction.

Admittedly, one can avoid difficult situations altogether, by having them occur off-stage, or having someone tell us what happened. But these are unsatisfying evasions.

Assuming that we have decided on a replica of the event, several methods are open to us:—

(i) A film of the actual scene or event.

(ii) **Direct imitation**—building an accurate replica (complete or partial).

(iii) **Indirect imitation**—through process work (back-projection, models, electronic-insertion, etc.), probably combined with built setting.

(iv) **"Cheated substitutes"**—juxtaposing carefully chosen shots of partial settings (see Chapter 8).

How do these methods compare?

(i) Filming is expensive, even when the subject and situation is available. Stock (library) shots are economical, but may not be exactly suitable.

(ii) Direct imitation is usually costly, and may require considerable skill to look convincing.

(iii) Indirect imitation through process work can provide complete realism, but may necessitate restricting subject or camera movement.

(iv) The cheated substitute provides interesting possibilities. Here a series of representative parts are intercut to convey a composite impression. Ingenious, imaginative and most economical. But this method fails if the component parts are not completely integrated.

A detailed example in Figure 18.5 shows how simply and effectively quite an elaborate sequence can be staged.

Non-associative settings (see Chapter 8)

Most television staging aims to suggest some definite atmospheric environment. For many programmes, however, non-associative, neutral backgrounds are less obtrusive and more suitable, and for these, plain, even-toned, surfaces are widely used.

Black backgrounds often find favour where complete subject-isolation is wanted. In a play using *Limbo shots*, the entire action may take place against a black background; augmented, perhaps, by essential furnishings, properties and built pieces (e.g. doorways). Enthusiasm for such treatment is not universal. Although frugal, the result is not altogether satisfactory, for its effect tends to vaccilate too easily between naturalism, fantasy and abstraction. Moreover, its sombre gloom can be out-of-key with lighter moods.

Black backgrounds present practical problems, too, for dark hair and clothing merge readily into the darkness. The eye tires with continued viewing, and many television receivers reproduce the blackness as varying values of grey.

White backgrounds, too, have had their vogue. They spatially isolate, appear ethereal, and reproduce well enough. Fashion parades have that glossy magazine look. But again, sustained viewing becomes tiring. And subject-tones may appear unduly darkened.

Crowds

Economies aside, limited floor-space and set-size prevent most studio crowd-scenes from using much more than a score of people. Recorded crowd-noises may augment our fragmentary group. We may even avoid the issue altogether, as with spectacle, by off-stage action or narration. But where a crowd is essential, various subterfuges have been tried, in order to conjure up a host from a handful of people. Each has evident weaknesses, but they can all be surprisingly convincing if wisely used.

Film in television studio production

Using high-grade telecine equipment, we can televise film with a quality equal to—often exceeding—that from the studio cameras. How widely we can use film productionally rests largely with the design of our apparatus. Some machines will not allow us, for instance, to run up to full-speed in vision, from a held-frame. So, if we need a still picture to burst into life, the film itself would have to be specially printed.

By means of film inserts, or by using sophisticated video-recording facilities, one can extend studio production techniques in various ways:

Enabling scenes and events remote from the studio to be included.
Providing illustrative material.
Increasing the extent of settings.
Reducing the number or size of studio sets required.
Increasing the number of shots available (where these camera set-ups are impracticable, or inconvenient in the studio).
Providing material that might be unsuccessful if enacted live (e.g. a critically-thrown knife).
For video effects (time lapse photography, reversed action, speed changes, appearances, transformations, etc.).
For animation sequences (i.e. cartooning, animated still-life, etc.).

Fig. 18.6.

CROWDS can be simulated by using selective viewpoints (*left*) and carefully positioned subjects crowded together along the lens axis (*centre*). Also by augmenting subjects with a background of a crowd scene (moving back-projection, photo-mural, painted cloth) or using dummy or cut-out foreground figures (*right*).

To include performers not otherwise available.

To allow time for costume, make-up, or setting changes.

To enable a performer to be in two places at once; for space/ time illusions.

Routine presentation

For many directors, their main preoccupation is in finding and organizing their programme-material. Its presentation to them is a comparatively routine matter. They have stock approaches, which they apply as convenient. For several types of programme, this is understandable, e.g. panel-games, interviews, talks, sporting events, —or where last-minute items have to have a fool-proof presentation method framed to meet them.

Unhappily, this attitude can spread further afield, and may be applied (unwittingly, perhaps), to theatre, variety, magazine programmes, even drama.

Creative presentation is rare in nearly all mechanico-artistic media. And perceptive creation even rarer. Still, despite one's artistic aspirations, we have to face the fact that certain subjects can only be presented in a very limited number of ways, short of eccentricity.

Supposing we are to show a pianist performing. How far can we reasonably vary our shots? Discouragingly little, if we want meaningful, attractive, pictures (Figure 18.7).

Again, when we have a single speaker, or an interview, the choice of shots is disappointingly restricted (Figure 18.8).

So, when we encounter what looks like routine presentation, we have always to ask ourselves how far it is avoidable, and how far it is inherent in that type of programme material.

When a motion picture director wants to tell us something, he usually does so through continual illustration; a commentator perhaps describing and discussing the situation we see before us. Live television has developed simpler, more economical, but less enticing approaches. People all too often talk about the subject instead. People sit and talk to us; or they sit together and chat; or one interviews another.

Unfortunately, the attraction of the human face is limited. So, unless he happens to be particularly animated, we shall seldom be prepared to concentrate interest on a speaker's face for any length of time, however much the viewpoint is changed for variety's sake. There are those who feel this visual boredom allows the viewer to concentrate upon the dialogue instead; but their hopeful confidence is seldom borne out in practice.

Opening routines

Introductory captions completed, we have now to open the programme. But how? There are half-a-dozen basic methods regularly used.

The formal welcome is amongst the commonest—and the dullest. After a "Good evening, ladies and gentlemen ...", or a friendly "Hello", a benign introducer tells us what he has in store for us. At least it gets on with the business in hand, and let us know what to expect.

The crash-start dismisses preliminaries and takes us straight into the programme, which probably appears to have started already. In dramatic form, the crash-start puts the viewer on the edge of his seat from the onset: a brick shatters the shop window, a hand grabs at the tray of jewels, an alarm-bell shrills, a police siren wails. The show has begun!

In the eavesdropping approach we join the action surreptitiously, perhaps by looking through a window, then going into the room. Watching a group round a television receiver, and tracking-in to its screen. We may look at a photo-album that comes alive.

Then there is the coy welcome; where we track up to someone who is supposedly preoccupied. Suddenly he realizes that we are there. His welcome ranges from the "So glad you could come" to

350

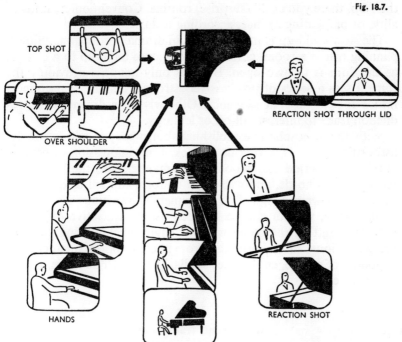

Fig. 18.7.

TOP SHOT

OVER SHOULDER

REACTION SHOT THROUGH LID

HANDS

ALONG KEYBOARD

REACTION SHOT

PIANO RECITAL. The basic range of pictorially attractive shots is limited when presenting a piano recital.

Fig. 18.8.

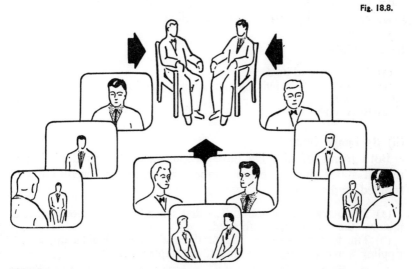

FORMAL INTERVIEW. In a formal interview the basic range of pictorially attractive shots is limited.

the "Oh, there you are!" (surprise) routine. Conventionless informality, or nauseating artifice, according to taste.

The slow build-up is mostly found in drama productions. Slowly panning round a room, curiosity or suspense build-up to a climax, as the door is thrown open. But this only happens providing the camera sees a series of progressively-stronger stimuli, that connect into a logical thought sequence. Otherwise, the drawn-out introduction leads to interest fall-off or lethargy.

With the atmospheric establishing shot, we open with strongly-associative symbols, giving us a quick, firm, impression of personality, place, period and mood. On a mantle-shelf, the brass telescope, ship in bottle, and well-worn uniform cap tell us quite a lot about their owner, long before the camera shows him.

Pictorially, we can choose between using associative objects of this kind, setting the scene with an environmental long shot, or using locational shots—probably from photo-captions or stock film.

Sound, of course, contributes similarly, through dialogue, mood-music, or effects.

Inter-scene transitions

One scene ends, we pass to the next. There are several methods available to us here to convey the idea of transition. Some of them have become unobtrusive conventions. Others, like the matched-dissolve, verge all too closely on the cliché.

(i) By editing

(a) THE FADE, and its derivatives.
(b) BY CUTTING—providing narration or action have heralded the scene-change.
(c) BY A WIPE.

(ii) By image control

Both forms of dissolve have become accepted as depicting a mental state: the act of remembering, dreaming, becoming befuddled, and so on.

(a) THE DEFOCUSED-DISSOLVE. The first scene becomes defocused, and mixes to the second, also defocused.
(b) THE RIPPLE-DISSOLVE. The first scene's image breaks up, as by rippling water, until indiscernible. (See Chapter 20.)

(iii) By intermediary elements. By interposing a suitable inter-

mediary link between two scenes, we can suggest a change of time, place, or mood, and indicate a relationship that a more direct transition would not give.

(a) BLACKOUT. A performer moves up to the lens, blocking it out. He (or another subject) moves away again, revealing a new scene.

The camera pans into deep shadow; panning out into the new scene.

(b) HELD SUBJECTS. A balloon is held to the lens. We mix to an identical balloon, which is burst to reveal the new scene.

The camera turns away from the main scene, to look at part of the setting; turning back to show that its appearance has changed. Time has passed.

(c) BRIDGING SUBJECT. Linking material forming an associative connection between two shots—e.g. turning car wheels, to imply travel.

Atmospheric transitions—e.g. swirling mist.

(d) DECORATIVE DESIGNS. Abstract forms—e.g. light reflections, shadows. Moving patterns—e.g. kaleidoscope, spirals, whip-pan.

(iv) **By matched shots.** Ending the first scene with a close-up and mixing-through to a matching shot of a similar subject in the new scene. Then tracking-back to begin the action.

Background music and effects

Where background music or sound effects are clearly motivated, they are accepted readily enough. Criticism generally comes when the viewer considers them as not factually justified, or as unnatural.

All art-media deviate from pure realism. That is part of their appeal. They use convention to underline, augment, or to circumvent lengthy explanations. But all conventions are liable to misuse and misunderstandings.

Background music and effects are normally misused when they are too loud, hackneyed (e.g. too familiar a theme or orchestration), over-obvious (e.g. "mickey-mousing", where every action has its musical imitation), obtrusive (surging into the slightest gap in dialogue), out-of-scale (e.g. over-scored music), or suggestive of wrong or misleading associations.

When discreetly applied, however, background music has very practical purposes, achieving a dramatic significance that natural sound-effects cannot obtain, providing an aural setting for visual subjects having no naturally-associated sounds (e.g. statues, flowers), evoking moods or mood-changes, building up a definite dramatic effect (Chapter 17), or providing a transitional bridge between differing moods, pace, etc.

19

GRAPHICS—TITLING—CAPTIONS

WHETHER written, printed, pictorial, diagrammatic, or sheer design, graphics have a place in almost every programme.

Titling

Good titling attracts, informs and excites interest. It is neither confusingly decorative, nor severely characterless. It is compact without being crowded. Successful lettering establishes atmosphere. A style suited to the dignity of a formal talk would be starkly misplaced in a gay, vaudeville background.

All captions need substantially 4 by 3 proportions to fill the screen. A surrounding safety-border prevents our cropping important material. The 12 in. × 9 in. title-card meets most needs, although more elaborate art-work may require caption cards up to twice this size; 35 mm. caption slides are more convenient but not readily recomposed (unless screen projected and shot by a camera).

For good legibility, 1/10 to 1/25 picture height is often recommended as minimum lettering size. Smaller lettering is still just readable (even to 1/60th), but lacks impact value. It may strobe badly.

Fig. 19.1.

TITLING can be varied in style to suit the occasion. Clarity and simplicity are the prerequisites of good titling, particularly against complex backgrounds.

Fig. 19.2.

CAPTIONS set the scene; but there are captions ... and captions. There is the half-informative opinion, giving us direct information, but only completely meaningful if we realize the significance of its data. There is the redundant caption, telling us the same thing twice. And the caption that tells us enough, without stressing the obvious; requiring us to interpret, and giving us imaginative pleasure in the process.

Explanatory captions

Time and location. Costume, architecture, set-dressing, etc., usually provide visual clues keying the viewer rapidly into the time and locale of an action. Occasionally, this may prove too long or too ambiguous, and a brief explanatory caption is obligatory.

Story captions. The bygone film practice of prefacing a story with a printed preamble dies hard. Still used occasionally, it has seldom aroused much audience enthusiasm. Where such explanatory introductions are essential, a background commentary is far more readily absorbed. Better still, the facts are revealed through dialogue and treatment.

Subtitles and head-titles. Explanatory subtitles enable us to supply supplementary information, without interrupting the visual flow of the main picture. When adding them (by superimposition, or electronic-insertion), it is best to choose an area of suitably contrasting tones, where they will not conflict with important subject-matter. An electronic "edge-generator" may sharpen lettering outline.

Diagrammatic captions. A good diagram has much to offer. It replaces detailed description, clarifies obscure information, makes the dull palatable, correlates, scales-down unimaginable proportions.

Fig. 19.3.

TITLE POSITIONS. Titling is normally localized to achieve the maximum impact.

355

On the other hand, our diagram may attract interest without the viewer appreciating its meaning. In aiming at simplicity, there is the danger, too, that we may over-simplify or distort.

Pictorial captions. Whether used as pure illustration, to broaden location, or to get those occasional shots that the studio cameras cannot reach, the pictorial caption has a wealth of applications. Many programmes are entirely compounded of such captions, covering subjects from children's stories to serious studies of arts and sciences.

Photographs of three-dimensional scenes need discriminating handling, however. If we pan or track to look around them, we are liable to find realism destroyed. The photograph's flatness becomes apparent, for we shall see no parallactic movement or proportional size changes as in reality. (Remember, this is one of the shortcomings of the zoom lens on a real scene.) Worse still, the whole sequence may degenerate into an ill-disguised lantern-lecture.

Types of caption

Static captions
Lettering. Variation in lettering-style gives us a considerable range of expression (try to avoid fine detail, thin lines, close pattern):

> Lettering can take many forms, being hand-drawn, printed, or derived from adhesive or rub-off sheets. Lettering can simulate sculpted relief.
> Displayed as shadows, silhouettes, reflections, stencils. ...
> Built-up in solid or outline, using transparent, opaque, or translucent materials, wood, plastics, metal-strip, wire, rope ... even toothpaste.
> Arrangements of inanimate objects—e.g. sticks, pebbles, toy bricks, thumb-tacks.
> Traced in sand, snow.

To avoid the need for skilled painting or printing, we can use pre-constructed lettering. Stuck, pinned, or clipped, as required, titling is made up quickly and economically. But unless the lettering-set is well designed we are liable to get rather dull, mechanical-looking layouts; a limitation of electronically-generated characters.

The backgrounds. The backgrounds for this lettering can take many interesting forms. For instance:

Fig. 19.4.

COMBINING LETTERING AND BACKGROUND. As we see here, lettering can be combined with its background in a number of ways. But keep good tonal contrast.

Plain, even-toned, or shaded.

Textured surfaces—e.g. wood, pebbles, brick, corrugated or embossed paper, ornamental glass sheeting.

Textiles—e.g. satin, printed cloth, wallpaper.

Foliage—e.g. grass, leaves.

Pictorial—e.g. studio setting, symbolic objects, water ripples, cloudscapes. ...

Fig. 19.5.

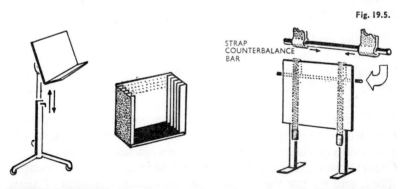

CAPTION HOUSINGS. We can support the caption-card on a tiltable shelf, which is self-aligning, but suitable only for small captions: in a slotted-box, also self-aligning, and intended for a series of pull-out captions; or on a strap easel, the weighted hang-over webbing straps of which can be adjusted to suit captions of all sizes—although they require individual re-alignment.

Captions in settings

Captions seem to have a particular attraction when they have been introduced as an integral part of the setting. They may simply be desk-charts, wall-maps, or projection screens set into a wall. Alternatively, we can introduce them as natural features: as titles on a book-shelf, book illustrations, suitcase labels, wall-posters; as street signs, or as drawings chalked on walls or pavements. They can also be attached to cushions, furniture, or clothing —in appropriate circumstances.

Fortunately, audiences accept these less formal methods of delivering caption information, without much hesitation, providing we take the precaution of avoiding its being seen accidentally, as a distraction in other shots during the course of the production.

Animated captions

Animation attracts. It brings the caption alive, awakens our interest. (If we are not careful, interest is captured by the ingenuity of the animation!) We can get movement by simply intercutting or superimposing different captions—or by panning or tracking the camera. Skilfully applied, a combination of zooming and camera movement can bring astonishing life to a detailed illustration.

Then there are visual effects such as wipes, spins and ripples writing the caption in vision, having snow, rain, or smoke pass in front of it, playing action in front of the caption (e.g. puppets), or even setting fire to it!

But that is only the start. In animation, ingenuity knows no bounds. Let us take a look at the animated caption devices one meets in most television studios.

The roller caption. The work-horse of daily production, the roller caption machine itself may be anything from a hand-operated "mangle", to a motor-driven, variable-speed, rotatable de-luxe robot. For all such *'travelers'*, an appropriate reading speed is essential.

ROLL TITLES: Give us a continuous, unbroken, stream of information. But as lengthy titling quickly becomes monotonous, to add visual variety, the principle of *crawl-lines* has been introduced.

CRAWL-LINES: Reposition the screen area in which the eye concentrates, and we are thereby helping to overcome feelings of visual restriction. Whether shown as an undulating decorative motif, or insinuated by the lettering's layout, crawl-lines increase audience interest, especially where tone or pattern are varied too.

THREE-DIMENSIONAL TITLE. Titling can be incorporated in a three-dimensional display, which the camera explores by trucking past its various sub-sections.

Fig. 19.6.

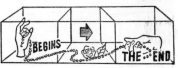

Fig. 19.7. Pt.1.

ANIMATED CAPTIONS. *Left:* The Slide provides a push-over wipe. *Centre:* Rotating boxes. *Right:* A loop of captions on a rotating box can be used either as a flop-over or vertical slide, depending on camera position.

Fig. 19.7. Pt.2.

Left: Multi-stage turntable. *Centre-left:* Rotating strips. *Centre right:* Drum captions *Right:* Superimposed on a graphic with a central subject, lettering moves round it.

Fig. 19.7. Pt.3.

Left: We can stand lettering on the drum. *Right:* Or make it part of a small set-up.

Fig. 19.8. Pt.1.

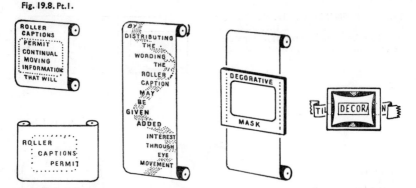

THE ROLLER CAPTION. *Left:* Roller titles. *Centre left:* Crawl titles. *Centre right:* Variations on the roller principle:—Proscenium title. *Right:* Decorative panel and pull-through strip.

Fig. 19.8. Pt.2.

A TRAVELLING PATH CAPTION. Can be devised from a horizontal roller caption.

Flips. Flips can provide the sharp transition of a cut, yet require only one camera for a complete set of captions.

Stencils (pierced lettering). Captions in which the lettering is pierced in stencil fashion provide the basis for several animation effects.

We can have the background seen through the lettering made by a moving pattern, e.g. a rotating disc of striped card, moulded glass, or interference patterns made by sliding together sandwiches of rippled-silk, wire-meshing, photographic etching screens.

With a stencilled caption, we can simulate unseen writing. Just stick a length of opaque tape along the path the pen follows in writing the letters, and rip off to animate (Figure 19.10).

Letter movement. We can animate the caption by having the actual letters move, having them slide-in, or pop-in; on springs; rotating; moving them with magnets, or forming the letters from movable subsections (e.g. a series of metal flaps).

360

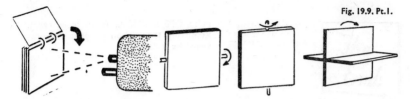

Fig. 19.9. Pt.1.

FLIPS. *Left:* In book form on ring-binders; as shown, flip-in (drop-in); or with binders at the bottom flip-out (drop-out). A version of this is the animated book, with a wire stay on the bottom of each page to facilitate turning the pages. *Centre left:* Flip-over: double faced caption with horizontal pivot. *Centre right:* Flip-round version has a vertical pivot. *Right:* Rotating-flip.

Fig. 19.9. Pt.2.

SHUTTERS. *Left:* Shutters flip-over individually or together to reveal a new caption, or a background beyond. *Right:* In a further version, a caption is seen through slots in a sliding shutter. This is then moved down, covering the first caption and uncovering the lines of a second.

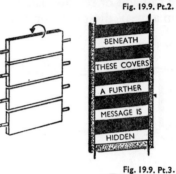

Fig. 19.9. Pt.3.

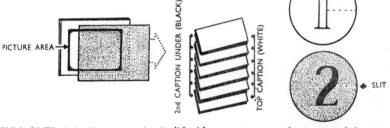

PULL-OUTS. *Left:* The top caption is slid aside, as an uncover-wipe, to reveal the next. *Centre:* One set of slats (white) is drawn up behind the other (black), covering up the first caption and uncovering the next, in a sectional wipe. *Right:* A split rotating flap disappears through a slit in the flap beneath, uncovering the next caption.

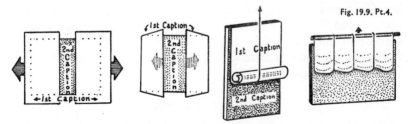

Fig. 19.9. Pt.4.

BREAK-AWAYS AND UNROLLED CAPTIONS. *Left:* Double pull-out. *Centre-left:* Double flaps. *Centre right:* Paper roll or spring-blind. *Right:* Curtains or venetian shade.

361

Fig. 19.10.

TEAR-OFF RIBBON
COVERING
STENCILLED LETTERING

PULL-OFF STENCIL. (seen from rear).
Removing the ribbon progressively reveals the lettering.

Multiplane captions. Here the caption is comprised of several transparent or cut-out layers, and by revealing or obscuring these layers, we can add or remove detail to order. So lettering, objects, design, can appear, disappear, or be transformed on cue.

Fig. 19.11.

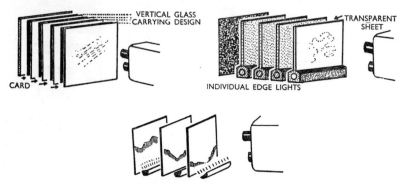

VERTICAL GLASS
CARRYING DESIGN

TRANSPARENT
SHEET

CARD

INDIVIDUAL EDGE LIGHTS

MULTIPLANE CAPTIONS. *Top left:* Multi-layer sandwich of painted glass and plain card. (Pulling out black background cards in order will reveal each stage underneath.) *Top right:* Similarly, as each transparent plastic sheet is lit (in any order), its drawn or engraved detail is revealed. By such means an acorn can be made to grow through sapling to forest giant. *Bottom:* The multiplane cut-out. The prominence of each cut-out can be varied by lighting.

Animation by lighting. Lighting can provide various animation effects.

Flickering light, or projected patterns, give the caption overall life. Changes in lighting colour can be used to make coloured detail appear and disappear, and to alter the size and shape of areas (Samoiloff effect). Flashing signs can be made up from separate lamps, or illuminated letters. But rear-lit stencils are equally convincing in most cases.

Constructing graphics

Graphics make statistics attractive and meaningful. We can show various complicated statistical facts in simple pictorial form, de-

Fig. 19.12. Pt.1.

ANIMATION BY LIGHTING. *Left:* Shuttering to localize attention. *Centre:* Illuminated spots to identify areas. *Right:* Shadows, to direct attention or create atmosphere.

Fig. 19.12. Pt.2.

The appearance of solid lettering can be altered by changing the lighting direction.

picting developments, proportions, relationships, so that the viewer can appreciate them quickly and easily.

Sometimes homely analogies are better than the most elaborate models. Consider the sun as a football, one planet the size of a pea a tennis-court's length away, and another a pin's head half a mile distant, and the enormous proportions actually involved really begin to mean something.

An endless range of materials can be pressed into service: coins, sand-piles, wooden blocks, etc. We have to be careful, though, to ensure that these analogies are strictly comparable. The same amounts of liquid in a tall, thin, jar and a short, fat vessel look entirely different.

The principles of animated captions

The exciting potentialities of animated film need no introduction. But film is relatively expensive for a once-only telecast, and does not lend itself to last-minute changes. What we often want, therefore, is to be able to obtain the effects of animated film by a cheaper and adaptable process. And animated captions have been developed in many studio centres that go a long way towards achieving these effects. (Video recorder single frame animation is not yet widespread).

They have had a variety of uses: moving maps, graphs, cartoons, diagrams showing the operation of machinery, parts of the body, scientific processes, to name just a few.

We have already met several of their constructional principles. The methods themselves are usually simple enough. The art lies in the skill and cunning with which these ideas are applied. Often we have several ways in which we can achieve the same effect.

What sort of things do we want to be able to do in animated captions? Well, fundamentally, we can summarize most of our needs as:

   (i) To remove or add information to a caption (detail, titling, etc.).

  (ii) To show movement (e.g. a wheel moving, a heart beating).

 (iii) To show flow (e.g. liquid down a pipe, goods down a conveyor belt).

 (iv) To show variation in flow (e.g. fluctuating pressure, density, etc.).

  (v) To show direction (e.g. the passage of a ship on a map).

 (vi) To direct attention to a particular detail.

(vii) To move detail about (e.g. showing a product passing through a machine).

(viii) To show changes in size (i.e. volume, amount, etc.).

 (ix) To show sub-surface detail, or sectional views (e.g. removing external casing to show machinery below).

  (x) To show enlarged portions of a subject.

There are more or less standard ways of going about each of these types of animation. And we can summarize them (cryptically, perhaps) as follows:

   (i) To remove or add information to a caption: **Rotate** part of the caption (boxes, drums, rollers). Drop solid or transparent flips over the original caption. Use pull-outs, stencils, multiplane, pop-ins, break-aways, unrolled-captions, lighting-animation, flaps. Manually attach new information by sticking, magnets, hooks, etc.

  (ii) To show movement: Have the moving surface planes hinged, pivoted, or in runners, on the background, and operated by tags, levers, strings, magnets.

 (iii) To show flow: Have the background cut-out in the areas requiring flow-treatment, and simulating movement by rollers, turning patterns, interference patterns, pull-outs.

 (iv) To show variation in flow: Change the density, tone, or speed of the flow-pattern.

  (v) & (vi) To show direction or to direct attention to a particular detail: Use stencilled pull-outs. Have the subject on a glass

364

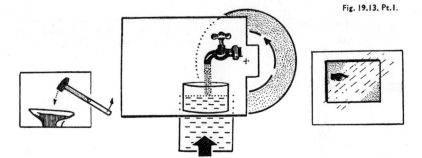

Fig. 19.13. Pt.1.

TYPICAL ANIMATION DEVICES. *Left:* Surface movement. *Centre:* Sub-surface movement. The falling water is simulated by a patterned disc appearing through a stencil, and the liquid rising in the bowl by a push-up card seen through a stencil. *Right:* Movable cover-glass. Pointer painted on movable glass sheet covering stationary background.

Fig. 19.13. Pt.2.

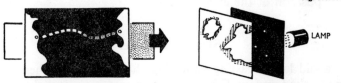

Left: Pull-out stencil. A pull-out, two-toned strip causes the stencilled route to become visible. This technique can also be used with lettering (see Fig. 19.10.) *Right:* Lighting Stencil. On the translucent map, stencil-marked areas appear bright. Individual areas can be operated by selectively blanking-off the stencil.

Fig. 19.13. Pt.3.

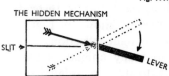

THE HIDDEN MECHANISM

The Pop-in. *Left:* An arrow suddenly appears in the target. *Right:* The arrow, affixed to the end of a rapidly-operated lever, pops-in instantaneously through a slit in the background.

sheet, which is slid over the background. Light movement. Rear magnet animation. Electronic insertion.

(vii) To move detail about: Use rear-magnets. Surface-glass. Tags or wires in background slots.

(viii) To show changes in size: See (i). Build the subject from a series of sub-divided parts, which are animated simultaneously. Intercut between captions showing progressive stages. Use coloured lighting changes.

(ix) To show sub-surface detail or sectional views: See (i). Remove the surface-area by unflapping, unsticking, sliding, etc.

Fig. 19.14.

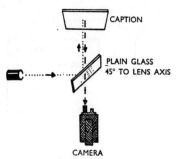

SHADOWLESS LIGHTING. Shadows on multi-layer captions can be avoided by using reflected frontal lighting.

(x) To show enlarged portions of a subject: Have a pop-in enlargement of the area. Electronic insertion, or superimposition of the enlargement.

Staging problems

To avoid layers casting confusing shadows upon others below, we need shadow-free lighting. Unfortunately, soft light sources seldom provide this. If we were able to place a point-source of light exactly where the lens is, the camera would see completely shadowless subjects. But in practice we cannot usually get the lamp close enough to the lens-axis. Instead, we can light the caption indirectly by the method shown in Figure 19.14.

Spurious light-reflections can mar multi-layer captions that use glass, so many animators use stretched gauze instead wherever possible, to support isolated surface areas.

Another well-tried device used to simplify animation is the glass table-top (Figure 19.15).

This table is made from a translucent sheet, or plain glass topped with a plastic, tracing-linen, or thin paper surface, upon which the background pattern is drawn. The details of any flat cut-outs laid upon it will be seen through on the other side, by the camera; while to make them disappear, all we have to do is pick them up.

Display screens

The display screen gives us an ever changing "blackboard" in the setting, upon which we can present all forms of graphical material. Still or moving back-projection on to a translucent screen in the setting wall, gives wide display opportunities.

SIMPLE ANIMATION TABLE. For correct reproduction, the mirror-image requires to be laterally reversed.

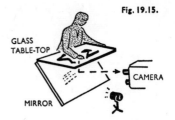

Fig. 19.15.

GLASS TABLE-TOP

CAMERA

MIRROR

Of lower intrinsic quality, but even more flexible, is the wall-housed television picture-monitor, for we can switch any video source on to its screen at choice. It is best to prevent the viewer from overseeing extraneous shots, film-leaders, and duplicate transmission pictures upon it. Otherwise his attention wanders. Nevertheless, this is probably the most convenient display method of all.

By suitable sound arrangements (hybrid circuits; audio switching; hearing-aids; beamed loudspeakers and mics.) we can have the studio speaker chat with others, via the monitor screen.

Whenever our television camera shoots on to a monitor's screen, certain unavoidable vagaries arise. (Aside, that is, from light-reflections, and tonal degradation from spill-light.) Inevitably, a horizontal light or dark shadow covers part of the monitor's picture, crawling vertically whenever we reposition the studio camera. One has to arrange it for minimum nuisance, and be content. Sometimes a photo-pulse effect is visible too. This is a faint ghost-image of the monitor's picture, superimposed over the whole shot. These defects may not be too serious—but forewarned is forearmed.

In colour systems, *chroma-key* (colour separation overlay) enables areas of other picture sources to be integrated into the master shot. A blue background or panel before the master camera will automatically insert telecine, animated maps, or work captions, into that area.

Colour Synthesiser

Colour can be generated "synthetically" in a television system, by controlling the relative proportions of the red, green and blue signals. Such colour can be used to convert monochrome captions and transparencies to coloured versions. With two-level systems, black reproduces as one selected hue, white as another. Switching produces instant colour changes (more elaborate versions provide half-tone switching for multi-hue transformations). Applied to normal colour or half-tone pictures, gimmicky "posterised" effects develop in the selected colours.

367

20

VISUAL AND AURAL EFFECTS

WE tend to expect effects to be spectacular or abnormal in some way. But effects devices are at their most exacting and interesting when we use them to recreate naturalistic illusions, and here, paradoxically enough, the more successful we are, the more self-effacing such effects become.

In this chapter we shall meet many methods and mechanisms, old and new. Some are widely-used tricks-of-the-trade, others are subject to patents, design registration and similar restriction.

Visual effects

Dry-ice—solid carbon dioxide. When highly compressed, carbon dioxide solidifies into a white, ice-like substance. This, in block form, is widely used as a coolant. Allowed to become warmer, the dry-ice evaporates slowly, giving off clouds of harmless white vapour as it returns to its natural invisible gaseous state. Quite cheap, and easily stored, the material is foolproof to handle. The only precaution necessary is to avoid prolonged skin contact, which causes blistering.

Small pieces dropped into water dissolve, bubbling with wreathing white "steam". Ideal for bubbling retorts, and magic potions, the effect is dramatic, but harmless. One can even drink the result; unlike the poisonous titanium tetrachloride often used for steaming liquids.

As the water becomes progressively colder, the dry-ice dissolves less vigorously. Hot water prolongs the action. But where we want continuous dense mist, the dry-ice is best placed on an electric hot-plate, or in a portable steam-oven. The heavy white vapour sinks rapidly and can be blown to represent swirling ground-mists, ethereal clouds around dancers' feet, smoke from burning rooms.

LYCOPODIUM POT. The lycopodium powder is blown through holes on to the burning wick.

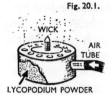

Fig. 20.1.

LYCOPODIUM POWDER

Because the vapour disperses fairly rapidly, continuous replacement is needed. This is just as well, for persistent clouds can be an embarrassment around the studio.

Fire. Although most mock fireplaces use perforated gas-pipes to provide flames, subterfuges are generally safer for more widespread fire-effects.

The lycopodium pot is a reliable safe controllable device, for its powder is only ignited when blown on to the wick. Sheets of flame can be generated on cue by regulating the air supply.

Wood-alcohol in jellied or liquid form enables torches, furniture, buildings, to burn fiercely; augmented, perhaps, with fragments of paraffin-wax or firelighters.

To avoid real flames altogether, we can superimpose filmed flames. Or, with care, smoke and flickering light alone will give the illusion. (Also see Figure 20.36.)

Water-tank effects. In a small water-tank we can contrive many strange and impressive sights. Substances dropped into the water (e.g. potassium permanganate), or stirred from its bottom, conjure up clouds, streaks of variegated tones, ephemeral patterns. With the picture inverted, whatever is dropped or squirted into the tank will seemingly rise from the ground—useful for explosions and volcanic eruptions. Water catastrophe, flooding, rushing torrents, are easily simulated. Lettering can dissolve to illegibility.

Glass partitions enable us to introduce several separately-controlled effects into the same tank.

Fig. 20.2.

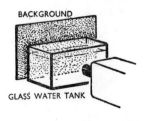

BACKGROUND

GLASS WATER TANK

THE WATER-TANK. Shot in big close-up the water-tank provides a number of visual opportunities.

369

Glass panels. An unbreakable glass panel in front of the camera will protect it against subjective assaults! It can effectively prevent such discomforts as thrown water, fruit, landslides, collapsing buildings, engulfing camera and cameraman.

Lecturers have used glass panels as invisible blackboards; standing behind, painting, facing the camera. The result is stuntish, and becomes bizarre and confusing unless suitably presented; for the panel-writing and the studio-scene can merge indecipherably.

Turntables. Ranging from hand-turned 6 in. miniatures to electrically-driven revolving stages, turntables have numerous applications. Horizontal turntables are invaluable where we want to display sculpture, models, etc., from various angles. Vertical turntables can spin titling, decorative shapes, or stills (e.g. to suggest crashing aircraft dizziness, and the like).

Passing backgrounds. To create the illusion that a background is moving past our stationary subject, there are several devices. A continuously-moving roller-cloth may suffice, providing it is soft-focused and without obvious detail.

We can make use of moving back-projection, electronically-inserted moving backgrounds. Walking in front of a moving background is usually dummied by marking-time. Treadmills or moving-floors are most effective, but seldom noise-free.

For night scenes, the illusion of passing backgrounds may be created by moving lights alone, using a cyclodrum or mirror-drum (Figure 20.13). A scenic projector (Figure 20.36) may also be used.

Fog and dense smoke. Smoke superimposed from film, or a smoke-filled glass tank (Figure 20.2), is convenient and controllable; but unconvincing where people walk, or the studio camera moves. Live smoke is invariably unpleasant, and liable to hang over into following scenes.

Special blow-torches releasing crude-oil vapour through heated tubes generate controllable clouds of dense smoke.

Fig. 20.3.

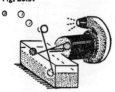

BUBBLE MACHINE. For a continuous stream of bubbles, the wire loops rotate slowly through a soap/glycerine solution. The film formed across them is blown into bubbles by the air-stream from the blower.

THE FLASH-BOX. (1) Low-current fuse wire and powder; (2) terminals; (3) asbestos board. Electric current melts the fuse wire, igniting the powder.

Fig. 20.4.

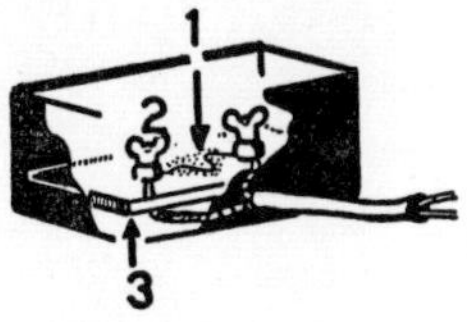

CABLE TO
ELECTRICAL SUPPLY

Smoke-producing pyrotechnics are ignited by slow-fuse or electrical contacts, but they burn persistently until exhausted.

Dry-ice provides innocuous "smoke", which disperses readily.

Smoke from compounds such as metafuel cubes (spread on to a hot-plate), or exposed titanium tetrachloride, is sometimes suggested, but is highly distressing to breathe.

Flash-boxes loaded with naked-ignition smoke-powder and magnesium flash-powder, produce bursts of smoke on cue.

Wind. For realistic wind, an assembly of vari-speed generating devices are available, from domestic fans to huge, encased airplane-type propellers. Most wind machines are bulky and highly directional. Moreover, by the time their breezes are strong enough to register, dust and paper are airborne, and microphones rumble. Localizing the wind may reduce these hazards. Often better still, we simulate the effect instead. Fanning with a hand-held board, or attaching cottons to waft drapes or blow-away hats, can work wonders.

Cobwebs. A special gadget, allowing rubber-solution to leak in a fine jet on to rotating fan-blades, produces cobwebs that would satisfy a spider. The thrown filament is normally spun on to a framework of stretched black cotton threads. A poorer substitute is devisable by squeezing rubber-solution between two boards, and drawing them apart. Dusting with Fuller's earth makes the cobwebs visible.

Gunfire and bullet-holes. Real gunfire being unpopular, bullet-holes are cunningly contrived by pre-drilled holes, plugged with dowel-rod and dust-packed, strings or springs remove them on cue or else by electrically-ignited charges in dust-packed holes.

Ground impact-spatter may be simulated by buried charges or, more conveniently, by a compressed air system (Figure 20.5).

Partial disappearances. If we place a subject before a black background, any part we then cover in black will disappear.

371

Fig. 20.5.

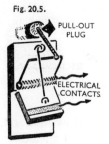

BULLET-HOLES. *Left:* When tripped, the mousetrap pulls the plug from the preformed bullet hole and makes contact with a metal strip, firing electrically-ignited
charges. *Right:* For ground-spatter effects, compressed-air is connected to a buried
perforated pipe. When released, air puffs through each hole in turn, giving a realistic
surface spatter. Rubber branch-pipes allow more widespread effects.

In this way we can produce headless bodies, disembodied limbs,
gravity-free ballet (assisted by black-clad partners), and kindred
mysteries.

Vignettes. Simply a black cardboard stencil placed in front of the
camera, to mask-off or isolate parts of the scene.

Lighting effects
Sparkles. Glitter, stars, stardust and similar scintillations, are
emulated by sequins, diamanté, or flake-mica. Glass chips are often
stuck to scenery for glitter effects, but, remember, they cause painful
abrasions if handled.

Metallic dust (used for sparkling coiffures) can be dropped into a
light-beam to produce a shimmering column of light. So, by cutting
from Mephistopheles to such a cloud, we have him vanish in a puff.

Reflected shapes. The decorative and dramatic possibilities of light
reflections have been to date little used.

The reflection can be thrown directly on to the scene (rear or

Fig. 20.6.

PARTIAL DISAPPEARANCES. Parts or complete subjects set-up before a black background will disappear, either by dressing subjects in black, or using black screens.

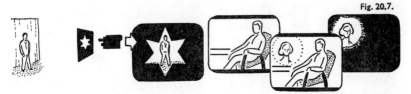

Fig. 20.7.

VIGNETTES. *Left:* The vignette used as a masking effect. *Right:* Vignetting to obtain localized superimposition.

forward), or superimposed upon another's shot. We can cover or paint the reflector with a stencil, and so lettering, patterns, or symbols, can be flexed and manipulated *ad infinitum*. By covering up, we can make any part of the reflection disappear. Varying the finish of the reflector's surface (e.g. greasing, varnishing) changes the relative brightness, clarity or texture of parts of the image. Foil, sequins, etc., can be used for localized reflections. Furthermore, by arranging patterns at either end of a reflector, their respective images can be juggled, superimposed, interposed, simply by flexing.

Amongst the countless uses of reflected shapes we have that of simulating stars, nebulae, planetary-motions, clouds, lightning, making mobile patterns—for abstractions, dream sequences, mental states, producing magic and supernatural effects, or animated titling.

Ripple reflections. To imitate the rippling-light reflected from nearby water, we have the apparatus illustrated in Fig. 20.9.

Apparatus employing moving patterned glass or metal stencils is used in theatres to project water-ripple effects, but the result may not satisfy the scrutiny of the television camera.

Firelight. Real firelight is too feeble to illuminate the scene, or to register satisfactorily on camera, so light from a powerful lamp on the ground is flickered instead. Waving linen streamers or leafy branches can produce quite convincing illusions—when skilfully handled. The motion picture practice of using smoke or gas-jets to animate the light is generally less convenient. There are automatic fire-flicker devices, but most give somewhat mechanical results.

Fig. 20.8.

REFLECTED SHAPES. Polished material, held in a sharply-focused light-beam, projects its image on to nearby surfaces. Bending the reflector distorts the reflected shape.

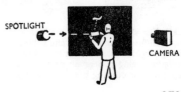

SPOTLIGHT

CAMERA

Fig. 20.9.

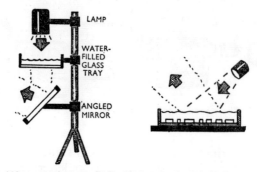

RIPPLES. *Left:* Water ripple-tray. *Right:* Mirror ripple-dish (fragments of mirror at the bottom of a water-filled dish).

Flashes. To get a light-flash suggesting sheet lightning, explosions, etc., we can momentarily uncover a bright light source, strike an arc between scissor-held carbon trims, or ignite a flash-box charged with flash-powder.

While for forked lightning there are several possibilities such as stock film effects, superimposed animated graphics, reflected patterns, electronic insertion, or stencils in scenic projectors.

Projected shadows. Most shadow effects we encounter will be locational—implying time, weather, environment; dramatic—

Fig. 20.10. Pt. I.

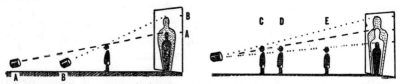

SHADOW SIZE, SHARPNESS AND INTENSITY. The shadow's size can never be smaller than the subject. *Left:* Size increasing as lamp-subject distance lessens. Close lamp positions give greater size-changes and perspective exaggeration. *Right:* Size increasing as subject-background distance increases.

Fig. 20.10. Pt. 2.

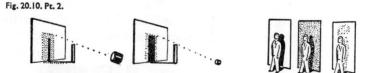

Left: The shadow's sharpness increases with smaller area light sources; decreases with light-diffusion, and also as the subject gets further from the background or closer to the lamp. *Right:* The shadow's intensity depends upon the tonal value and finish of the background (being poorer on dark-toned, smooth or patterned surfaces), and how much spill-light is diluting the shadow.

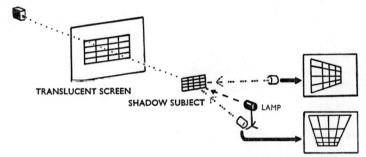

Fig. 20.11.

CAMERA

TRANSLUCENT SCREEN

SHADOW SUBJECT   LAMP

SHADOW DISTORTION. Unless the camera and lamp are in line, and the shadow-screen and shadow-subject at right-angles to it, there must be some degree of distortion. Tilting a background relative to the camera, or the lamp's direction relative to the background, will introduce distortion.

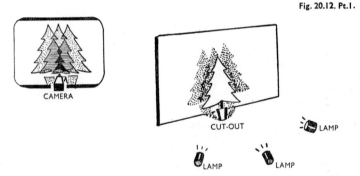

Fig. 20.12. Pt.1.

CAMERA

CUT-OUT   LAMP

LAMP   LAMP

MULTIPLE SHADOW EFFECTS. Two or more lamps casting shadows on the same background give deep shadow where their respective shadows coincide, half-tone shadows (penumbra) where the shadow from one lamp is diluted by the light of another, and the background lit by their combined light.

Coloured lamp filters can supply subtle colour mixing. By fading between lamps with individual cut-outs, we can change the background shadows at will.

Fig. 20.12. Pt.2.

Several lamps on a rotating wheel cast multiple shadows that weave from side to side, continuously changing in tone.

375

Fig. 20.13. Pt.1.

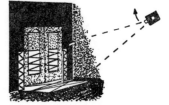

MOVING LIGHTS AND SHADOWS. *Left:* Cyclodrum. A rotating circular framework fitted over a lamp, with stencils or shadowing objects (e.g. branches) mounted on it. *Centre:* Mirror-drum. Focusing a spotlight on a drum set with pieces of mirror suggests passing lights (street-lamps, vehicles). In either case we have to avoid needlessly repetitious patterns. *Right:* The Boomlight. The counter-balanced spotlight can be used to provide a swinging key light (e.g. to suggest the slow roll of a ship).

Fig. 20.13. Pt.2.

Lift-door Shadows. *Left:* Lift-gate shadows fall across the waiting passengers as the masking card is raised. *Right:* A tilted slide-projector provides a moving shadow-pattern on the translucent lift-doors.

It may be sufficient to have a sliding sheet or roller blind behind translucent windows.

Fig. 20.14.

LOSING BACKGROUND SHADOWS. To avoid background shadows and display a subject as if isolated in space, we can either place it on a glass panel, away from its background or arrange it on an illuminated translucent panel.

e.g. a tell-tale silhouette on a window-blind; or decorative—tracery of grilles, lattice-work, etc.

We can get these shadows from optically-projected stencils or slides, by placing cut-out shapes in a spotlight's beam, or throwing shadows of actual staging (foliage, windows, people, etc.).

Multiple shadows. By multi-source lighting, we can turn a single cut-out's silhouette into a complex multi-tone pattern. This gimmick has been used for moving decorative patterns, silhouettes and abstracted devices (Figure 20.12).

Moving lights and shadows. To give the impression that passing lights or shadows are being cast upon a subject, it may be sufficient to move diffusers, branches, etc., past a key-light. But for more extensive effects, the *cyclodrum* is invaluable. In some cases we can use a scenic projector (Figure 20.36) for this purpose.

Background shadows. Although diffused light obviates background shadows, it also reduces surface modelling and texture (Figure 20.14).

Black-light (U.V.—ultra-violet light). Anything lit by ultra-violet light alone, remains invisible to our eye and camera, the exception being fluorescent materials, which glow with vivid colours under its rays. When costumes and settings are treated with fluorescent paint, and lit only with black-light, they glow attractively in otherwise dark surroundings. This idea has been utilized in advertising, stage-spectacle and ice-shows.

Unfortunately, its value in television is negligible. Despite intense U.V. sources, the camera receives insufficient light, and picture quality is very poor. Colour-appeal is lost entirely. Nevertheless, directors periodically try to use this effect—and fail. (Remedy: Try white-detailed design on all-black costumes, staged before black drapes. Dancing skeletons, for instances, are easily devised this way.)

Mirrors

The versatile mirror crops up regularly in visual effects, so first a few reminders about how mirrors behave.

The mirror can alter the camera's or the subject's virtual position.

To find the area seen, draw the lens-angle limits up to the mirror, and reflect each at an angle equal to its angle of incidence.

Fig. 20.15.

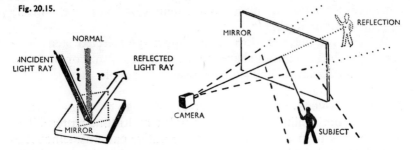

BASIC LAWS OF REFLECTION. *Left:* Light is reflected at an angle (r) equal to the angle of incidence (i). A line perpendicular to the mirror surface from where the beams meet is the normal. The incident and reflected rays and the normal lie in the same plane. *Right:* The reflected image appears laterally reversed and as far behind the mirror, as the subject is in front.

Fig. 20.16. Pt.I.

CHANGING THE VIEWPOINT WITH MIRRORS. Using a single mirror. *Top left:* Top shot, with image inverted and laterally reversed. *Right:* Similar reverse-angle shot. *Bottom left:* Very low shot.

Fig. 20.16. Pt.2.

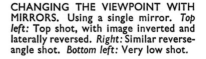

Using two mirrors. *Left:* Top shot. *Right:* High-angle shot.

Fig. 20.16. Pt.3.

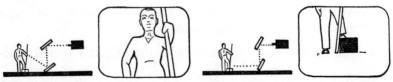

Left: Low-angle shot. *Right:* Shot along floor. The mirrors may be mounted separately, fixed in a portable periscope stand, or attached to the camera mounting.

378

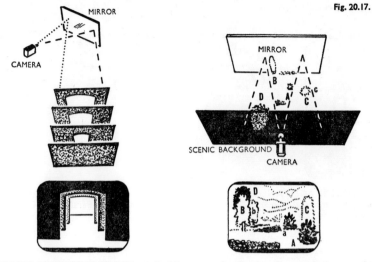

MIRRORS TO EXTEND SPACE. *Left:* The camera's effective distance is increased to obtain a longer shot than studio space permits. *Right:* Looking into a mirror through a peep-hole, the camera sees a reflection of a scenic background, apparently twice the background-mirror distance away.

Focus on the subject image, not on the mirror surface.

The more obliquely we shoot the mirror, the larger it must be.

Tilting an upright mirror from the vertical causes the reflected image of vertical subjects—

for downward-tilt—to lean forward, putting the subject higher in the mirror;

for upward- tilt —to lean backwards, putting the subject lower in the mirror.

Let us now look at further common ways in which mirrors are used in productions.

Half-silvered mirrors (transparent mirrors). Being only very thinly coated with reflective silvering, these special mirrors behave either as transparent or reflecting surfaces, according to the relative brightnesses of subjects on either side.

A useful facility for beauty demonstrations; and convenient sometimes as a peep-hole for otherwise inaccessible shots. In either case, the camera shooting through it gets less than 1/4 of the scene's reflected light, so we must compensate accordingly by using a larger lens-aperture, or more intense lighting.

Fig. 20.18. Pt.1.

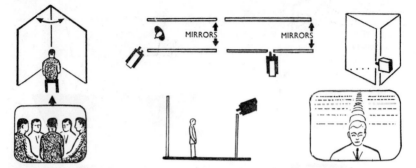

MULTIPLE IMAGES AND SPLIT-SCREEN EFFECTS. *Left:* The smaller the mirror-angle, the more the images. Number = 360 ÷ angle. *Centre:* Parallel mirrors. *Right:* Triple mirrors.

Fig. 20.18. Pt.2.

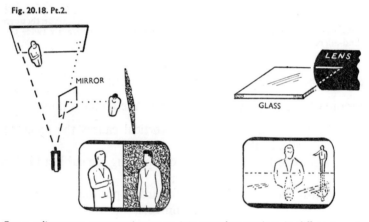

For a split-screen conversation or to join together settings in different parts of the studio. *Right:* For a mirage or water-reflection effect.

Fig. 20.19.

THE MIRROR-SHOT, in which we combine the reflection and the directly-seen subjects.

380

Fig. 20.20.

HALF-SILVERED MIRRORS. Set in a wall, we can watch the performer full-face (1), and from behind, together with his reflection (2).

Duplexing. By this arrangement, one camera can handle two captions or small displays, or service two slide or film projectors Extending this idea, we can multiplex between numerous picture sources. But only with added complication and image-degradation.

Pepper's ghost. An old theatrical device; the only apparatus needed is a sheet of clear glass angled before the camera. Simultaneously we see the solid scene beyond and a ghostly superimposition reflecting an off-stage subject. By adjusting their relative brightness, we can change their respective prominences. Using a half-silvered mirror instead of plain glass, we get a brighter, more distinct reflection.

The Pepper's Ghost principle is not restricted to *Hamlet* and "Things that go bump in the night". We can show where a component part (in solid) fits into its complete installation (a ghost car, perhaps) (Figure 20.22).

In the *shadow-box*, this idea is applied to caption presentation. We can dissolve, cut, superimpose, just by lighting-changes—and yet use only one camera.

Fig. 20.21.

DUPLEXING. The camera sees graphic (A) via a central movable mirror, and a fixed mirror compensating for lateral reversal. When the movable mirror turns or slides, the camera now sees along another path instead, to graphic (B); the picture changing with a soft-edged wipe.

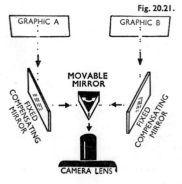

381

Fig. 20.22.

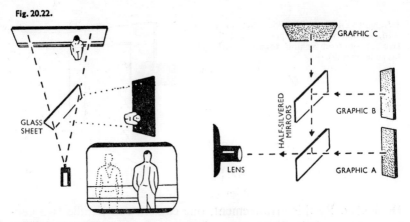

PEPPER'S GHOST. *Left:* The out-of-shot figure is reflected into the lens as a transparent ghost beside the subject. *Right:* The Shadow Box. As each graphic is lit, it becomes visible, (A) directly; (B) reflected correctly; (C) reflected laterally reversed.

Unseen caption-drawing. Seeing a caption form itself as we watch, gives it a persuasive fascination. Drawn-while-you-wait techniques hold our attention. More important, when we have a lot of information to introduce, we can do so gradually, rather than face the viewer with a completed graphic. Some directors use unseen drawing to make a little visual material go a long way; as pictorial padding to a lengthy discourse. But this usage—in story illustration, for example—can be very tedious.

Fig. 20.23.

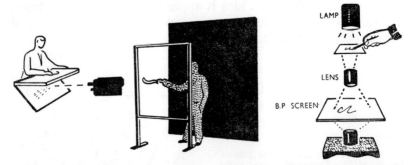

UNSEEN CAPTION-DRAWING. *Left:* Drawing on a flat glass table. The image reflected in a mirror is seen by the camera. *Centre:* The artist, dressed in black, stands before a black background and paints in white on a clear glass panel (guide outlines may be lightly sketched in beforehand). *Right:* The design is drawn on a translucent sheet (or smoked-glass slide), and focused on to a back projection screen, being picked up by the camera on the reverse side.

Fig. 20.24. Pt.1.

TILTING THE PICTURE. Tilted shots from elevated and depressed angles, using canted mirrors. (A vertical 45° mirror on the lens cants the scene as the camera tilts.)

Fig. 20.24. Pt.2.

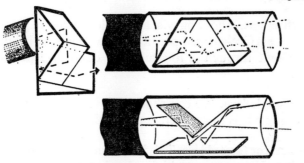

Left: The Porra prism (a pair of reversing prisms). Rotating one of the prisms causes the image to rotate. A similar device can be constructed from two sets of double mirrors. *Top right:* Image inverter prism (straight-through reversing prism). Rotating the assembly causes the picture to revolve. *Bottom right:* A similar device uses a three-mirror assembly. Lateral reversal takes place.

Inverting and reversing the picture. Wherever we need to invert or laterally reverse (mirror) the picture, we can do so optically or electronically. To correct a mirror-shot, or for an effect (ceiling-walking, perhaps), we can switch circuits in the camera-equipment (changing the camera's line-scanning direction for mirroring, its field-scanning direction for inversion). Mirrors or reversing-prisms are effective, too, where circuitry-switching facilities are not fitted.

Tilting the picture. Several mirror and prismatic lens-attachments are available to enable us to tilt or rotate the picture. By mixing a tilted subject (by superimposition or electronic insertion) with a normal scene from another camera, we can simulate floating in free space, flying, cliff-climbing and similar prodigious feats—the easy way.

Distorting the picture

There are times when we want to distort the picture in some characteristic fashion for comic or dramatic effect.

Treated glass. Shooting through a sheet of clear glass that has been lightly smeared with grease, petroleum jelly, liquid paraffin, oil, etc. Moving the glass will animate the distortion. Grading the treatment, we can vary the distortion until the picture finally becomes indecipherable. Nor need we distort the entire picture. We can localize the effect, or use it to introduce halation or diffusion.

Moulded glass. Of the countless varieties of moulded glass, many with dimpling, ribs, or prismatic surfaces, transmit clear but distorted images: to obtain ripples or multiple images.

Metal reflections. Shooting via a flexible mirror (e.g. chromium sheeting, or surface-silvered plastic), we can curve, twist, distort the picture, shoot through a foil tube for decorative peripheral reflections.

Very wide-angle lens. We can shoot a close subject on an abnormally wide-angle lens to exaggerate proportions (Chapter 3).

The prismatic or faceted lens. This lens-attachment comprises a small multi-prism drum, or a multi-faced prismatic lens. It gives us a number of juxtaposed upright images, merging at their borders.

Rotated either remotely or by a small handle, the images can be made to revolve around their centre while still remaining upright.

Picture quality adjustment

As we saw in Chapter 4, the effective brightness and contrast can be modified electronically within limits, to augment lighting and scenic effects.

Tonal reversal (phase reversal)

In monochrome television systems, it is possible to reverse the video polarity and hence make the picture negative. Thus we can;

Transmit negative film (so saving print expense).

Reverse caption tones (to improve a superimposition of titling, perhaps).

Change the tones of stock film of smoke, clouds, etc. (dark or light, simply by switching).

Achieve comic effects (e.g. transforming the appearance of scenery).

Achieve dramatic effects. Some subjects appear strangely sinister in negative (e.g. black lightning in a white sky; black highlights glistening on a white sea; white shadows).

Fig. 20.25.

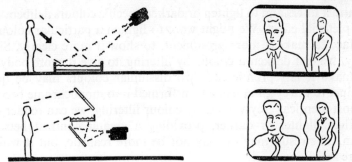

DELIBERATE DISTORTION. *Left:* Water Reflections. The camera shoots into a shallow mirror-bottomed water tank, giving an inverted image, or a glass tray is interposed in the path of a mirror periscope. Ripples can be made to flow across the picture, finally breaking it up altogether. Pouring an opaque liquid into the tray blots out the picture with a fluid wipe. *Right:* Electronic Ripple. A vertically travelling S-distortion, obtained electronically (by varying the line trigger timing pulse).

Fig. 20.26. Pt.1.

MULTIPLE IMAGES. The prismatic lens. The number and position of these images is determined by the facets on the lens. The peripheral images rotate as the lens is turned, but remain upright.

Fig. 20.26. Pt.2.

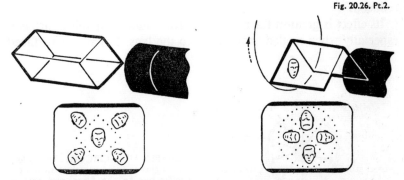

Left: The Kaleidoscope. Shooting through three-or four-sided mirror-tube. Turning the tube rotates the images round the central subject. *Right:* Two angled mirrors. The number of images equals 360 divided by angle of mirrors. Surface-silvered mirrors are needed to avoid spurious ghost images. In neither case do the images stay upright.

Filters

Contrast filters aim to lighten or darken specific colours deliberately, for pictorial effect. We might want to lighten a particular coloured surface to make it more prominent, to show it more clearly. Similarly, we can darken a colour by filtering, to give it more body, or to make it less prominent. Two dissimilar colours may be quite distinct to the eye, but when transformed into monochrome become almost identical grey-values. By colour filtering, we can render one of them lighter or darker, providing a greater contrast between them. The result may or may not be more realistic, but it will be more satisfying.

Fundamentally, a colour filter causes subjects of its own colour to be reproduced of lighter tone than normal, while subjects of complementary colour will be darkened. Remember that most surfaces reflect impure hues that are a mixture of colours spread over part of the spectrum. Filters, too, cover a spectral band. This means that filtering will affect quite a number of scene tones, apart from the most obvious ones. The deeper the filter, the more pronounced the change.

When will we use filters in practice? To get more effective contrast in delicately coloured fabrics, packaging, reproducing faded documents, visual effects, etc. This is not possible with colour systems.

Polarizing filters have the unique property of being direction-conscious towards light rays. Non colour-discriminating, the polarizing filter can be used in monochrome and colour systems, to cut down the light reflected from shiny surfaces. By rotating the filter in front of the lens, particular reflections can be selected and reduced.

Its effect is greatest for reflections from smooth or shiny surfaces, especially when angled around 35° to the lens. For rough-surfaced

Fig. 20.27.

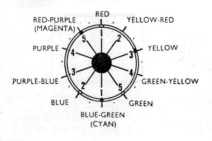

MUNSELL HUE CIRCUIT. The colour filter lightens tonal values of surfaces of similar hue to the filter, and darkens tonal values of surfaces of complementary hue to the filter, lying opposite on the colour-circle.

materials, little change will occur. Of limited use on a moving camera, a polarizing filter on a static shot can enable us to subdue strong flares on water, glass, metal surfaces, etc. or to see through transparent media without showing any surface reflections (e.g. to look into water, windows, etc., without any surface kick-back).

Unfortunately, as with all filters, a light loss is inevitable; in this case only allowing about a quarter of the overall light to pass.

Special filters are:
HAZE FILTER.—An ultra-violet filter reducing haze-blur.
NIGHT FILTER.—Simulates night or evening effects, for scenes shot in bright light. Frequently graded neutral density filters.

In monochrome: dark red, to darken blue and green shades; exaggerate the pallor of skin-tones.

In colour systems: filter imparting a bluish tinge to the pictures.
FOG FILTER.—Provides an effect ranging from slight mist to dense fog, with strongly haloed highlights.
STAR EFFECT FILTERS.—Close-ruled clear filter creates pattern of 4, 6, 8 ray star pattern (increasing with closer spacing) around scene highlights.

Diffusion discs. Strictly speaking, these are glass discs with moulded ribs or dimpling, that impart a soft, diffused look to the picture. Sometimes coarse muslin net, or nylons held over the lens will serve as a satisfactory substitute. Whether the effect is ethereal, or just woolly and irritating, depends upon the occasion. Facial wrinkles are obscured; point sources of light develop glowing haloes.

Samoiloff effect. In monochrome, an actor in scarlet make-up who is shot with red filtering or lighting, will look natural enough. But change to a green filter (or green lighting), and his face will reproduce as black. This Samoiloff effect can be used to achieve sudden

Fig. 20.28

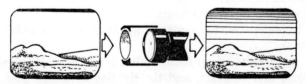

SKY FILTER (graded filter). Graded from deep yellow at the top, to clear glass in the lower half, it increases contrast in the upper portion of a shot without increasing foreground exposure. Thus a plain, uninteresting sky becomes more dynamic. It can be used wherever the upper shading does not obscure the main subject.

ageing or rejuvenating, comic face-transformations and, to some extent, to change coloured backgrounds or graphics. While coloured lighting can introduce localized changes, a filter necessarily affects the entire shot.

Decorative coloured lighting in monochrome television is value-less, for the visual effects are not transmitted. Coloured light simply alters tonal rendering of coloured objects. Similarly, viewers watching monochrome receivers lose the entire colour impact.

Miniatures (model shots)

Miniature settings are comparatively seldom used, for to carry conviction, they require a very scrupulous attention to detail and finish.

So model-making is necessarily a skilled, lengthy, and expensive business.

Within limits, moving model people and vehicles can be introduced.

Indirect signs of life (smoke, flashing street signs, running water, etc.) help to suggest reality, but each situation presents its own problems.

There are problems, too, when shooting models: depth of field tends to become unnaturally limited in close shots.

Speed of movement will usually require scaling down, also, to appear realistic, particularly for such violent disasters as collisions, explosions, hurricanes and cliff-plunges.

Slow-motion filming, at a rate around the square root of the size ratio between real subject and model is often recommended for this purpose.

So the time/energy scale becomes slowed-down during normal speed projection.

Partial model shots are occasionally blended with the studio setting, These enable elaborate scenic features, such as vaulted roofs, arches, to be built in miniature at much lower cost than a full-size replica, and lit to match the studio scene.

They have the same fundamental weakness as glass and Schufftan processes.

They require very deep focus. What is more, scale, perspective, tonal range and contrast, and the lighting of model and setting, need to match. The camera is immobile; at most we can only zoom to alter shots. People's movements must be restricted to specified areas, to remain visible.

Fig. 20.29.

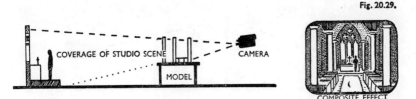

COMPOSITE EFFECT

MINIATURES AND MATTES. A miniature foreground can provide scenic elaboration cheaply in restricted space. Alternatively, we can have a two-dimensional foreground painted on glass or cut out (matte), through which the studio scene appears. Their main drawback is the great depth of field required.

Glass shots; camera mattes; gobo shots

Whether we use a vertical painted glass sheet, a painted or photographed matte, held before the camera, these devices all follow the same principle. The studio scene is shot through a pictorial foreground *stencil*. Moving away from the cut-out areas, people disappear behind the foreground-piece.

Schufftan process

Seldom used, the Schufftan process involves a mirror angled at 45° to the lens axis. The reflected model (laterally-reversed) is set to one side, while the studio scene is visible through holes scraped in the silvering.

Back projection (B.P.; rear projection)

Perhaps one of the most far-reaching innovations introduced into motion picture making was the advent of back projection. Despite its simplicity, it offered considerable flexibility and economy, and television naturally inherited this valuable facility.

The screen itself is all-important, for its specially prepared cellulose skin, containing minute quartz crystals, is designed to provide evenly-illuminated pictures over a wide viewing angle. Screen size ranges from a few square feet to stretches 20 feet or more wide. Wherever the screen's edges are likely to be seen, they are usually masked-off with strategically-placed scenery. A sloping or curved ramp (cove) may hide the join where the screen meets the floor.

Considering we can project pictures from the scenic to the decorative, from the microscopic to cartoon, we might expect that this would mean the end of built scenery in the studio. Quite ambitious presentations have been staged with B.P. alone. But, as we shall see, its various disadvantages prevent its being used exclusively.

One B.P. screen can supply the backgrounds to many scenes in a

Fig. 20.30. Pt.1.

BACK PROJECTION. A slide or film projector throws its image on to one side of a matte translucent screen. Action on the opposite side of the screen appears as if within the background scene.

Fig. 20.30. Pt.2.

We can use back projection in several ways. *Left:* As the complete backcloth, or supplemented by pieces of built set, furniture, etc. (see Figure 8.14). *Centre:* Used to supplement a built set. *Right:* To extend a built set.

single production. The studio floor is marked out to locate the positions of performers, scenic pieces, and furniture for each set-up.

With ingenuity, we can press the projection principle to greater economies. For example, we can use one stage for two consecutive scenes, by cross-cutting to an adjoining wing-flat while the projected slide is changed.

Moving B.P. The high quality image that glass transparencies can provide meet most B.P. applications, particularly where the background scene would contain little or no movement. But many subjects—seascapes, for example—look unreal when "frozen". To carry conviction, we need to use moving projection.

The projection of moving pictures poses several technical difficulties. Unless both the film camera and the projector are precision instruments, the projected image will weave vertically and horizontally, and this will be especially obvious behind stable foreground scenery. We saw, too, in Chapter 7, how one must synchronize the picture projection-rate with that of the television system, otherwise a horizontal black phase-bar will travel up or down the background picture. Even despite synchronism, it may be marred by a dark hum-bar when shot with some types of camera-tube.

Whereas a still-slide can be held indefinitely,[1] a moving back-

---

[1] Former arc-luminants have given way to quartz–iodine ($2\frac{1}{2}$–5 kw.) or xenon sources, needing no periodic adjustment. Blown-air and oil cooling prevents slides overheating.

ground requires a continuous supply of film so, unless we happen to be using film-loops, we have to guard against running-out during a scene.

Normally, the background film is photographed from a static camera. If there is a cut in this background sequence, the foreground performers will be transferred instantaneously to the new location. Unless we are supposed to be in a vehicle, tracking or zooming during photographing can make the live performers appear to move, or even to change scale. Panning or tilting in the background picture will cause foreground subjects to slide. These phenomena are useful enough devices when we need them, otherwise they constitute hazards for the unwary.

Fig. 20.31. Pt.1.

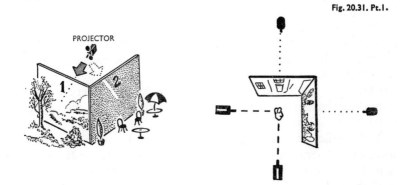

MULTI-SCREEN APPLICATIONS. *Left:* One projector can double between two equi-distant screens, turning between them as needed *Right:* Two screens at right angles show two viewpoints of the same scene, or a character in two consecutive scenes to simulate filmic space and time.

Fig. 20.31. Pt.2.

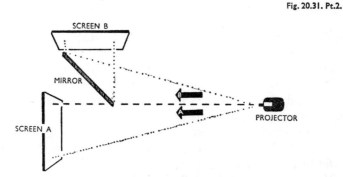

Half the projected image falls on each screen, providing backgrounds for two separate settings. With a split slide, two different scenes can be projected simultaneously.

Back projection problems. The benefits of B.P. are many, but in production it presents several staging problems, most of which have been collected together here.

(i) STUDIO SPACE. The size of the background image depends eventually upon the projector's lens-angle, and upon the projector-to-screen distance (throw). A 7 in. projector lens with a slide of 3 in. × 2 in. picture area, gives an 18 ft. × 12 ft. picture with a 42 ft. throw. Even a 5 ft. × 3 ft. screen needs a 12 ft. throw. A wide-angle projector lens allows us to use shorter distances, but at the expense of optical-distortion, and falling off of light at the edges of the screen.

As the screen is enlarged, the projector's light output must be increased considerably to maintain a suitable screen brightness. This is only practicable where the projector's luminant provides sufficient reserve power; difficult for colour TV's high intensity needs.

For large-screen television projectors (fed directly with a video-signal) the light output is seldom sufficient—especially for dark backgrounds.

(ii) OPTICAL DISCREPANCIES. Ideally, the studio camera should be placed in the same relationship to the screen as the photographic camera had to the original scene.

The studio camera's lens-angle and set-up should match the location camera's exactly, but that would mean restricting our studio camera to one static position. Few directors would accept such handicaps, for in live, continuous production, this would necessitate either semi-static camerawork or innumerable different backgrounds. Happily, we can deviate considerably from the ideal with many background subjects, without the knowledge of the viewer.

Whenever we move the studio camera, the projected background shows no spatial or proportional changes, such as the true scene would produce. This falsity can reveal the background's flatness and destroy any illusion of depth.

Shooting at an angle to any flat picture will encourage distortion, and an overall reduction in its brightness. Where there are scenic foreground pieces intended to blend in with the projected picture, further perspective discrepancies can arise.

If the location camera is tilted up, the picture's horizon (eye-line) will lie below the centre of the B.P. frame. If tilted down, it will be above picture-centre.

We can compensate partially for incorrect studio camera elevation by adjusting the slide's height-position on the B.P. screen, but the

FOLDING THE LIGHT PATH. By folding the light-path with a mirror, floor space can be saved. Using a second mirror even greater compactness is possible, but with a further loss of clarity.

Fig. 20.32.

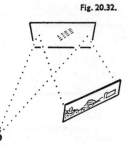

Fig. 20.33.

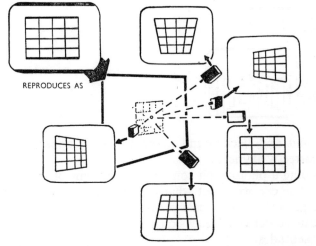

REPRODUCES AS

BACKGROUND DISTORTION. Unless the studio camera is perpendicular to the flat two-dimensional background picture, distortion will occur. Similarly, the projector must be dead square to the B.P. screen to avoid distortion.

Fig. 20.34.

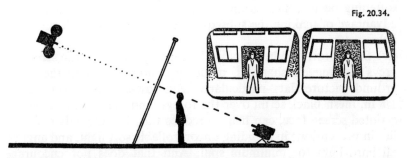

THE EFFECT OF TILT. If the location camera is tilted, the studio set-up should be similarly angled (ideally, the screen being tilted to keep it at right angles to the lens axis) otherwise, the foreground subject will lean off-vertical.

393

Fig. 20.35. Pt.1.

MATCHING EYE-LINES. If the television camera's eye-line does not coincide with the eye-line of the background, the foreground subject will not match its background's perspective.

Fig. 20.35. Pt.2.

Background eye-line above subject eye-line.

Fig. 20.35. Pt.3.

Background eye-line below subject eye line.

foreground and background floors will only appear continuous when correctly aligned.

Providing a photograph contains visible signs of scale and perspective, one can work out what the original set-up was, and with patient geometry, even scale plans and elevations of the original scene can be derived. Television staging to date generally ignores such exactitudes.

(iii) DEPTH OF FIELD. Because the background picture is flat, the studio camera focuses it as a whole. In longer shot it will be sharp, but for closer views, depth of field is more restricted, and overall softening occurs. This is an unnatural effect, although it has the advantage of making the subsequent perspective errors less noticeable.

(iv) SPILL-LIGHT. Where studio lighting spills on to the screen, it greys the image's shadows, reducing contrast, making the background picture flat—particularly with low-key backgrounds. A 1/8 in. mesh black scrim over the foreground face of the screen, or a tinted screen face, can help to reduce this, but the only real cure lies in prevention; in avoiding uncontrollable soft light, and angling all hard light for minimum spill. And that does not encourage good lighting techniques. Keeping people over 6 ft. from the screen will help, but lighting treatment is still a compromise. With continually varying background scenes, accurate matching may become impracticable.

To be convincing, the direction and contrast of the foreground lighting must be suited to that of the background scene. This is easier said than done, for slide-quality is variable and, anyway, its lighting treatment may not match the productional needs of the foreground action.

(v) HOT-SPOTS. Looking from the camera position at the picture projected on to a single-skin screen, we see that it does not have even overall brightness. A brighter central area falls off to lower intensity at the screen edges (how markedly, varies with the screen material). This hot-spot's position moves as we re-angle our viewpoint.

A double-skinned screen, using two close parallel sheets, reduces this unevenness, but does not eliminate it entirely.

Wide angle lenses aggravate the situation, either on the projector or camera.

We can reduce hot-spots for a given camera position by holding a small, irregularly-shaped piece of filter-medium some 2 ft. in front of the projector's lens. A cross-wired frame is a useful accessory here; the localized dimming produced can be adjusted to suit the circumstances.

The same idea can be used to take-down an overbright sky, or a glaring scenic surface in the background picture, and so allow improved exposure of the rest of the image.

(vi) ACOUSTICAL PROBLEMS. Back projection poses difficulties for the sound mixer, too, for there are limits to how far we can change the studio sound's acoustical character to match the background scene.

Despite dialogue-equalizers, deadened studios, artificial reverberation, filtering, etc., it is often impossible under typical studio conditions to simulate matching sound quality. Sound for exteriors invariably has a studio resonance about it, even in motion picture making—although background effects may disguise this artificiality to some degree.

Further uses for the B.P. screen. We can use the B.P. screen (or a stretched-cloth equivalent) without a projector, in several ways: for adjusting its tone (locally or overall) by lighting treatment—for skies, neutral backgrounds, etc., for rear-projecting silhouettes of people or scenery, or for displaying projected light patterns, shadows, clouds, etc.

Back projected graphics. The most modest studio can run to a junior B.P. set-up. A small screen 2 to 3 ft. wide enables us to project slides or filmstrips, drawings, captions, microscope-slides, etc., for a multitude of purposes. (It can even provide backings for small windows, backings for model shots, etc.)

Two forms of projector are used here, possibly combined into one instrument: a transparency projector (Diascope) with manual or remote controlled frame change, and an opaque projector (Episcope, Balop). In this form, a flat subject (photograph, drawing, coins), strongly lit, can be focused on to the screen.

Front projection

We can project still or moving images on to backgrounds, or over the acting area, for decorative or scenic effect.

However, especially in colour systems, direct front projection has major disadvantages: it gives lower image-brightness than with back projection; spill light is more troublesome; tonal contrast is more restricted; background surface-texture may glare through.

Reflex (axial) front projection (Fig. 20.36) produces a bright, integrated image; but only for a single static camera.

Because the projector is often necessarily angled relative to the background surface, we get keystoning and overall-focus snags. Projected window shadows inevitably have this shortcoming.

With scenic projectors, one can provide simple moving images of such varied subjects as snow, sea-swell, clouds, water-ripple, flames, twinkling stars, and fireworks. Their realism may be challengeable, and their low intensity may restrict their use, but they are convenient where space or cost preclude motion picture type projectors.

Electronic picture insertion

There are two basic methods of electronic insertion, enabling us to take subjects from one picture and integrate them into another.

(A) An optical-mechanical system requiring special masks to actuate the process. This is known as: inlay, keyed-insertion, or static electronic matte.

(B) A solely electronic process, needing a marked tonal difference between the inserted subject and its surroundings. This is brightness separation overlay, combined montage amplifier, or moving electronic matte.

Fig. 20.36.

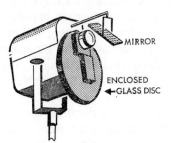

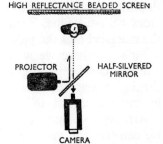

HIGH REFLECTANCE BEADED SCREEN

MIRROR

ENCLOSED
◄—GLASS DISC

PROJECTOR

HALF-SILVERED
MIRROR

CAMERA

FRONT PROJECTION. *Left:* The Scenic Projector. A 2 ft. diameter glass (or mica) disc has painted or photographed patterns on it. This is rotated at an adjustable speed, by a small motor. The whole is attached to the front of a spotlight or arc. A section of the disc is projected on to the background to simulate moving clouds, water, snow, etc. *Right:* Reflex (axial) Front Projection. The background scene is projected along the lens axis, via a half-silvered mirror, on to a highly directional beaded screen. The camera sees subject and reflected scene integrated. Actor's lights swamp the image projected over him.

(A) Inlay—keyed-insertion—static electronic matte. The apparatus consists basically of a picture-tube displaying a plain white TV picture (raster) on its screen, which is seen through a transparent mask-plate or *cel* by a lens focused on to a photo-cell (photo-tube). This operates a switching amplifier (see Figure 20.37). The whole is usually arranged vertically for operational convenience.

One makes the matting-mask empirically, looking at its effect on an adjacent picture monitor. Thin card, or modelling clay are typical materials. Alternatively, it can be painted on to the transparent cel, held positioned by registration-pins.

By adjusting the height and width of the raster on the *inlay-tube*, we can alter the masked area's size and proportions, either to aid fitting, or to change or animate its shape (e.g. from circular to oval).

Because we only use as much of the scene as corresponds with the masked area of the picture, inlay is economical in studio space. The rest of the shot—whatever it contains—is suppressed.

As to restrictions: one cannot insert the whole of the one picture into a smaller area of the other. Nor shall we usually be able to move either camera without untoward effects.

When part of the background is inlayed care must be taken that an actor will not cross over to that area, because he will disappear. This effect can be used, however, for magical purposes.

On the credit side: numerous kinds of *wipes* between pictures are obtainable, simply by passing a mask over the raster on the inlay-tube. We can use mechanical masks, stencils, etc., manually or push-

397

button operated, their speed and direction being readily adjusted. Pouring an opaque fluid over the inlay-tube (via a glass tray) provides a liquid wipe. Furthermore, by interposing a prism between the mask-cel and the pick-up lens, we can rotate the wiping action.

Finally, where we want even further elaboration, the masking patterns can be photographed on cine film. This film is then run through a telecine projector, its video signal being used to switch the insertion apparatus in lieu of live masks.

Although we have been talking here of camera pictures, clearly we can use any video source (e.g. telecine, caption-scanners, electronically-generated signals). Herein lies the flexibility of this device. We can even omit the subject picture altogether, and use masks laid on the inlay-tube to "punch" black (or white) silhouettes into the background picture.

Let us try to summarize the main applications of this facility.

Summarizing. We can use inlay (keyed-insertion; static electronic matte).

(i) To INSERT a particular area from one picture into a corresponding area of the same size, shape and position in another picture (Figure 20.37).

(ii) To OBLITERATE all or part of one picture, revealing another (wipes; split-screen).

(iii) To INSET—

    (a) Static insets. Static decorative masks (mattes) insetting material obviously abstracted from another picture.

    (b) Stencil vignettes. Mattes punching symbolically-shaped stencil in which action appears in otherwise blank screen. (For keyhole effect, etc.)

    (c) Moving insets. Moving arrows—produced by a cut-out moved around the mask-cel.

        Placing a pattern before one of the cameras will impart a texture to the moving arrow.

    (d) Titling

        Similarly we can draw on the cel in "spotted" or "striped" paint.

        Ready-made lettering can be moved about the cel to give animated captions.

        Prepared titling on cells or film-strips can be used to provide static or roller-captions.

(iv) To ALTER PICTURE FORMAT (Figure 15.25). By blanking-off areas of the picture, we can alter the screen's overall shape, the change

398

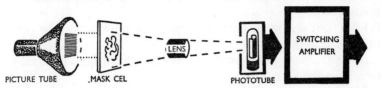

Fig. 20.37, Pt. I.

INLAY—KEYED INSERTION—STATIC ELECTRONIC MATTE. The layout of components. Alternatively, a light-panel–cel–TV camera layout can feed the switching amplifier.

Fig. 20.37. Pt.2.

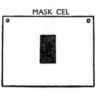

We select part(s) of the scene (*left*) and make a mask (any shape and number) to cover this part of the inlay-tube's raster (*centre*). The insertion apparatus then automatically blanks out that area of the picture. Wherever its photo-cell "sees" the raster, it allows the scene (camera 1) to be transmitted, but mattes-out during the masked area, switching instead to another picture source, called here camera 2. This camera is looking at the subject to be inserted (*right*). Only as much of the subject-shot as corresponds with the mask area is picked out by the insertion apparatus.

Fig. 20.37. Pt.3

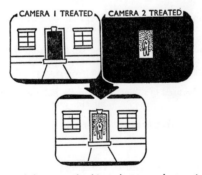

When the treated scene and the treated subject shots are electronically mixed together, a composite picture results.

Fig. 20.38.

INSETS. *Left* and *left centre:* Static insets. *Right centre* and *right:* Stencil vignettes.

being surprisingly unobtrusive if introduced during camera movement.

(v) TO OBTAIN APPEARANCES AND DISAPPEARANCES, by switching insertion in or out.

(vi) TO TELEVISE MULTI-TONE TRANSPARENCIES (slides or film; positive, negative, or colour).

The transparency is introduced in place of the cel. The constant brightness flying-spot which traces the raster on the inlay-tube is "seen" through this transparency by the lens and photo-cell; so the spot's intensity appears to vary as it scans, according to the transparency's tones. The photo-cell generates from this fluctuating light, the video signal. (This set-up is identical with the *flying-spot scanner* apparatus used for televising films and graphics.)

TABLE 20.1

THE MAIN ADVANTAGES AND DISADVANTAGES OF INLAY

Advantages	Disadvantages
The apparatus is simple to operate.	Masks may have to be made accurately for each set-up.
Reliable and definite in its action.	These can take time and skill to make, adjust and replace.
There is no spurious break-through between pictures.	The studio set-up during transmission must duplicate the rehearsal version accurately, if the mask is to match.
Any tonal values may be used in subject and background.	Normally neither camera should move. Any telecine must be free from weaving.
Only limited studio-space is required.	Foreground and background scenes must match—in proportion, perspective, lighting, etc.
Inlaid titling overcomes the mutual tonal-degradation that superimposition involves.	We cannot inlay the whole of one picture into part of another.
It replaces a studio-camera set-up for graphics.	
Subjects can be made to disappear behind foreground scenery, by suitable masking.	
Permits numerous visual effects: wipes, split-screen, transformations, appearances. disappearances, unseen writing.	

Fig. 20.39.

BRIGHTNESS SEPARATION OVERLAY.
Set before a plain white background (the
switching-tone), the background shot is
suppressed electronically wherever the
subject appears in its treated picture.
The subject shot is suppressed electronically wherever the switching-tone
appears. The two treated pictures faded
up together produce a composite shot.

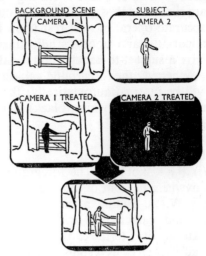

(B) **Brightness separation overlay—combined montage amplifier—
electronic moving matte.** This is an automatic electronic-switch which
necessitates placing the foreground subject in front of a plain white
(black[1]) backing. All the subject's tones must be considerably
darker (lighter[1]) than this backing for the switching-circuits to
operate effectively. Wherever the *overlay apparatus* "sees" about
50 per cent darker (lighter[1]) tones in the foreground subject than
the background, the foreground picture is selected; while wherever
the apparatus "sees" the extreme backing-tone, it switches to the
background picture source. The combined result is the foreground overlaid solid upon the background.

Unfortunately, the overlay process is far from fool-proof, and is
liable to let the background break through into parts of the fore-

[1] We use black-separation instead of white when subjects are light-toned.

Fig. 20.40.

LIMITATIONS OF THE PROCESS. *Left:* Typically unsuitable subject tones. *Centre:*
The effect of using a white switching-tone. This is liable to occur in most normal scenes
in highlights, reflections, fair hair. *Right:* Using a black switching-tone, the result would
be this. In normal scenes this can occur in shadows (e.g. open mouth, armpits) and
dark hair.

ground. This spurious switching occurs wherever foreground tones are similar to those of the background. The difference necessary for accurate operation will rest upon the selectivity designed into the apparatus, and upon its adjustment; but it is never advisable to have a subject-to-background tonal difference of less than 2 : 1.

Wherever the foreground subject moves in the picture, its outline will be inserted automatically; the only sign of the overlay action usually being a white (black) slightly-ragged border to the foreground subject.

Uses of overlay are:
(i) BACKGROUNDS. The foreground subject and scenery can move anywhere in front of the background scene.
(ii) WIPES. Sliding a card of white (black) in front of the foreground camera will wipe to the background picture. Or, by panning over to the card, we get a sort of push-over wipe. Stopping half-way gives us a split-screen.

This idea can be extended to use a model-theatre curtain, drapes, shutters, doors, etc., as an interlinking wipe-device between scenes.
(iii) APPEARANCES/DISAPPEARANCES. Fading either the foreground or background picture in or out, we have appearances or disappearances to order.
(iv) TRANSFORMATIONS. A vase of roses can change into a bowl of goldfish if the latter, placed in corresponding position before the foreground camera, is faded up on overlay. We can even replace one person's face with another's on the same body—if need be.
(v) GROWTH AND SIZE. On tracking its camera, the overlaid subject's size will increase or decrease relative to the background scene and *vice versa*. Useful for *Alice in Wonderland* situations.
(vi) SPECIAL APPLICATIONS. By using overlay in a less orthodox fashion, many interesting abstract and decorative effects become

Fig. 20.41.

NATURAL WIPES. The subject camera looks at the window shutters, which conceal an area of switching-tone. Opening the shutters will automatically wipe this area, revealing the background camera's shot. Close them (wiping out that background), cut to another background camera, and upon re-opening the shutters, we now see a different scene outside the window.

possible. For instance, supposing we use a foreground picture containing a wide range of tones, and employ a white tone to actuate the overlay process, the result will be a strange image in which lighter tones have become solid black, while all lower grey-scale values are reproduced normally. Mixed with a background camera—of a white or textured surface, perhaps—these solid black areas now appear filled. Had we used a black to switch on, the darkest tones of the subject would have been stencilled, while lighter values were reproduced normally.

By mixing the stencilled picture from a multi-tone subject with the original untreated subject picture (now applied as a background-source), we get strange solarization effects.

The possibilities of such trickery are unbounded: for bizarre representations of people (nightmares, the unknown); strange transformations of everyday scenes; animated titling.

In colour television, chroma-key (colour separation overlay) has increasing applications. A blue background is used as the master-

TABLE 20.11

THE MAIN ADVANTAGES AND DISADVANTAGES OF OVERLAY

Advantages	Disadvantages
Its operation is automatic.	Similar to the last four disadvantages of inlay.
No masks are needed.	There must be considerable tonal separation between the subject and its backing, for clean overlay.
The subject's movements are not confined.	Even then spurious break-through and ragged fringing can arise.
Subjects do not disappear or become mutilated accidentally.	Normally an area must be cleared before the subject camera, sufficient for its entire coverage.
A foreground vignette of switching-tone can provide limitations where necessary; or enable people to move behind background subjects.	
Permits wipes; split-screen; appearances; disappearances; size-changes; transformations.	
Various visual effects are obtainable; unachieved by other means.	

Fig. 20.42.

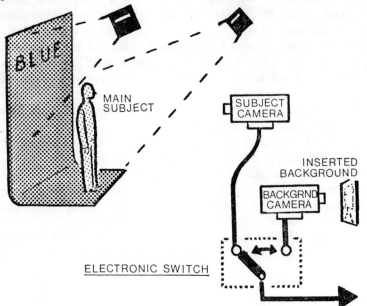

CHROMA-KEY (COLOUR SEPARATION OVERLAY). A development of the bright-ness separation overlay system, this method achieves automatic subject isolation by keying from a particular hue. Blue is usually chosen, as most subjects contain relatively low blue-content; consequently all blue must be excluded from the required main sub-ject (otherwise spurious break-through of the background scene arises in these areas). The subject to be isolated is positioned before a bright blue backdrop. Wherever this master camera sees a blue area it switches (or rapidly mixes) to a separate background picture source (actually a "blue minus luminance" signal is used).
Colour reflected from the blue backing may cause accidental triggering.

camera switching-hue (excluding blue from the subject scene). These blue areas are electronically replaced by the full-colour background picture from another source.

Special effects amplifier. A cousin to the electronic insertion apparatus, this device generates a selection of electrical wave-forms of various shapes (square, triangular, etc.), which are used to switch automatically between two video sources. Its main purpose is to obtain simple wipes and split-screen effects at the push of a button.

Convertible backgrounds

Occasionally, we want to change the entire appearance of a back-ground while still watching the foreground subject. Behind the homesick singer, city rooftops are transformed into his native countryside. We can accomplish this metamorphosis by several means:

404

(i) By gauzes—frontally or rear-lit.

(ii) By projection devices—back or front projection; Pepper's Ghost, etc.

(iii) By overlay—mixing to another background camera.

(iv) By double-faced screens—revealing one scene when frontally-lit, and another painted or shadow-effect on its reverse side when rear-lit. Usually only successful for simple changes.

(v) Multi-layer B.P. slides—the projector is focused in turn upon each slide in a multi-layer sandwich, so we get a sharp picture of that slide and a blurred, indistinct image of the others. (Photographic or painted slides, embossed glass, may all be used here.)

(vi) Utilizing the Samoiloff effect, with scenic hues.

Filmed effects

For certain visual effects we need the help of motion picture magic. For fast-motion, slow-motion, suspended-animation, reversed-motion, stop-motion, cyclic-motion, extended/contracted space or time, complex appearances, disappearances and transformations. (Video disc provides limited speed and stop-motion facilities.)

Fast-motion. Speeding up subject movement and time-scale.

(i) To exaggerate energy for comic effect, or dramatic force (e.g. fast-moving clouds).

(ii) To shorten an action's duration.

(iii) To reveal the development of a slow process (e.g. plant growth).

To achieve fast-motion, we film at a speed below normal (e.g. from below 24 f.p.s. to 1 frame per hour or so), reproducing this film at normal speed (24 frames per sec.).

Slow-motion. Slowing down subject movement and time-scale.

(i) To exaggerate movement form (e.g. of dance-movements).

(ii) To allow rapid movements to be discerned (e.g. a hummingbird's wing-beats).

(iii) To lay emphasis to a movement, giving it force and importance (e.g. emphatic body-movements).

To achieve slow-motion, we film at a higher rate than normal (e.g. 64 f.p.s.), reproducing this film at normal speed.

Stretch (skip-frame) printing slows action by repeat-printing selected frames (e.g. duplicating alternate frames).

Suspended animation (freeze-frame). Stopping action during its course, to show development (e.g. a high diver "frozen" in mid-air to demonstrate body positions). Achieved either by stopping the projector (telecine) at that point or by having the appropriate frame repeat-printed during processing, for the required duration.

Reversed-motion. Causing a movement to be reproduced backwards, for comic, magical, or explanatory effect. So a fallen chimney rises and pieces itself together; goods wrap themselves up; scattered lettering arranges itself.

Achieved by having the film printed with its frames in reverse order; feeding the film into the projector tail-first, and electronically-inverting the picture; photographing the scene upside-down and subsequently inverting; or running the film in reverse through the projector.

Stop-motion photography (single-frame animation). Here we shoot a series of pictures one frame at a time, rearranging the subject a little for each. On reproducing them in quick succession (at any speed), these shots are given a group significance, usually creating the illusion of subject movement. Thus one animates cartoons, stringless puppets, still-life, etc. Edited video-tape is sometimes used.

So-called *photo-animation* is a dodge to cheapen animation processes. One photographs the subject in a series of stills. Then, on enlarged prints of each, we add any drawing, titling, retouching, double-exposures, etc., required. Finally, these doctored prints are re-photographed in a frame-by-frame sequence. Projection in quick succession gives them photo-animation. Innumerable process-effects are economically managed this way, including wipes, spins, phantom-sections and zooms.

Repeat (cyclic) motion. Where a process is recurrent or best illustrated by repetition, film-loops are invaluable. But for only a few repeats, one simply joins several identical copies, running them straight through the projector.

Extended and contracted space and time. By judicious film-editing one can suggest:

    (i) That two adjacent areas are nearer or further apart than they are.

    (ii) That a room is larger or smaller than it actually is.

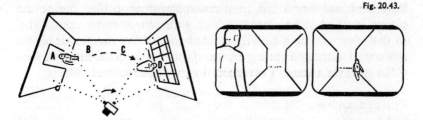

Fig. 20.43.

MANIPULATING TIME AND SPACE. *Left:* Panning the performer throughout his walk gives a fairly accurate idea of the room size. Omitting part of the walk by cutting from pan A—B to pan C—D contracts time and space. The room seems smaller. But unobtrusively repeating, by cutting from pan A—B—C to pan B—C—D, extends time and space. *Right:* Cutting from a walk down a corridor, to a repeat walk up the corridor, suggests scenic continuity.

(iii) That a stretch of scenery (e.g. corridor, stairway) is longer than it is.

Complicated appearances, disappearances and transformations. All become possible by stopping the camera during filming; modifying the scene; and continuing. Rose bushes can burst into bloom. Man can change into mouse with the greatest of ease.

Aural effects

Adjusting sound character

Occasionally we shall turn from our normal "high fidelity" target, and deliberately modify the sound quality; for dramatic, novelty, or environmental effects. We can do this by filtering, distorting, controlling reverberation, or speed adjustment.

Filtering (using equalizers, tone-controls, audio-filters). Filtering enables us to increase or decrease bass or top with varying severity. Emphasizing or suppressing selected parts of the audio-spectrum can make the sound seem shrill, edgy, thin, tubby, hollow, harsh, brittle, etc. Most audio consoles include comprehensive filter arrangements.

Distortion. Distortion devices help to simulate the sound of short-wave radio pick-up, contort sounds for dramatic effect, or to produce synthetic noises.

Using a carbon microphone, or holding a studio microphone near a midget loudspeaker or head-set, will often do the trick. Adjusting an audio-amplifier's biasing conditions can produce distortion *ad nauseum*. Try manipulating audio-tape during replay.

407

Reverberation. For dead (non-reverberant) acoustics, the sound source itself must be placed in highly absorbent surroundings (i.e. in the open air, or a heavily-padded room). We cannot deaden a live sound, although cutting bass and top may help the illusion.

To increase a sound's reverberation, we have several devices:

An echo-room; echo plate; multi-speaker reproduction (ambiophony).

A multiple reproducing-head tape recorder (Figure 6.17).

Playing a disc with two or more slightly-displaced pick-ups simultaneously.

Electro-mechanical arrangements utilizing springs or liquids, between their reproducer and pick-up heads.

Of these, magnetic tape has greatest flexibility for, by selectively coupling the tape recorder's head, we can achieve many effects, including duplicate repeats, slap-back, rapid multiple-echoes (stutter), and the multi-path reception of public address loudspeakers.

We can contrive further interesting variations by mixing a proportion of the pure, filtered, or distorted sound with a reverberant version—itself pure or treated.

Fig. 20.44.

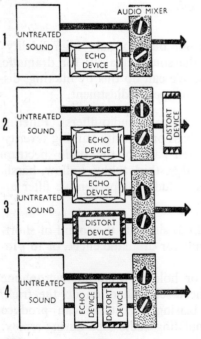

ECHO EFFECTS. The original sound is made more reverberant, the extent depending upon the relative proportions of untreated and echoed sound mixed together.

The original sound is mixed with an echoed version as before, the result being distorted by a further device.

Here the original sound has been distorted, while an undistorted version of the original is echoed. They can be combined in any proportions.

The untreated original sound is mixed with an echoed and distorted version.

Speed adjustment. As we all know, playing a recording faster than the speed at which it was made, will raise its pitch and tempo, shortening its duration. The sound reproduced will lack bass, and have excessive top. We get striking subjective effects, too. For slight speed increase, one feels that the recording is running fast; but speeded further, its complete character changes. Around $1\frac{1}{2}$ to double speed, sounds become strangely miniature. They can appear perky and comic, often unassociated with the original. Here we enter the audio-land of singing mice, insects, household utensils, etc., of cartoon. By articulating correspondingly slower during recording, we get natural-speed speech and song in midget voices on speed-up. (Rotating multi-gap reproducing-heads on tape recorders even enable us to alter pitch, without changing the actual tape-speed.)

Reproducing sounds at speeds lower than they were recorded produces lowered pitch and tempo, and increased reverberation. Quality becomes bass-heavy, and lacks highest frequencies. Most sounds take on a deliberate, sinister, character. Slow-speed reproduction of xylophones, bells, thunder and gongs, for example, is strikingly powerful, but still recognizable. Other sounds lose their identity. Moaning winds become suggestive of subterranean rumblings—the earth in turmoil. Rippling streams are transformed into the ominous gurglings of engulfing flood waters.

Reversed sound. Playing sounds backwards may modify sound character in several ways. Sustained sounds are often changed little. Parts of Schubert's *Unfinished Symphony*, for instance, reproduce backwards recognizably as intelligible music. Percussive or transient sounds all have a characteristic indrawn-breath quality when reversed.

Many everyday noises can be reversed, and when filtered, reverberated, or pitch-changed, manage to satisfy the most extreme imaginative needs, from Martians to Monsters.

Repeated sound. Repeating a sound insistently over and over again, we may heighten tension, imply hysteria or mental derangement (over-used, we may harass our audience, too). Short tape or film-loops are more reliable than a repeating disc groove for this purpose.

Fading and distortion. Playing two identical recordings very slightly out of step produces fading and distortion reminiscent of short-wave radio reception, as their relative phasing varies. Try misguiding tape.

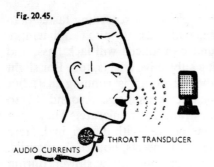

Fig. 20.45.

ARTICULATED SOUND EFFECTS. The sound to be treated is fed into a throat-attached device. Actuated by the audio currents, a small plunger vibrates against the operator's throat. Mouthing the words, the articulated effects come from the operator's mouth, and are picked up by the studio microphone.

AUDIO CURRENTS

THROAT TRANSDUCER

Articulated sounds. In the fantasy of audio effects, speech is not only a human prerogative. We can have a locomotive-whistle cry "All aboard"; an organ can sing its own lyrics. There are two basic arrangements enabling us to articulate non-human sounds. Both have their merits and shortcomings, but both open up a fascinating field of opportunity. The commonest uses a special throat attachment to produce sound which is articulated by the mouth.

An all-electronic method also exists, which permits even further wizardry, by automatically modulating one audio signal with another.

Dissociated sounds. A great many sounds lose their identity when divorced from their original context. Some are unfamiliar anyway, like those of straining pack-ice, or roaring gas-flames. Others carry something of their original significance with them. A jet engine's sound can accompany typhoon scenes, to suggest incredibly destructive forces. Some sounds take on an entirely new significance (e.g. crushing a match-box near a microphone simulates the crash of a falling building).

Many musical instruments have astonishingly elastic propensities, especially when abnormally played: the disembodied waverings of the musical-saw, the violin producing creaking noises (doors, wind-jammers), squeaks, whines, the jew's-harp with its twangy-spring effects, the squeaks, waverings, metallic sounds of the electric guitar, as well as the organ, musical glasses, clavioline, harmonica, Swannee whistle, etc.

To the discriminating director, these are much more than audio-stunts. They are the means of stimulating the viewer's imagination, of improving on actuality, of giving a voice to the inarticulate, of materializing the Unknown.

21

COLOUR TELEVISION

The nature of colour

COLOUR is such a part of our natural world that we tend to accept it
in the television and photographic picture as a most obvious and
unarguable development. A step nearer, or so it seems, to complete
realism; away from the artificiality of the black-and-white image.
But behind the "obvious" can lie new problems, and the need for
revised techniques. These are best considered separately from earlier
principles.

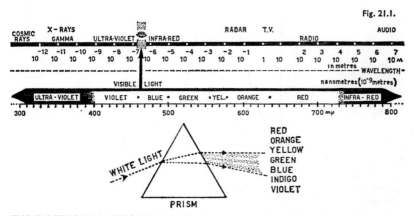

Fig. 21.1.

THE ELECTROMAGNETIC SPECTRUM. Electromagnetic radiation (naturally or artificially generated) has many applications; its character changing gradually with wavelength. Visible light occupies only a tiny segment of this range.

The spectral colours a particular light source contains is revealed when a thin light
beam is passed through a prism. As each hue is diffracted from its path at a different
angle from others when passing from one medium to another (e.g. air to glass), the
component hues become spread in a spectral band.

Visible light wavelengths are variously quoted in millimicrons ($m\mu$) or nanometers
(10^{-9} metre); Ångström units (10^{-10} M.).

Colour reproduction can be a many-splendoured, but very per-
verse, business. We may find we prefer one picture that bears little
resemblance to the original scene, while dismissing another as lacking visual appeal, although technically it renders a scene exactly. At

411

times our brains believe what our eyes should detect as quite false. Clearly, the more we understand the peculiarities and potentialities of colour, the better our opportunities to use it with cunning.

When Newton took a prism and used it to split a beam of sunlight he demonstrated one of the most important fundamentals of colour. "White" sunlight was shown to be made up of a continuous progression of spectral hues; from red, orange, yellow, green, blue, indigo, to violet. It is possible to detect radiation beyond these extremes (in the form of infra-red and ultra-violet light), but these are outside the visible spectrum our eyes can discern.

Our eyes do not respond equally well over the entire spectrum. More sensitive in the yellow-green region, they would require higher light intensities at other hues to accept them as of equal brightness. But our eyes' response, however uneven, is nevertheless the one by which we assess the "naturalness" of the world around us. This unevenness, however, does affect the quantity of light required at extreme hues (e.g. blue) to register strongly to the eye (Fig. 21.2).

Fig. 21.2.

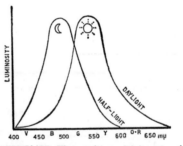

THE EYE'S COLOUR RESPONSE. The eye's sensitivity to colour varies throughout the visible spectrum. Under half light conditions it changes (Purkinje shift) from photopic to scotopic (twilight) vision; thus losing red-orange vision and gaining greater blue-green sensitivity.

Exactly how our eyes detect colour is still not really known. But it has long been established that the light-sensitive receptors or nerve-cells forming the retina at the back of our eyeball take two forms. The majority are rod-shaped and extremely light-sensitive, but cannot discriminate fine detail or any colour differences. They provide our unsharp peripheral vision. The fewer, cone-shaped receptors, on the other hand, are concentrated in the eye's central *fovea*. Although less light sensitive, they acutely detect detail, and are responsible for our colour vision. One convenient theory suggests that our *cones* fall into three groups: those responsive from violet to blue; green to yellow; orange to red.

412

It has been found in practice that most natural colours can be approximately matched by mixtures comprising three *primary* colours (how well, depends upon the actual primaries chosen).

Fig. 21.3.

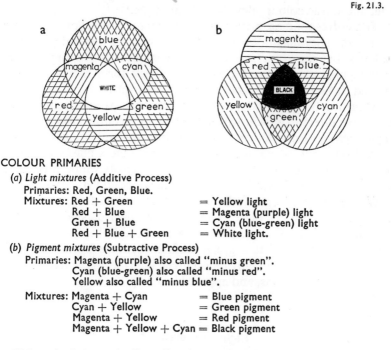

COLOUR PRIMARIES

(a) *Light mixtures* (Additive Process)
  Primaries: Red, Green, Blue.
  Mixtures: Red + Green   = Yellow light
            Red + Blue   = Magenta (purple) light
            Green + Blue   = Cyan (blue-green) light
            Red + Blue + Green = White light.

(b) *Pigment mixtures* (Subtractive Process)
  Primaries: Magenta (purple) also called "minus green".
             Cyan (blue-green) also called "minus red".
             Yellow also called "minus blue".

  Mixtures: Magenta + Cyan   = Blue pigment
            Cyan + Yellow   = Green pigment
            Magenta + Yellow   = Red pigment
            Magenta + Yellow + Cyan = Black pigment

This principle underlies all colour television and film processes. We shall meet two sets of primaries:

those for coloured *light* (additive mixing),
and those for coloured *surfaces* (subtractive mixing).

The reason for the two sets of primaries is simple. We see the sensation of *light* as a direct effect. As in Fig. 21.3, one coloured light always *adds* to another to produce a new colour light mixture. But when we look at a coloured *surface* (i.e. painted, dyed, inked, etc.) what we interpret as its colour is in reality only the residue of colour reflected from it after this surface has absorbed most of the incident light. Thus, when white light falls upon a "blue" surface most of the spectral energy is lost within it as heat, and only a proportion of the light's blue content is returned to us, and seen as "surface colour". So the action is described as *subtractive*. Total absorption would mean no reflection, so the surface looks black.

413

When the *coloured light* primaries of red, blue and green are mixed in equal proportions they cover the entire visible spectrum, and produce "white" light. Varying their proportions will alter the resultant hue. We shall meet this *additive mixing* in coloured lighting, and in the colour television picture tube.[1]

The corresponding primaries for pigments are yellow, magenta and cyan. They may not be as familiar as our pre-blended paint-box colours, but they represent the basic components offering a wider colour-mixture and brightness range. They form the basis (supported by black) of much commercial colour printing.

Magenta, being a mixture of the spectrum's extreme limits of red and blue, does not appear within the true spectrum, and so is termed *non-spectral*. Certain other apparently non-spectral colours actually arise from contrasting normal but low-brightness hues with much brighter surroundings. Thus grey is a low-intensity white; brown is a dim orange or yellow. Pink is a red with white added.

Where coloured light falls upon a coloured surface, again subtractive mixing takes place. The resultant effect depends upon the colours involved. For example, in a pure red light a white or red surface looks red. But a green surface (because by its nature it absorbs all red and can only reflect green light) would reflect nothing under these conditions, and appear black.

Light measurement

In specifying and measuring colour, three aspects are normally examined:

Firstly, the predominant sensation of colour—more properly called *hue*; i.e. whether it is red, green, blue, etc.

Secondly, how strong it appears to be; how far the basic colour has been "diluted"—called its *saturation, chroma, intensity* or *purity*.

[1] Further colour sources:

Light scatter. Atmospheric light scatter from water molecules and dust (blue skies, sunsets).

Interference. Light reflected from inner and outer surfaces of thin films mutually add/subtract, creating spectral colours (oil films, bubbles, "Newton's rings").

Diffraction. Spectral colour arising from fine, closely scribed grooves (iridescence on beetles, diffraction-gratings).

Dispersion. Spectral light arising from "prismatic dispersion" (rainbows, prisms).

Certain low-intensity sources are ineffectual for colour television: *Fluorescence* (page 377); *Photo-luminescence* (materials excited by electric fields); *Phosphorescence* (materials that glow after their excitation source is removed).

Thirdly, its apparent brightness—called variously *brightness, luminance* or *value.*

The group designations normally used are:

hue, chroma, value; or hue, saturation, brightness.

We may meet further terms, too: *tint* (hue diluted with white); *tone* (a greyed hue); *shade* (hue mixed with black). Further descriptions, such as brilliant, pastel, deep, pale, vivid, although descriptive, are rather subjective, as they involve both saturation and brightness of a colour.

Achromatic or grey-scale steps represent intermediate stages between black and white. "Monochrome", although widely used to mean the same thing, strictly means "single colour", and therefore could refer to varying brightnesses of *any hue.*

Just as a series of synonymous terms has evolved to describe aspects of a colour, so various methods exist to assist us in matching or specifying a particular colour.

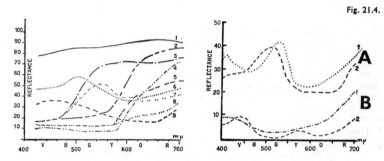

Fig. 21.4.

COLOURED SURFACES. Surfaces are seldom coloured in pure spectral hues. Most colours spread over a wide portion of the spectrum. According to relative proportions of hues, we interpret subjects as having a particular colour.

1. White card	6. Pale Blue
2. Orange	7. Light Green
3. Canary Yellow	8. Maroon
4. Red Rose Petal	9. Sky Blue
5. Signal Red	

Sometimes colours that match under one light (e.g. daylight) may appear dissimilar under another (metametric), due to marked differences in extreme red and blue content. Thus in (A) two green plastic materials differ and in (B) two blue-green paints differ in spectrophotometric curves.

Sample cards (as for paint-matching) are the most familiar. They may be accompanied by a *spectral analysis curve*, as in Fig. 21.4. Of these matching-chip methods, the most sophisticated and widely used notation is the *Munsell system.*

As we see in Fig. 21.5, it is based on the idea of a series of colour chips showing all the hue variations normally encountered. From an eleven-step grey-scale, each of 100 hues in turn is progressively built up to its fully saturated strength for successive steps in the grey scale (in practice, 40 hues up to 12 chroma steps and 9 values are used). So, for a given grey-scale value, colour purity of successive steps increases along the saturation axis. Looked at on a "monochrome" film or TV system, colours of similar value will appear identical in monochrome.

Fig. 21.5.

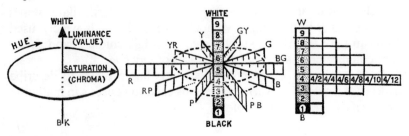

MUNSELL SYSTEM OF COLOUR NOTATION. Devised to classify a wide range of hues, for varying degrees of saturation. Formed from a series of pages evenly spaced around a central pivot (*left*) each analysing a slightly different hue. Between each of the ten principle hues (Fig. 20.27), each scaled as "5", lie sets of four subdivisions (2·5, 5, 7·5, 10), giving a total of 40 constant hue charts.

Each page contains a series of sample "chips" (*right*). At the inner margin "pivot" these form a vertical grey-scale (steps 1–9). For each grey value a horizontal branch is developed, of progressively increasing colour purity to a maximum saturation (chroma) for that value. Steps (total 12) are evenly numbered.
To specify a colour, quote:

hue deviation—value (luminance)—chroma (saturation) steps, e.g. 7·5 G. 7/10
       (a slightly bluish-green, of light, strongly saturated character).

For yellow hues, full saturation can take place at light grey-scale values (e.g. step 8), while for purples full saturation is seen only in low-luminance steps (e.g. steps 3–5).

We can tabulate a wide range of colours by analysing the proportion of red, green and blue light involved. Examining the scene through separate red, green and blue filters, we detect the proportions of these colours and their derivatives present—remembering the very wide colour range possible from their mixtures. (Thus, yellow will reflect some red and some green light; white reflects all three primaries, and so on.) The coloured image can then be reconstituted by mixing light in similar ratios.

The three primaries appear at the corners of the colour triangle. Its centre-of-gravity represents white light. The triangle reveals,

416

incidentally, that not only can white light be produced from these *tristimulus* primaries but equally well by appropriate blends of any complementary colours: magenta/green, or cyan/red, or blue/yellow light.

Fig. 21.6.

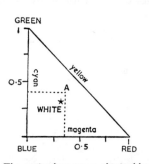

THE COLOUR TRIANGLE. The entire hue range obtainable with three given primaries can be plotted, knowing any two. Using *trichromatic* (T) primaries (i.e. proportioned to add to unity), white is derived from thirds. A colour that is 0·4 g, 0·35 r, (0·25 b), is shown at "A".

As by definition one makes $R + G + B = 1$ (i.e. unity), then we need only to specify the proportions of any two colours (e.g. red and green) to deduce the amount of the third present.

Mixtures of any of the primaries will produce a colour represented by a point falling within the triangle. Any colours that cannot be exactly matched by these primaries will be reproduced as noticeably less saturated than the originals.

The international C.I.E. system (Commission International de l'Eclairage) derives from this concept. Whereas the colour triangle indicates in broad terms the results of mixing primaries, the *C.I.E. chromaticity diagram* (Fig. 21.7) provides a reference system for precisely defining a hue. The "horseshoe" boundary plots the entire range of spectral hues. So any primaries we choose, being less pure (i.e. desaturated), must lie within its confines.

The psychology of colour

To be able to describe, measure and specify colour is essential. But ultimately we are concerned with its impact upon our audience. The psychological aspects of colour must necessarily concern both artist and engineer alike. One's assessment of colour depends on many arbitrary factors; some physiological, some illusory.

Initially our impressions depend upon our own colour vision.

One in ten people has colour blindness, while "normality" is itself variable.

Viewing conditions modify our colour evaluation. The act of putting a border around the coloured scene can create colour enhancement. Bright hues tend to look more vivid, even gaudy, viewed

Fig. 21.7.

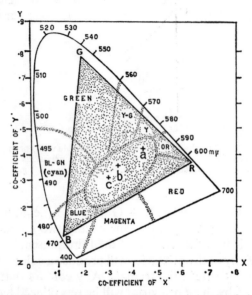

C.I.E. CHROMATICITY DIAGRAM. All visible hues can be classified as points within the curved spectral locus. Colour TV primaries (N.T.S.C.) are plotted here as R.G.B., and televised colours lie within the triangle formed. Subject hues falling outside it are inaccurately reproduced (pure greens and cyans are rare in nature).

A line from a white point to a spectral hue depicts all its possible values (saturations). If two hue points are joined any proportional blend falls along this line. Standard "whites" are shown:

Luminant A = 2850° K. The whiteness of tungsten-filament lamps.
Luminant B = 4800° K. The whiteness of standard daylight.
Luminant C = 6600° K. Colour TV picture-tube "white".

in dim surroundings. Room lighting modifies our eyes' adaptation; warm-toned surroundings make the screen colours look cooler. Room furnishings can be reflected in the screen.

Within the scene itself, the colour-quality of the incident light will affect apparent hues—especially if it varies. Pigments that match under certain lighting may not match under another (Fig. 21.4).

The texture of a surface, too, can affect its apparent brightness and vividness. Identical hues in smooth and rough-textured finishes will be subjectively different; the former looking more saturated.

418

Our camera's viewpoint will also modify this effect. A smoother surface will change in brightness as the angle of the light upon it and/or the camera views it more obliquely. The hues of surfaces lit with hard light will appear brighter and more saturated than when only diffused light is used.

As the real, or effective, size of coloured detail is reduced, our eyes can less readily detect colours. They seem to pale, some losing their identity before others. Yellow tends to light grey, while blue is indistinguishable from dark grey. Similarly, bright areas of green, blue-green and blue may merge. Eventually even the hue-difference between reds and blue-greens is lost, and only brightness variations left (this conveniently obviates the need to transmit tiny colour detail in colour television).

Colour becomes illusorily associated with concepts of temperature and distance. We find that warm colours (red, yellows, browns) normally appear nearer and bigger than cool colours (blue, green). Similarly, darker and saturated colours seem nearer than light or desaturated ones.

Once we juxtapose coloured surfaces in a scene, many interesting —often frustrating—things are liable to happen. For we are establishing both a physical and an artistic relationship between them.

One surface may reflect on to its neighbour—e.g. a bright green bush tingeing one side of a bystander's face to a sickening hue. The more saturated the colours, the greater will be their mutual effect.

The illusion of *simultaneous contrast* goes further, so that a green dress against grey drapes will not only look greener but the grey will take on the complementary hue, and appear warmer, even pinkish. Similarly, if a white area is lit with both white and a coloured light a shadow cast by the latter appears in complementary hue.

The same turquoise object can appear lighter, darker, bluer, greener, according to its background colours. When a strong colour appears against a pale version of itself the latter appears greyer than it would against a strong complementary colour.

A light colour appears stronger backed by white than by black. Conversely, a dark colour is seen more strongly against a black background.

Colour composition involves an extension of the principles discussed in Chapter 15. Although coloured areas may in reality be quite separated, they can appear close together in a flat picture, and so mutually interact. Warmer tones appear heavier than cool, while strong, saturated colours look heavier than duller, desaturated ones. It is preferable, therefore, to keep strongly saturated colours to small areas, for they will balance large desaturated areas.

Attention is held most readily by bright colours. Particularly by scarlets, bright yellows and orange. But this can prove an embarrassment, for, especially when defocused, we are likely to become distracted by them. Desaturated colours do not pose these problems, but in excess lack vigour and visual appeal.

A disharmonious tension can occur between strong, saturated colours of equal area, particularly when they are not complementary. Far better to have one or both desaturated (i.e. greyed-off, less vivid, pastel). A colour will appear brighter and stronger than its complement.

Above all, we have to remember that in the monochrome picture all colours are reduced to grey values. Discordant or distracting colour relationships are overlooked. Dissimilar colours can have identical grey values. An advantage we have in monochrome is the opportunity to merge or separate subject-tones at choice; whereas in colour they have distinct interactions.

Colour harmony is a popular but elusive topic. Broad principles are recognizable. Near-complementaries are quoted as harmonious, as are *triads* or a trio of hues equally spread round the Munsell colour cycle (Fig. 20.27). But harmony depends so much in practice upon image-proportions, relative saturations, application, social criteria, fashion, etc., that colour complexes are not reliably tabulated.

Approximate colour constancy is a mental phenomenon that continually affects our colour evaluation. With many familiar subjects of a known colour we have a predisposition to see a colour reproduction as "correct", even although it deviates considerably from a true representation. Having fixated on this, we judge adjacent colours accordingly. So we accept the mail-box as being of correct colour, but believe other scenic hues to be inaccurate, even when the reverse is the case. This happens with "white" areas on the screen, which we may interpret as such, although they are really quite blue, or yellow, in fact. Paradoxically, we are equally liable to scorn the colour rendering of certain scenic hues, although they may be quite accurate. We may discount the idea of blue puddles on the sidewalk, although they truly reflect a blue sky. Perversely, we can be wildly gullible in accepting misrepresentations of unfamiliar subjects. Our judgments are very relative. A "black" object in bright sun may actually be reflecting more light than a dimly lit "white", although our brain, making a comparative judgment, refuses to believe this.

After looking at a coloured surface for some time, one is liable to experience a brief colour "hang-over" affecting our impression of the next picture. Following exposure to red, we shall see on a black screen a blue-green ghostly *after-image*. Orange gives a peacock blue after-image; yellow gives blue; green gives purple; blue gives orange-yellow. Bearing this in mind, one would expect some interaction between a succession of colours. Such *successive contrast* causes white to appear rather blue-green after our scrutinizing a red screen. Similarly, yellow appears bright green; blue appears more intense and greenish; while repeated red has a greyed appearance.

These are, admittedly, extreme situations, but colour-distortion on quick cuts, and the visual fatigue that follows over-exposure to a strong colour are not to be disregarded.

Colour and emotion are inextricably interlinked. *Colour associations* are legion. For example:

Red with warmth, anger, crudity, excitement, power, strength.
Green with Spring, macabre, freshness, mystery, envy.
Yellow with Sunlight, the Orient, treachery, brilliance.
White with snow, delicacy, purity, cold.

Staging for colour

Scenic presentation for colour TV falls into two broad groups: those in which the excitement, vigour and persuasive potentialities of colour can be fully explored, and those in which colour, unless carefully contained, can introduce distraction, false associations, tawdriness, exaggerated glamour or an inappropriately vivid background to action. In excess, colour defeats its purpose and becomes visually tiring to watch. Too subdued, the picture has vagueness, and lacks definition.

Beyond all else, colour should have a purpose. It should be used where it is needed, remembering that it readily dominates a shot. Subjectively, one has the overwhelming impression that the televised image exaggerates (even caricatures) the coloured scene. In reality, this effect may be quite untrue, and purely illusory. But it is the viewer's *impression* that counts, and one can compensate only by anticipation in avoiding undue colour emphasis when arranging staging.

A good general adage is to arrange staging by working from the unalterable objects in the scene when determining the surrounding décor. As with so many maxims, this is easier said than done.

Corrective remedies when we discover a particular colour unduly prominent or distracting are varingly effective. Removal or substitution are obvious, but not always possible. Spraying with white, grey or black may improve matters, although one is liable to lose surface detail and character in the process. Suitably coloured lighting may sometimes save the day, or one may change light intensities falling upon it. Occasionally one may cover a surface with gauze.

Strong colour is easily achieved—perhaps rather too easily. Subtle but effective colour without simply resorting to neutrals requires interpretative skills. One can have real problems when trying to simulate drab, sombre or slummy surroundings in colour. Reddish browns (e.g. mahogany furniture) appear markedly red, while dark browns may lose detail and contour.

Lighting, unless entirely soft and flat, will always extend effective scenic contrast, so that scenic tonal values should be restricted to around a contrast of 10:1. (4–60% reflectance is sometimes quoted.) Overlight tones will bleach out; overdark tones will simply merge into dim masses. Higher contrast should be confined, therefore, to small areas only. (We are, of course, concerned with *brightness contrast*, not with *colour contrast*, which, being partly subjective, may not reveal the brightness range the camera is being asked to handle.) If the entire scene is of high tones these can be accepted, providing the overall contrast range remains suitably low.

Large areas of colour should generally be avoided. Subtlety is the keyword. Equally, one should avoid large areas of a single unrelieved tone, breaking it up with lighting, set dressing or similar devices.

We become rather more aware of the effects of surface texture in colour. At some angles a green material may look grey, or blue. Materials with a pile (e.g. velvet, velours) appear much darker than smooth surfaces of identical hue. The specular reflection from shiny surfaces will appear as blobs the colour of the incident light, rather than as desaturated patches. The same distraction arises with glossy coloured floors.

As in monochrome television, horizontal planes (floors, table tops and their associated papers, table-cloths, etc.) catch back-light and reproduce over-light. But, in addition, light reflected from them is liable to illuminate in colour other, nearby subjects. (The adage "reduce the backlight for colour" is not entirely meaningful in a complex show, where one camera's backlight may be another's keylight.)

Colour emphasis is achieved in staging through the use of colour in costume, key "props" (e.g. set dressings and ornaments) and furnishing, while backgrounds tend to neutrals. Wallpapers definitely require camera testing before use. Fine pattern may "strobe" (line-flicker) or change appearance with angle and distance. Strong background hues easily dominate.

Scenic opportunities exist in using coloured light with translucent surfaces and with plain grey cycloramas (page 160). But where large areas of blue are required it is less economical to use blue light (as the lamps' blue colour-media pass little light) than to provide blue-coloured surfaces.

Foliage gives problems in colour. When dead or withering it looks pathetic; when fresh it looks over-green. Plastic versions can be even worse. Any coloured lighting or electronic hue-changes introduced to compensate can affect other subjects also.

Realistic scenic backings (e.g. views beyond windows) are not achieved as simply as one might suppose. Underplayed colour, such as a dye-retouched monochrome photo blow-up, has looked more convincing (and is cheaper) than a high-quality colour reproduction.

Graphics in colour require special care. Background hues should not dominate (white, yellow, green, neutrals are suitable), while small lettering is always best avoided. If faces cannot be excluded from a caption, remember that they should be of good colour quality, and will become the subjective reference-hue for the picture. Glossy graphics can be defaced with light reflections.

Finally, in most networks the scenic designer must remember that a compatible monochrome picture is all that much of his audience will see. Effects based upon colour relationships alone can become meaningless to them. Only *tonal* distinctions will convey atmospheric illusion, and this, in the time available, is a challenge indeed!

Lighting for colour

Lighting for colour is often, mistakenly, imagined as simply a stepped up version of monochrome television lighting. Some have attempted to make use of overall soft-light and rely upon colour-differentiation for pictorial effect. As we shall see, these are over-simplifications.

Light intensities for colour television are necessarily higher than for monochrome systems; some twice to nearly three times. Lighting must be more even, that is, of more constant intensity from various camera viewpoints. One is less able to compensate for lighting variations by adjusting lens-apertures, black-level, gamma and channel-gain, for these can create hue changes. To achieve lower-contrast even lighting which is pictorially attractive, and still effective in monochrome, requires perceptive skills.

Lighting contrast modifies our impression of hue. It is worth repeating that under soft light colours appear paler (desaturated), while under hard light colours appear bright and purer (saturated). Low-intensity lighting will darken hues, while over-lighting will pale-out colours.

Lighting proportions between the foreground subjects (actors, costumes) and the background demand care. An overlit foreground

Fig. 21.8.

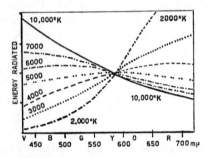

COLOUR TEMPERATURE. Measurement of a light's colour quality. The spectral distribution of the source under discussion is compared with the heat required to cause a non-reflective black-body to emit a similar spectral colour-spread. Values of 1000K. (yellowish-red) to +20,000K. are encountered (0 K = −273° C.). The colour temperature unit, once known as the Kelvin degree is now a kelvin (K).

Colour temperature is also measured in *mired values* $\left(\dfrac{\text{one million}}{\text{Kelvin temp.}}\right)$. Curves here show relative energy distributions over the spectrum at different colour temperatures.

will necessitate stopping down the camera lens, consequently the dim background becomes dimmer and more colourful. An underlit foreground produces converse results.

It follows that lighting contrast and intensity must directly modify facial tones. A "white" face may be made to look pallid, sickly, ruddy, according to its lighting. Under high contrast lighting it may be bisected to look pale on its brightly lit side and swarthy in its shaded planes. Brown skin-tones may appear yellowish when overlit, brown-black or terracotta in shadow. Consequently, a compromise must be achieved between causing false skin rendering through excessive contrast and losing the illusion of solidity that contrast creates in both colour and monochrome. Doubly lit, or shadowy areas, need careful handling, therefore.

Whereas the setting and costume, once correct, can remain satisfactory for an entire show, the impact and technical problems of lighting change with each camera angle. As we saw in Chapter 5, even slight repositioning of subject or camera can alter the lighting

Fig. 21.9.

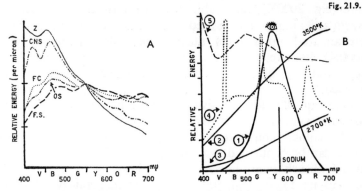

WHITE LIGHT. Many luminants are accepted as "white" light, although their actual spectral ranges differ considerably.
(A) Variations in colour quality of daylight (Northern Hemisphere)

    Z   —Clear zenith sky
    CNS—Clear northern sky
    FC —Full sun with clear sky
    OS —Overcast sky
    FS —Full, direct sun

(B) Variations in typical luminants compared with the Eye's response (I).

    (2) High colour temperature incandescent lamp (e.g. overrun lamp, studio-type lamps).
    (3) Low colour temperature incandescent lamp (e.g. domestic 40–75-watt type).
    (4) Fluorescent lamp, showing peaks in spectrum lines (from mercury vapour filling) within smooth spectrum from fluorescent coating.
    (5) Carbon-arc projector.

illusion. So, wherever one intercuts mobile action between varying viewpoints, the function of lamps will continually change, and ingenious forethought is necessary to achieve high-grade "filmic" lighting.

In the colour studio one soon hears reference to *colour temperature* (Figs. 21.8 and 21.9). Colour temperature refers to the colour quality of "white light". Light of a warm reddish-yellow quality (e.g. candles, very dim lamps) is said to be of a low colour temperature; while cold bluish light-sources (e.g. sunlight, over-run lamps, arcs) have a high colour temperature. As the colour temperature of our light source changes (i.e. its proportion of each spectral colour), so will the apparent brightness and hues of the scenic colour range alter.

The camera channels are adjusted for a balance in which equal r.g.b. proportions produce a "white" of a specific colour temperature. The studio lighting standard may, for instance, be 2850 kelvins ($\pm$100K) to suit this balance. If studio lamps are dimmed considerably their light quality will fall to a lower colour temperature (e.g. 2000K), and subjects lit with them will appear overwarm and akin to having used a reddish overall filter. If we should use an arc-lamp with its much higher colour temperature (over 5000K) its subjects' hues will all appear modified by this comparatively blue light, unless the camera channels are re-aligned.

One can alter the effective colour temperature of any lamp by placing a suitable corrective colour filter over it. A bluish filter upgrades colour temperature to a higher value. Conversely, a high value can be brought down with a warm filter. Corrective colour media absorb light, however. Dimming any incandescent lamp reduces its colour temperature, so that filament switching (in a multi-lamp fitting) or a neutral wire-filter may be preferable to provide light reduction.

Colour in television immediately suggests the opportunity to use coloured light. In practice, though, it must be used with discrimination. In Variety and Dance spectaculars colour can abound—under control. But for serious Drama one seldom encounters more than a glimpse of bluish or orange-yellow light to enhance a mood, for it too easily looks artificial and blatant.

Coloured light may enable us to ring changes on neutrally toned scenery; to modify unsatisfactory scenic hues; or to vary atmospheric effects. Any white light spilled or reflected on to surfaces will dilute these projected colours as it would similarly degrade any other form of projected image.

Make-up for colour

Make-up for colour film and television has required a reassessment of techniques and materials that had proved successful for monochrome media. Not only must mechanics be more subtle but the ingredients and colours need to be more consistent. More time is required for treatment.

Face tones fall into three broad groups: brown tan; pink-tan to pink; olive-tan to beige. These have been found to cover a wide

426

range of make-up conditions for Caucasian, Negro and Oriental skin-tones. Former "panchromatic" shades have proved unsuitable, due to their orange, yellow or brownish trends. While it is agreed that the best reproduction is for skin-tones ranging from olive-tan to beige, avoiding pinker finishes, it is essential not to have unnatural over-uniformity.

To make-up without the subject appearing inappropriately overtreated is the general dilemma. Skin blemishes and natural colour must normally be covered, although the aim is to use thin applications of increased pigmentation, rather than to obscure with heavy mask-like coatings. Whereas formerly *background* performers needed little or no make-up beyond, perhaps, their own colour, television more often demands that they be treated. Arms, hands, necks, ears will similarly need application, blended to an even tone, whereas hitherto these may have been left.

In general, skin quality modifies the make-up used, heaviertexture faces providing more definite modelling, while finer skins tend to possess veining or blotches that the camera accentuates. Ears readily appear reddish and translucent, complexions become flushed with exposure to heat (or hospitality), and make-up gets disturbed (e.g. lip colours "eaten off") in the course of action. Nor do hazards end here, for bleached hair can exhibit alarmingly green shading, while blue-tinted white hair looks startlingly overcoloured, and both require corrective rinses.

Make-up will generally be designed to suit close-up shots (for, like Lighting, readjustment to the bolder, stronger treatment for longer shots may be impracticable). Under this scrutiny, detail, texture and colour are therefore most clearly seen. Hair work, which must be of good quality, has to be blended on to natural hair without hard edges. Pseudo-contours by shading are far less convincing in colour, often looking like grime. Opinions vary as to the best substitute; greyed colours are preferred by some experts, while others use blue or green-grey media. Reds and browns are definitely out. In any case, shading or highlighting must be softly blended and of values nearer to the foundation (this tends to make extreme ageing a more difficult procedure). Otherwise, character make-up follows similar principles as for monochrome.

Highlighting is achievable by colour accents. Rouge, if used, is preferably of a brown-orange hue, although some experts contend that any rouge can inadvertently reproduce as shaded areas in monochrome.

Powdering (to "set" a foundation, to reduce shine, to absorb perspiration) has presented problems due to the tendencies of many products to change colour in use, or obscure base-colouring. Shine is liable to reproduce as distracting white blobby areas, whether owing to perspiration, grease, spirit-gum or surface sheen. When they shine through make-up fresh treatment will often be the only remedy—unless, of course, the effect is appropriate for the character concerned.

Lip colour is suited to the complexion, costume and role, but usually brownish or light clear reds have proved successful. Pinks are rather too strong, whereas standard monochrome materials have shown an excess of orange, yellow or blue ingredients that are emphasized in colour.

Eye shading in grey or brown (replacing the familiar blues and greens) overcomes the suggestion of red eye-sockets, but again must be subtle. Eyebrow pencils and lining colours are as for monochrome.

Prosthetics of all kinds, wax, modelling clay, latex, etc., are best camera tested, to ensure colour and textural suitability.

Colour make-up for men follows conventional form, save that it is easily overdone. Neutral complexions appear too pink, and textural differences very noticeable. One must be careful, though, to avoid a routine universal tan. Some experts prefer complete make-up treatment, with a base covering the whole face, while others reduce the demands by localized treatment (ears, beard, mop-up).

For women, colour make-up follows lines akin to that for monochrome, but with the important differences already outlined, although perhaps rather lighter than normal.

Costume for colour

We shall usually find it preferable to concentrate strongest colour in costumes, leaving the scenery as a less colourful foil to the performers. One must remember though, that colour tends to be exaggerated. A drab dress may look vividly over-emphasized; apparently grey suits can have a bluish hue. Colour matching that looks good to the eye may reproduce with considerable differences— e.g. "reds" differing due to their having dissimilar brown or blue proportions. Colours that have a strong, bold appeal in long shots can seem harshly crude in closer views. We can find, too, that closely patterned materials (e.g. small checks) show changes in appearance with length of shot, giving rise to flicker, colour break-up or hue changes. For these reasons it is preferable to use correct or closely representative costumes for all rehearsals.

Materials that are too light-absorbent (e.g. serge, velours) are liable to appear over-dark, and lose draping and detail as tones merge. Conversely, over-light materials can look pale and lose modelling, unless heavily textured.

Reflective materials (e.g. glazed finishes, satin, plastics) will look brighter and more saturated than matte versions, but more readily develop white blotchy detailless highlights (blooming); reflect colour on to faces (neck, jaw, cheeks); and prove visually distracting. Broken-up reflections from sequins, lamé and metallic cloths are effective in colour, but may give rise to "comet-tails" (lag, smearing, after-image).

Perhaps the most exciting costume opportunities in colour are the ways in which the emotional significance of hue can be drawn upon. With a neutral setting one can change mood from sadness to gaiety through costume alone. But again we must not forget the audience watching monochrome receivers!

Directing colour production

To the discerning Director, colour production has many quite distinct differences from monochrome television. Particularly when a compatible monochrome service has to be considered, the dilemma can be a real one, for the impact, significance and relationships of colour will be entirely lost.

Co-ordinated, planned team effort is an essential for colour if high-grade results are to be achieved in the time available. Where routine quizzes, panel games, etc., are involved, this may be marginal, but more ambitious productions demand close liaison with Lighting, Design, Wardrobe and Make-up, so that efforts combine to a complete whole.

Many of the opportunities and hazards that colour offers the Director we have examined in the foregoing sections. But there are a few further aspects that we might well bear in mind:

In monochrome excess detail distracts and confuses unless it is defocused; then planes merge to form an unobtrusive background. In colour one's attention can be drawn to unsharp coloured detail, as we seek to decipher it. One is rather more conscious, therefore, of the available depth of field in colour, and focus-pulling is over-dramatized.

Exact *matching* between different picture sources may be hard to achieve, so that direct comparative cuts between studio, film, B.P., tape are best avoided.

The *cutting-rate* in colour usually needs to be slower than for monochrome, for varying background colours that would appear as simple grey changes in monochrome now become distracting "colour hops".

Mixing between different colours will produce additive effects during the transition. Hence superimpositions of different colours will result in a combination hue.

The electronics of colour

Although the *image orthicon* camera-tube has long been used for colour television, its relative insensitivity and electronic quirks (redistribution, varying black-level, etc.) result in variable colour quality. Currently, the *vidicon's* relation, the lead-oxide camera tube (Plumbicon, Leddicon) provides simpler, smaller, more stable colour cameras.

Various combinations of camera-tubes exist in the colour camera. (Three tubes may give red, green, blue images, the brightness—luminance—being derived from their combination. Or tubes may provide luminance, red and blue images—green being matrixed from white. A fourth tube may provide luminance for the RGB hue tubes. Even two-tube cameras exist.) Each has its own technical advantages, for one problem is to keep unrelated in the video signals the hue and brightness aspects of the scene.

Although the Plumbicon has many merits, it has certain weaknesses to date. Insensitive at the far red end of the visible spectrum, its rendition of red hues and colour mixtures containing this colour region can be defective. It is liable to produce a coloured *trailing* (lag) behind fast-moving high-contrast subjects. It lacks the edge-definition to fine detail that the image-orthicon provides spuriously. Nevertheless, technical progress has improved or circumvented these drawbacks and resulted in a camera-tube of considerable capability.

Whether other camera-tubes under development will prove rivals, only time will reveal. Currently, electronics are developing so rapidly that today's innovation is all too likely to be succeeded by tomorrow's discoveries.

In previous discussion of colour we have indicated that the contrast range of the studio scene (after lighting) must be confined to limits of some 20:1 overall, unless we are prepared to accept tonal crushing. We must not forget, though, that many colours are, in fact, mixtures. So if, for example, a surface presents strong green, medium red and medium blue signals when we let the mixture "crush out" (e.g. beyond video circuit clippers) we shall find ourselves compressing the strong green component before the red and blue reach this limit. Hence we see a spurious hue change towards red/blue (magenta). Adjusting the camera-channel's black-level (sit) can similarly modify the saturation and hues within the picture.

It is important in any colour reproduction process to keep the analysed component primaries in the same proportions as in the original subject's colour mixture. Any change in their relative strengths (or in their individual spectral coverage) will create a hue change. Ideally the "tracking" should be perfect for the entire brightness range of that combination. But, in reality, we shall encounter mis-tracking to varying degrees, anywhere in the chain, from camera-tubes to picture-tube. (Colour film and colour printing have like problems.) So, nominally accurate colour reproduction at one brightness level can deviate at other brightnesses.

Picture monitors and receivers present this dilemma. One can find that when primary proportions are adjusted for good colour quality they may not provide a pleasingly tinted monochrome picture, and vice versa.

Where the overall colour quality is strongly unbalanced, tingeing all colours towards a particular hue, it is said to have a *colour cast*. We may introduce this deliberately to enhance the "warmth" or "coolness" of a picture. Some organizations, indeed, use a *"hue control"* which, part of the vision-control facilities, will deliberately introduce a colour bias, either for effect or to counteract a bias in the studio scene. This requires discretion, however, for *all* picture hues will be affected, including face-tones. Furthermore, the viewer may subconsciously accommodate to the colour cast, seeing it as "normal", and then interpret the following shot as having an abnormal complementary cast.

The colour camera-channel is lined-up to produce balanced colour output for a particular lighting colour-temperature. As tungsten lamps normally produce light quality of around 3200K., colour studio cameras are often aligned to "Illuminant A", i.e. 2850K. (standards differ between organizations). This permits lamps' brightnesses (and colour temperature) to be changed within about ± 100K.

Colour monitors are typically designed for a colour balance of the order of 6600K. (Illuminant C) to 6500K. (Illuminant D); although this may prove very nominal in the home receiver.

Colour film presents certain problems for colour television. The cinema audience sits in near darkness and is able to adapt to any colour imperfections. The home viewer, lacking such adaptation, should be able more quickly to detect variations in exposure, print grading and colour balance, so that these must be more evenly matched. The gamma and contrast-range that can be accommodated when a print is directly projected in the cinema are higher than for colour television. The contrast possible on a colour shadow-mask

Fig. 21.10.

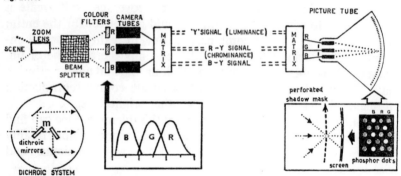

COLOUR TELEVISION PRINCIPLES. All modern colour TV systems (i.e. PAL, SECAM) are developed from the N.T.S.C. system.

Light from the focused scene is split by dichroic mirrors or prismatic devices into three colour bands which, after filtering, cover red, green, blue regions of the spectrum. Camera-tubes produce pictures for each. These R.G.B. signals could be used directly to form a colour picture. But to transmit *compatible* colour television, a complex interim process is necessary. This involves "matrixing" to derive a *luminance* and two types of *chrominance* signal.

The transmission process varies with the system used. The N.T.S.C. system possesses certain drawbacks (e.g. dot patterns, reception colour errors) which later systems partly improve upon.

In the receiver the R.G.B. signals are restored by further matrixing. These are applied to the picture-tube's guns, each shooting through mask perforations on to its particular set of phosphor dots (three types, one for each primary). The colours from these close, intermingled dots merge to appear as colour mixtures. When red, green, blue phosphors are activated the screen appears grey to white.

tube is about 30:1, whereas cinema-projection can meet a 160:1 contrast. Thus, a film made for one medium need not look at its best on the other.

Various imperfections arise in film colour fidelity. These derive initially from film dyes being affected spuriously by colours other than their own specific hue-range. Dye masking improves this situation, but causes certain colours to grey-off (lose luminance). Laboratory grading is more critical.

Fortunately, *electronic masking* compensation can considerably improve televised colour quality. Gamma and video-gain adjustments to the individual red, green and blue channels (TARIF) provide good correction for errors in the red and blue components (although rather less for cyan defects.).

Analysis errors arise from the additive synthesis of television picture-tube display of film materials designed for direct optical projection (subtractive synthesis). This leads to grey-scale errors (e.g. a magenta cast), particularly in low-luminance areas. Electronic masking corrects this, but not saturation of luminance defects.

Telerecording of colour is possible by both video-tape and film methods. Video-tape has achieved very high colour standards, so that third-generation recordings (i.e. dubbings of dubbings) are widely transmitted, following editing processes. Direct film (Kine-scope) recording from the face of a colour picture-tube still offers difficulties. Colour film-stock being relatively insensitive, the most practicable approach is to record the separate red, green, blue images on monochrome film-stock, and then recombine them in film laboratories (e.g. by imbibition process) into a single colour print.

TABLE 21.1

TYPICAL COLOUR TEMPERATURES

	K.
Standard candle	1930
* Household tungsten lamps (25–250 watts)	2600–2900
* Projector bulbs	3200
* Studio Tungsten lamps { 500 watts / 1000 watts }	3000
* Studio Tungsten lamps 2000 watts	3275
* Studio Tungsten lamps 5000, 10,000 watts	3380
* Tungsten halogen lamps	3300–3400
* Overrun tungsten lamps	3400–3500
High-intensity arcs	6000
Sunset, sunrise	2000–3000
Sunless daylight	4500–4800
Midday sun	5000–5400
Overcast sky	6800–7500
Hazy sky	8000
Clear blue north sky	10,000–20,000

* Run at their correct, full voltage.

BIBLIOGRAPHY

Adams, C., Producing and Directing for Television, *Holt, New York.*
Alton, R., Painting with Light, *Macmillan, New York.*
Amos., S. W., Birkenshaw, D. C., Bliss, J. L., Television Engineering, *Iliffe, London.*
Amos, S. W. and Godfrey, J. W., Sound Recording and Reproduction, *Iliffe, London.*
Arnheim, Rudolph, Art and Visual Perception, *Faber & Faber, London.*
Bálazs, Béla, Theory of the Film, *Dobson, London.*
B.B.C. Engineering Training Staff, Microphones, *Iliffe, London.*
Bethers, Ray, Composition, *Pitman, London.*
Bettinger, Hoyland, Television Techniques, *Harper Bros., New York.*
Bretz., R., Techniques of Television Production, *McGraw-Hill, New York.*
Cameron, Ken, Sound and the Documentary Film, *Pitman, London.*
Chinn, Howard A., Television Broadcasting, *McGraw-Hill, New York.*
Creamer, J., and Hoffman, W. B., Radio Sound Effects, *Ziff-Davis, New York.*
Curran, C. W., Handbook of Television and Film Techniques, *Pellegrini & Cudahy.*
Dome, R. B., Television Principles, *McGraw-Hill, New York.*
Eisenstein, S. M., The Film Sense, *Faber & Faber, London.*
Eisenstein, S. M., Notes of a Film Director, *Lawrence & Wishart, London.*
Ennes, H. E., Principles and Practice of Telecasting Operations, *Sams & Co. Inc.*
Evans, Ralph M., An Introduction to Color, *Wiley, New York.*
Fink, Donald G., Television Engineering, *McGraw-Hill, New York.*
Frayne, J. G., and Wolfe, H., Elements of Sound Recording, *Wiley, New York.*
Graves, Maitland, Design Judgement Test, *Psychological Corpn., New York.*
Halas, John, and Manvell, Roger, The Technique of Film Animation, *Focal Press.*
Hoddap, W. C., The Television Manual, *Farrar, Strauss & Young, Inc., New York.*
Hubbell, R., TV Programming and Production, *Chapman & Hall, London.*
Huntley, John, and Manvell, Roger, The Technique of Film Music, *Focal Press, London.*
Kehoe, Vincent J-R., The Technique of Film and Television Make-up, *Focal Press.*
Lindgren, Ernest, The Art of the Film, *Allen & Unwin, London.*
McMahan, Harry Wayne, Television Production, *Hastings House, New York.*
Nurnberg, W., Lighting for Photography, *Focal Press, London.*
Nurnberg, W., Lighting for Portraiture, *Focal Press, London.*
Olson, Harry F., Musical Engineering, *McGraw-Hill, New York.*
O'Meara, C., Television Programme Production, *Ronald Press, New York.*
Pieron, Henri, The Sensations. Their Functions, Processes and Mechanism, *F. Muller.*
Pousette-Dart, Nathaniel, Art Directing, *Hastings House, New York.*
Pudovkin, V. I., Film Technique, *Newnes, London.*
Read, Herbert, The Meaning of Art, *Penguin Books, Harmondsworth.*
Reisz, Karel, The Technique of Film Editing, *Focal Press, London.*
Research Council of the Academy of Motion Picture Arts and Sciences, Motion Picture Sound Engineering, *Chapman & Hall, London.*
Rotha, Paul (Ed.), Television in the Making, *Focal Press, London.*
Royal, J. F., Television Production Problems, *McGraw-Hill, New York.*
Skilbeck, Oswald, ABC of Film and Television Terms, *Focal Press, London.*
Smith, Paul, Creativity, *Hastings House, New York.*
Sposa, L. A., TV Primer of Production and Direction, *McGraw-Hill, New York.*
Spottiswoode, R., Film and its Techniques, *Faber & Faber, London.*
Spottiswoode, R., A Grammar of the Film, *Faber & Faber, London.*
Stanley, T. S., The Technique of Advertising Production, *Prentice Hall, New York.*
Stasheff, E., and Bretz, R., Television Programs, *Wyn, New York.*
Strenkowsky, S., The Art of Make-up, *Dutton & Co., New York.*
Turnbull, Robert B., Radio and Television Sound Effects, *Rinehart, New York.*
Vernon, M. D., Visual Perception, *Cambridge University Press, Cambridge.*
Wade, Robert J., Designing for Television, *Pellegrini & Cudahy, New York.*
Willis, D., A Source of Gestalt Psychology, *Ellis, New York.*
Wills, Edgar E., Foundations in Broadcasting, Radio and TV, *O.U.P., New York.*

INDEX

435

439